3D Printed Microfluidic Devices

3D Printed Microfluidic Devices

Special Issue Editors

Savas Tasoglu
Albert Folch

MDPI • Basel • Beijing • Wuhan • Barcelona • Belgrade

MDPI

Special Issue Editors
Savas Tasoglu
University of Connecticut
USA

Albert Folch
University of Washington
USA

Editorial Office
MDPI
St. Alban-Anlage 66
4052 Basel, Switzerland

This is a reprint of articles from the Special Issue published online in the open access journal *Micromachines* (ISSN 2072-666X) in 2018 (available at: https://www.mdpi.com/journal/micromachines/special_issues/3d_printed_microfluidic_devices)

For citation purposes, cite each article independently as indicated on the article page online and as indicated below:

LastName, A.A.; LastName, B.B.; LastName, C.C. Article Title. *Journal Name* **Year**, *Article Number*, Page Range.

ISBN 978-3-03897-467-3 (Pbk)
ISBN 978-3-03897-468-0 (PDF)

Cover image courtesy of Nirveek Bhattacharjee and Albert Folch.

Contents

About the Special Issue Editors

Savas Tasoglu joined the University of Connecticut (UCONN) in 2014 as an Assistant Professor in the Department of Mechanical Engineering. He received his Ph.D. in 2011 from UC Berkeley, with a research focus on transport phenomena and the pharmacokinetics of anti-HIV microbicide drug delivery. Dr. Tasoglu held a postdoctoral appointment at Harvard Medical School and Harvard-MIT Division of Health Sciences and Technology until he joined UConn in 2014. His current research interests are point-of-care diagnostic devices, bioprinting, magnetic focusing, microfluidics, and regenerative medicine. He serves on the editorial board of several journals including *Bioprinting (Elsevier), Scientific Reports (Nature), Journal of 3D Printing in Medicine (FSG),* and *International Journal of Bioprinting.* Dr. Tasoglu's achievements in research and teaching have been recognized by fellowships and awards including Chang-Lin Tien Fellowship in Mechanical Engineering, Allen D. Wilson Memorial Scholarship, and UC Berkeley Institute Fellowship for Preparing Future Faculty. Dr. Tasoglu has published 50+ co-author articles in journals such as *Nature Communications, Nature Materials, Advanced Materials, PNAS, Small, ACS Nano, Chemical Society Reviews, Trends in Biotechnology,* and *Physics of Fluids.* His work has been featured on the cover of *Advanced Materials, Small, Trends in Biotechnology,* and *Physics of Fluids* and highlighted in *Nature, Nature Physics, Nature Medicine, Boston Globe, Reuters Health, and Boston Magazine.*

Albert Folch received his BSc in physics from the University of Barcelona (UB), Spain, in 1989. In 1994, he received his PhD in surface science and nanotechnology from the UB's Physics Dept. During his PhD, he was a visiting scientist from 1990–91 at the Lawrence Berkeley Lab working on AFM. From 1994–1996, he was a postdoc at MIT developing MEMS. In 1997, he joined Mehmet Toner's laboratory at Harvard. He has been at UW BioE since June 2000, where he is now a full Professor, accumulating over 7,600 citations (averaging >82 citations/paper). His lab works at the interface between microfluidics, 3D-printing, and cancer. In 2001 he received a NSF Career Award, and in 2014 he was elected to the AIMBE College of Fellows. Albert Folch is the author of four books, including "Introduction to BioMEMS", a textbook now adopted by more than 78 departments in 18 countries. Since 2007, the lab has run a celebrated outreach art program called BAIT (Bringing Art Into Technology), which has produced 7 exhibits, a popular resource gallery (>2,000 free BioMEMS-related images), and a YouTube channel (>133,000 visits since 2009).

Preface to "3D Printed Microfluidic Devices"

Three-dimensional (3D) printing has revolutionized the microfabrication prototyping workflow over the past few years. With recent advances in 3D printing and 3D computer-aided design (CAD) technologies, highly complex microfluidic devices can be fabricated via single-step, rapid, and cost-effective protocols as a promising alternative to the time-consuming, costly, and sophisticated poly(dimethylsiloxane) (PDMS) molding and traditional cleanroom-based fabrication. Microfluidic devices have enabled a wide range of biochemical and clinical applications, such as cancer screening, micro-physiological system engineering, high-throughput drug testing, and point-of-care diagnostics. Using 3D printing fabrication techniques, the alteration of design features is significantly faster and easier than with traditional fabrication, enabling agile iterative and modular design, 3D printing services and marketplaces, and rapid prototyping. Biocompatible resins for 3D printing are now available that, contrary to PDMS, feature very low drug absorption and are thus very good material candidates for building complex 3D organ-on-a-chip systems in the future. These advances will make microfluidic technology more accessible to researchers in various fields and will accelerate innovation in the field of microfluidics. Accordingly, this Special Issue showcases 14 research papers, communications, and review articles that focus on novel methodological developments in 3D printing and its use for various biochemical and biomedical applications. The papers of this special issue explore the following aspects of 3D printing and how it pertains to microfluidic devices: (1) fabrication methods and materials, (2) applications for equipment and tools, (3) biological applications, and (4) biological compatibility.

Savas Tasoglu, Albert Folch
Special Issue Editors

micromachines

MDPI

Editorial

Editorial for the Special Issue on 3D Printed Microfluidic Devices

Savas Tasoglu [1,2,3,4,5,*] **and Albert Folch** [6,*]

1   Department of Mechanical Engineering, University of Connecticut, Storrs, CT 06269, USA
2   The Connecticut Institute for the Brain and Cognitive Sciences, University of Connecticut, Storrs, CT 06269, USA
3   Department of Biomedical Engineering, University of Connecticut, Storrs, CT 06269, USA
4   Institute of Materials Science (IMS), University of Connecticut, Storrs, CT 06269, USA
5   Institute for Collaboration on Health, Intervention, and Policy (InCHIP), University of Connecticut, Storrs, CT 06269, USA
6   Department of Bioengineering, University of Washington, Seattle, WA 98195, USA
*   Correspondence: savas.tasoglu@uconn.edu (S.T.); afolch@uw.edu (A.F.)

Received: 19 November 2018; Accepted: 19 November 2018; Published: 21 November 2018

Three-dimensional (3D) printing has revolutionized the microfabrication prototyping workflow over the past few years. With recent advances in 3D printing and 3D computer-aided design (CAD) technologies, highly complex microfluidic devices can be fabricated via single-step, rapid, and cost-effective protocols as a promising alternative to the time-consuming, costly, and sophisticated poly(dimethylsiloxane) (PDMS) molding and traditional cleanroom-based fabrication. Microfluidic devices have enabled a wide range of biochemical and clinical applications, such as cancer screening, micro-physiological system engineering, high-throughput drug testing, and point-of-care diagnostics. Using 3D printing fabrication techniques, the alteration of design features is significantly faster and easier than with traditional fabrication, enabling agile iterative and modular design, 3D printing services and marketplaces, and rapid prototyping. Biocompatible resins for 3D printing are now available that, contrary to PDMS, feature very low drug absorption and are thus very good material candidates for building complex 3D organ-on-a-chip systems in the future. These advances will make microfluidic technology more accessible to researchers in various fields and will accelerate innovation in the field of microfluidics. Accordingly, this Special Issue showcases 14 research papers, communications, and review articles that focus on novel methodological developments in 3D printing and its use for various biochemical and biomedical applications. The papers of this special issue explore the following aspects of 3D printing and how it pertains to microfluidic devices: (1) fabrication methods and materials, (2) applications for equipment and tools, (3) biological applications, and (4) biological compatibility.

1.   Fabrication methods and materials: Fuad et al. [1] characterized additively manufactured molds fabricated via stereolithography and material jetting, as well as the positive replicas produced by soft lithography and PDMS molding. They showed that stereolithography provides finer part resolution with no toxicity observed in the corresponding positive replicas. Beauchamp et al. [2] demonstrated a custom 3D printer which, via optical dosage control, provides very high-resolution printing capabilities down to ~ 30 μm and ~ 20 μm scales for positive and negative surface features, respectively. The custom printer was used to fabricate and optimize various microfluidic particle traps. Kim et al. [3] also developed a novel printing scheme: specifically, a sequential stereolithographic co-printing process which utilizes two different molecular weight poly(ethylene glycol) diacrylate (PEG-DA) resins to produce microchannels with embedded porous barriers. The semi-autonomous fabrication process reduced the processing time, manufacturing costs, and eliminated complications with assembly. In addition to fabrication methods, Kotz et al. [4] presented work on materials

formulated for 3D printing, namely highly fluorinated perfluoropolyether (PFPE) methacrylates, for which 3D printing has seldom been demonstrated. The developed formulations, printed and cured using stereolithography, exhibited high light transmittance and high chemical resistance to organic solvents.

2.   Applications for equipment and tools: Brennan et al. [5] utilized the capabilities of 3D printing technology to create an open source design for micropipettes which can be assembled from 3D-printed parts and a disposable syringe. The open source design exemplifies scientific tools that can be produced via 3D printing as inexpensive alternatives to commercial products. Oh et al. [6] also developed 3D-printed laboratory equipment to measure the viscosity of fluids, which is applicable, for example, to the quality assurance of liquid products and for monitoring the viscosity of clinical fluids. They designed 3D-printed capillary circuits, with graduations to serve as a flow meter for easy readability and a syringe modified with an air chamber to generate pressure-driven flow, to provide equipment and calibration-free viscosity measurement of Newtonian and non-Newtonian fluids. Van den Driesche et al. [7] presented methods for a wide variety of applications to design and fabricate microfluidic chip holders with integrated fluidic and electric connections, such as fluidic sealing by O-rings and electric connections by spring-probes without glue or wire bonding. Microfluidics can also be applied to 3D printing technologies, as demonstrated by Serex et al. [8]. For application to bioprinting, Serex et al. demonstrated the integration of micromixers, micro-concentrators, and microfluidic switches into the tip of the print heads for extrusion-based 3D printing, thereby enabling new prospects in 3D bioprinting.

3.   Biological applications: There are countless biological applications for 3D-printed microfluidic devices, a few of which are included in this Special Issue. Lim et al. [9] reported an automated platform for a colorimetric malaria-Ab assay, assembled from stereolithographic-printed elastomeric reservoirs, fused deposition modeling-printed framework, plastic tubing, servomotors, and an Arduino microcontroller chip. Kim et al. [10] demonstrated a 3D-printed millifluidic platform for bacterial preconcentration and genomic DNA (gDNA) purification, by immunomagnetic separation and magnetic silica-bead-based DNA extraction, to improve the molecular detection of pathogens in blood samples. The platform was verified for preconcentrating E. coli in blood, suggesting that the platform is a useful tool for lowering limitations on molecular detection. In addition to these research articles, Sharafeldin et al. [11] wrote a review on the applications of 3D-printed microfluidic devices in biomedical diagnostics and on how 3D printing enables low-cost, sensitive, and geometrically complex devices. Three-dimensional printing can be used for the fabrication of microfluidics, supporting equipment, optical and electrical components, in addition to 3D bioprinting which can incorporate living cells or biomaterials into diagnostic systems.

4.   Biological compatibility: In their work on connections and holders for microfluidic devices, for bioanalysis applications, van den Driesche et al. [7] addressed the possible cytotoxicity of cured 3D-printed resin by introducing a surface coating of parylene-C. Carve et al. [12] reviewed commonly used vat polymerization and material jetting materials with respect to the materials' biocompatibility, in addition to discussing methods to mitigate material toxicity to promote the application of 3D-printed devices in biomedical and biological research, such as for monolithic lab-on-chip devices. In addition to biocompatibility with cells, interactions with biomolecules such as protein have been studied by Lepowsky et al. [13]. They demonstrated a simple cleaning chip design with an integrated cleaning procedure to study the long-term cyclic biofouling burden on 3D-printed microfluidic devices, and verified the cleaning chip for urine sampling handling for a protein assay. Lepowsky et al. [14] also provided a perspective on traditional and emerging anti-fouling methods as applicable to enabling the greater reusability of 3D-printed microfluidic devices for biomedical applications.

We wish to thank all authors who submitted their papers to this Special Issue. We would also like to acknowledge all the reviewers for dedicating their time to provide careful and timely reviews to ensure the quality of this Special Issue.

Micromachines **2018**, *9*, 609

Conflicts of Interest: A.F. declares no conflict of interest. S.T. is a founder of, and has an equity interest in mBiotics, LLC, a company that is developing microfluidic technologies for point-of-care diagnostic solutions and QRfertile, LLC, a company that is developing technologies for fertility testing. S.T.'s interests were viewed and managed in accordance with the conflict of interest policies. The authors have no other relevant affiliations or financial involvement with any organization or entity with a financial interest in or financial conflict with the subject matter or materials discussed in the manuscript apart from those disclosed.

References

1. Fuad, N.M.; Carve, M.; Kaslin, J.; Wlodkowic, D. Characterization of 3D-printed moulds for soft lithography of millifluidic devices. *Micromachines* **2018**, *9*, 116. [CrossRef] [PubMed]
2. Beauchamp, M.J.; Gong, H.; Woolley, A.T.; Nordin, G.P. 3D printed microfluidic features using dose control in X, Y, and Z dimensions. *Micromachines* **2018**, *9*, 326. [CrossRef] [PubMed]
3. Kim, Y.T.; Castro, K.; Bhattacharjee, N.; Folch, A. Digital manufacturing of selective porous barriers in microchannels using multi-material stereolithography. *Micromachines* **2018**, *9*, 125. [CrossRef] [PubMed]
4. Kotz, F.; Risch, P.; Helmer, D.; Rapp, B.E. Highly fluorinated methacrylates for optical 3D printing of microfluidic devices. *Micromachines* **2018**, *9*, 115. [CrossRef] [PubMed]
5. Brennan, M.D.; Bokhari, F.F.; Eddington, D.T. Open design 3D-printable adjustable micropipette that meets the ISO standard for accuracy. *Micromachines* **2018**, *9*, 191. [CrossRef] [PubMed]
6. Oh, S.; Choi, S. 3D-printed capillary circuits for calibration-free viscosity measurement of Newtonian and non-Newtonian fluids. *Micromachines* **2018**, *9*, 314. [CrossRef] [PubMed]
7. van den Driesche, S.; Lucklum, F.; Bunge, F.; Vellekoop, M.J. 3D printing solutions for microfluidic chip-to-world connections. *Micromachines* **2018**, *9*, 71. [CrossRef] [PubMed]
8. Serex, L.; Bertsch, A.; Renaud, P. Microfluidics: A new layer of control for extrusion-based 3D printing. *Micromachines* **2018**, *9*, 86. [CrossRef] [PubMed]
9. Lim, C.; Lee, Y.; Kulinsky, L. Fabrication of a Malaria-Ab ELISA Bioassay Platform with Utilization of Syringe-Based and 3D Printed Assay Automation. *Micromachines* **2018**, *9*, 502. [CrossRef] [PubMed]
10. Kim, Y.; Lee, J.; Park, S. 3D-Printed Microfluidic Platform Enabling Bacterial Preconcentration and DNA Purification for Molecular Detection of Pathogens in Blood. *Micromachines* **2018**, *9*, 472. [CrossRef] [PubMed]
11. Sharafeldin, M.; Jones, A.; Rusling, J. 3D-Printed Biosensor Arrays for Medical Diagnostics. *Micromachines* **2018**, *9*, 394. [CrossRef] [PubMed]
12. Carve, M.; Wlodkowic, D. 3D-printed chips: Compatibility of additive manufacturing photopolymeric substrata with biological applications. *Micromachines* **2018**, *9*, 91. [CrossRef] [PubMed]
13. Lepowsky, E.; Amin, R.; Tasoglu, S. Assessing the Reusability of 3D-Printed Photopolymer Microfluidic Chips for Urine Processing. *Micromachines* **2018**, *9*, 520. [CrossRef] [PubMed]
14. Lepowsky, E.; Tasoglu, S. Emerging Anti-Fouling Methods: Towards Reusability of 3D-Printed Devices for Biomedical Applications. *Micromachines* **2018**, *9*, 196. [CrossRef] [PubMed]

micromachines

MDPI

Article

3D Printing Solutions for Microfluidic Chip-To-World Connections

Sander van den Driesche [1,2,*]**, Frieder Lucklum** [1,2]**, Frank Bunge** [1,2] **and Michael J. Vellekoop** [1,2]

1   Institute for Microsensors, -actuators and -systems (IMSAS), University of Bremen, 28359 Bremen, Germany; flucklum@imsas.uni-bremen.de (F.L.); fbunge@imsas.uni-bremen.de (F.B.); mvellekoop@imsas.uni-bremen.de (M.J.V.)
2   Microsystems Center Bremen (MCB), University of Bremen, 28359 Bremen, Germany
*   Correspondence: sdriesche@uni-bremen.de; Tel.: +49-421-218-62652

Received: 2 January 2018; Accepted: 3 February 2018; Published: 6 February 2018

Abstract: The connection of microfluidic devices to the outer world by tubes and wires is an underestimated issue. We present methods based on 3D printing to realize microfluidic chip holders with reliable fluidic and electric connections. The chip holders are constructed by microstereolithography, an additive manufacturing technique with sub-millimeter resolution. The fluidic sealing between the chip and holder is achieved by placing O-rings, partly integrated into the 3D-printed structure. The electric connection of bonding pads located on microfluidic chips is realized by spring-probes fitted within the printed holder. Because there is no gluing or wire bonding necessary, it is easy to change the chip in the measurement setup. The spring probes and O-rings are aligned automatically because of their fixed position within the holder. In the case of bioanalysis applications such as cells, a limitation of 3D-printed objects is the leakage of cytotoxic residues from the printing material, cured resin. This was solved by coating the 3D-printed structures with parylene-C. The combination of silicon/glass microfluidic chips fabricated with highly-reliable clean-room technology and 3D-printed chip holders for the chip-to-world connection is a promising solution for applications where biocompatibility, optical transparency and accurate sample handling must be assured. 3D printing technology for such applications will eventually arise, enabling the fabrication of complete microfluidic devices.

Keywords: 3D printing; stereolithography; microfluidics; chip-holder; fluidic and electric connections

1. Introduction

3D printing technology has evolved remarkably the last couple of years. Instead of a plastic melting machine that creates low-resolution structures with a high surface roughness, now, liquid resin-based 3D printers with a resolution of tens of micrometers and a surface roughness of a few micrometers have become affordable [1–7]. With two-photon polymerization-based 3D printers, it is even possible to attain a resolution down to 100 nanometers [8,9]. 3D printing has become an attractive tool to fabricate measurement setup components. It enables the fabrication of sample holders for various applications such as electron paramagnetic resonance measurements [10], surface plasmon resonance spectroscopic measurements to align various illumination angles [11] and for single-plane illumination microscopy [12]. It allows the construction of cartridge components to align the optical setup containing a smart-phone, battery, cuvette and electrical circuitry for fluorescence measurements [13], a smart-phone adapter including fluidic reservoirs [14], a connector to realize the electrical connection to a digital microfluidic chip to conduct droplet size analysis by color measurements [15] and lab ware [16]. In microfluidics, 3D printing is also becoming an accepted technology. There is a series of 3D-printed microfluidic devices described in the literature, including micromixers [17–19], flow channels [20] and valves [21,22]. Besides the printing of such structures, also the integration of fluidic and electric connections can be realized [23]. A 3D-printed holder was presented that compartmentalizes a glass microscope slide [24]. In this holder, recesses were made to fit an electrochemical sensor and PEEK tubings to make fluidic connections. 3D-printed microfluidic circuits that handle multiple fluid streams were presented that consist of integrated chip-to-world interconnects [25]. A fluidic interconnect based on a 3D-printed clamping structure yielded a maximum sealing pressure up to 416 kPa [26].

Apart from the many advantages, there are several limitations in 3D printing for direct fabrication of microfluidic devices [2,4]. The removal of support material, structures that are required to temporarily reinforce the design to prevent its collapse during the printing process, of small fluidic geometries requires careful handling. Furthermore, it is hard to print microfluidic channels with a diameter of less than several hundred micrometers. The surface roughness of constructed designs, material selection and biocompatibility to biological samples are also not optimal to realize devices that from a functional point of view can compete with silicon and glass chips [2,27,28].

Microfluidic devices constructed from hard non-cytotoxic materials, like silicon and glass, have many advantages for the analysis or actuation of biological samples. The chips can be fabricated by standard clean-room processes. The channel geometry ensures a closed and thus controlled environment that can be optimized for the sample [29]. However, the connection of microfluidic devices to the outer world by tubes and wires is an underestimated issue. The diversity in microfluidic chip dimensions, the fluidic inlet and outlet amount, geometry of the channel, bonding pad sizes and number to connect the electrodes, as well as optical measurement window geometry make it infeasible to design a universal chip holder. This means that almost every designed chip requires a specialized holder. The gluing of bonded-port connectors directly on the chip is a possible solution [30]. However, these connectors have a relatively large footprint that might not fit on the chip. In addition, they incur considerable expenses, and gluing is a process with limited reliability for fluidic interconnects. A method is needed to quickly construct and easily redesign holders for different microfluidic chips. The following design criteria should be taken into account: (i) the holder should have integrated fluidic and electric connections; (ii) easy exchange of the chip; no gluing; (iii) the dimension of the holder should be as small as possible. A useful overview of fluidic connections and electrical interconnects in microfluidics is given in the Design Guideline for Microfluidic Side Connect [31] and the Guidelines for Packaging of Microfluidics: Electrical Interconnections [32]. Both documents are part of an initiative to standardize microfluidic devices and component interfaces [33,34].

Injection molding and micro-milling are existing techniques to construct customized microfluidic chip holders [35]. However, these techniques require specialized expensive equipment or molds. A separate mold is needed for every design. Furthermore, these techniques do not give many degrees of freedom for the design of geometric channel structures within the holder. This makes them impractical when multiple fluidic and electric interconnects have to be accommodated. 3D printing has several advantages compared to injection molding and micro-milling. It allows the integration of complex geometric structures such as curved and closed sub-millimeter channels and specialized fluidic and electric connector ports printed directly within the holder. Injection molding for mass production of simple chip holder geometries has certain advantages compared to standard 3D printing technology. However, when the fluidic chip-to-world connections consist of an inner channel structure with curves, 3D printing shows enormous advantages. In this work, we combine the best of two worlds: clean-room technology for the fabrication of silicon and glass microfluidic chips with submicrometer resolution and 3D printing to make chip holders with integrated electric connections and fluidic connections that do not require a very high resolution. Several designed and assembled microfluidic chip holders are presented, including fluidic and electric connections, to show 3D printing as the solution for chip-to-world connections.

2. Materials and Methods

2.1. Stereolithography

Stereolithography, a form of additive manufacturing that is based on layer-by-layer photopolymerization of liquid resins, is a technology to create high resolution complex 3D-printed microfluidic geometries. The design of 3D-printed devices can easily be created and adapted in CAD software such as Inventor (Autodesk, San Rafael, CA, USA) or SolidWorks (Dassault Systèmes SolidWorks Corp., Waltham, MA, USA). The completed design is exported in the STL (Standard Tessellation Language) format at high resolution for pre-processing with the 3D printer software. Pre-processing the designs includes checking for and fixing any errors to create a completely valid STL file, orienting the part on the build platform and adding support structures for overhangs as necessary. Finally, the 3D printer software slices the STL file and supports into individual layers. In most cases, these pre-processing steps are automated and just takes a few mouse clicks to confirm the proposed changes.

There are many different additive fabrication technologies, each with its strengths and weaknesses [1,36]. Stereolithography technology is recommended to construct parts with a feature size on the sub-millimeter scale. It features a high lateral resolution and high surface smoothness. Microstereolithography with DLP (Digital Light Processing) micromirror array projection reliably yields feature sizes (such as channels, holes and walls) down to 100 µm [37].

In this work, a high-resolution microstereolithography DLP printer (Perfactory Micro HiRes, EnvisionTEC Inc., Dearborn, MI, USA) was used to fabricate the chip holders. A high-resolution, high-temperature, low viscosity acrylic polymer resin (HTM140 M, EnvisionTEC Inc., [38]) is used as the printing material. E-Shell 300 [39], a Class IIa biocompatible resin used for hearing aid shell manufacturing, was tested for cytotoxic effects on the growth of mammalian cells. Prior to printing, the stirred resin is filtrated into the carefully-cleaned printer bath for optimal printing conditions. The process is set to a lateral (x/y) resolution of 31 µm and a slicing thickness (z) of 25 µm. The exposure time is typically set to 3000 ms per slice. The maximum build size is 40 mm × 30 mm × 100 mm.

After printing, all constructed parts are removed from the build platform and rinsed with isopropanol in an ultrasonic bath. Then, the realized part is dried with nitrogen or air, and the support structures are removed. Finally, all parts are cured in a UV flood chamber for at least five minutes. The total processing time is primarily dependent on the height of the printed structure, typically ranging from one to a few hours, which highlights the rapid prototyping characteristics of this approach.

2.2. Cell Line for Biocompatibility Test

The influence on the viability of Madin-Darby Canine Kidney cells (MDCK, ATCC CCL-34, a gift from Dr. Manfred Radmacher, Institute for Biophysics, University of Bremen, Bremen, Germany) and HaCaT cells (a gift from Dr. Ursula Mirastschijski, Centre for Biomolecular Interactions Bremen, University of Bremen, Bremen, Germany) exposed to UV-cured 3D-printed structures has been investigated. The epithelial kidney cell line MDCK and keratinocyte skin cell HaCaT are well-established models used in many biological studies [40,41].

2.3. Cell Line Preparation

The MDCK and HaCaT cells were cultivated in Glasgow Minimum Essential Medium (GMEM) and 10% fetal calf serum, supplemented with 100 units/mL penicillin and 100 µg/mL streptomycin (all obtained from Sigma-Aldrich, Darmstadt, Germany). The cell culture was kept in a humidified atmosphere containing 5% CO_2 and 95% ambient air at 37 °C.

2.4. Coating of 3D-Printed Structures with Parylene-C

Parylene-C is a USP Class VI and ISO-10993-6 certified biocompatible material [42–44]. We have coated 3D-printed structures with parylene-C with a Labcoater series 300 (Plasma Parylene Systems

GmbH, Rosenheim, Germany) to investigate the inhibition of cytotoxic effects on mammalian cells. The deposition process of parylene-C is depicted in Figure 1.

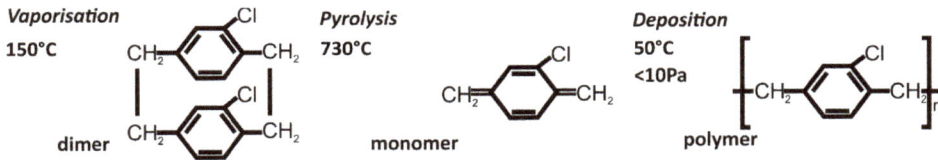

Figure 1. The deposition process of parylene-C. First, parylene-C dimer powder or granulate is vaporized at 150 °C, followed by a pyrolysis step at 730 °C to obtain parylene-C diradical monomers. At a temperature of 50 °C and a pressure of approximately 10 Pa, the monomer vapor condensates and forms a conformal coating on the substrate.

3. Results and Discussion

The cytotoxicity of printing material is discussed in Section 3.1; followed by Sections 3.2 and 3.3, where methods are described about how to realize reliable fluidic and electric chip-to-world connections, all based on 3D printing. Several chip holder concepts for on-chip cell growth and/or analysis are presented in Section 3.4.

3.1. Cytotoxicity Tests of MDCK and HaCaT Cells Exposed to 3D-Printed Structures

In the case of bioanalysis applications, a strong influence on cellular behavior of mammalian cells and bacteria exposed to 3D-printed structures realized from liquid resins was observed [29]. The resins commonly used for stereolithography 3D printing contain UV-sensitive photo-initiators, which exhibit a pronounced cytotoxic effect [45]. Biocompatible classified resins such as E-Shell 300 from Envisiontec [39] or Med610 from Stratasys [46] need a thorough post-processing, including UV-flood exposure to cross-link residual monomers. However, the 3D-printed microfluidic chip holders contain inner channel structures with regions inaccessible to UV light. When measurements are conducted with sensitive biological cells, an efficient and attractive method to prevent such cytotoxic effects is by coating the 3D-printed structures with parylene [47]. During the deposition process of parylene-C, reactive monomer diffuses into the 3D-printed channel, coating the inner structure. This assures that liquid sample, pumped from syringes into the chip, does not become contaminated with resin residues that are toxic for the biological sample.

In Figure 2, measurement results are depicted of MDCK cells grown in a multi-well plate for 24 h exposed to 3D-printed parts with and without a 10 μm-thick parylene-C coating. Two samples were prepared for each test.

The printed structures from the resins E-Shell 300 and HTM140 both show a strong cytotoxic effect on MDCK cells (Figure 2b,c). The cells did not attach to the bottom of the culture well within the 24 h of incubation. The parylene-C coating strongly inhibits the cytotoxic effect of the resin. The MDCK cells behavior in Figure 2d is similar compared to the negative control sample shown in Figure 2a (viability of respectively 84% and 85%), while the cells exposed to uncoated parts die. HaCaT cells yielded the same coated versus uncoated viability results as the MDCK cell measurements. For each HaCaT test, four samples were prepared, and the exposure time to 3D printed structures was 72 h. The control group and the parylene-C coated samples both yielded a HaCaT cell viability of >95%. The viable tests were conducted by trypan blue staining (Thermo Fischer Scientific, Darmstadt, Germany).

The measurements show that the coating of 3D-printed HTM140 structures with parylene-C is an easy and effective method preventing toxic resin components from diffusing into the biological sample.

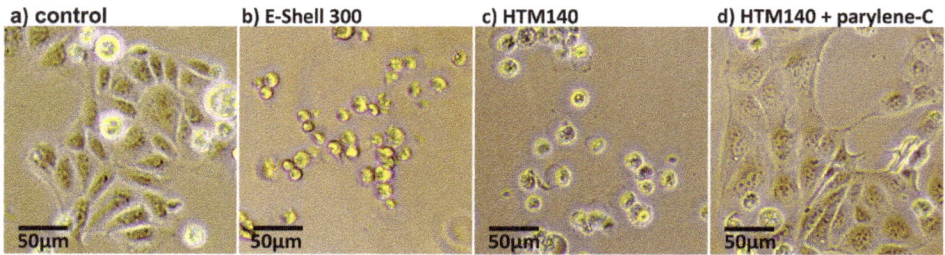

Figure 2. Cytotoxicity tests of MDCK cells exposed to 3D-printed structures built from E-Shell 300 and HTM140 resin and a 10-μm parylene-C-coated structure. The MDCK cells were grown in a multi-well plate for 24 h in Glasgow Minimum Essential Medium (GMEM) supplemented with 10% fetal bovine serum and antibiotics. (**a**) Negative control. MDCK cells attached to the bottom of a well plate show normal cell proliferation. (**b,c**) MDCK cells exposed to a 3D-printed part made from E-Shell 300 or HTM140 resin, respectively. The cells did not attach to the well-plate. (**d**) MDCK cells exposed to a 3D-printed part coated with a 10-μm parylene-C layer. The cell attached to the bottom of the well, showing similar behavior as the negative control sample.

3.2. Fluidic Chip Holder Connections

In microfluidic applications where a continuous flow of sample and/or carrier liquids is required, the chip needs to be connected to syringe pumps. In Figure 3, a method for fluidic connections is depicted where the holder acts as an interface between the chip and the syringe pumps.

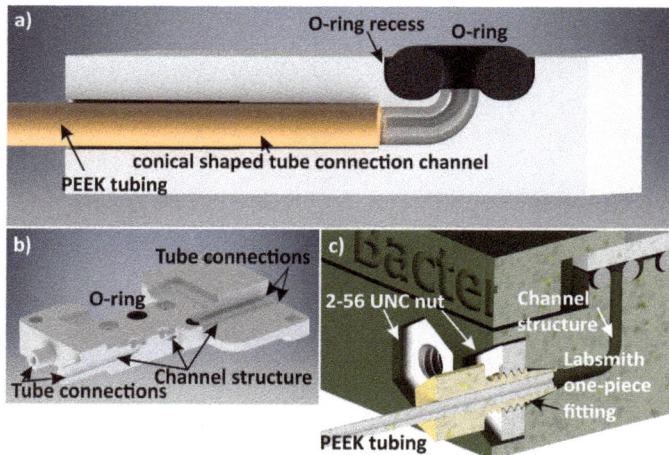

Figure 3. Methods to realize the fluidic connection between microfluidic chips and syringe pumps. (**a**) A schematic of an O-ring positioned in a recess and a tube fitted in a conically-shaped channel geometry; (**b**) a 3D rendered image visualizing the inner channel geometry; (**c**) a Labsmith one-piece fitting connection integrated in a chip holder.

The fluidic connections between the microfluidic chip and 3D-printed holder are sealed by O-rings. By creating half doughnut-shaped recesses in the 3D-printed holder, the O-ring positions are secured, which simplifies the assembly of chip and holder. These recesses are designed in such a way that 80% (height) of the O-rings are located within the holder. The width of the recesses are 110% of the O-ring diameter, allowing the rubber material to be pressed by the chip, yielding an air- and liquid-tight chip-to-holder connection (Figure 3).

The fluid connections from the holder to syringes are realized by fixing PEEK tubing with an inner and outer diameter of 250 μm and 800 μm, respectively, directly into the holder (Figure 3a,b). A channel structure with a diameter of 0.7 mm starting at the O-ring recess is directed within the holder to the tube connection, fitting the PEEK tubing. This connection is achieved by mechanically clamping the tubing in a conically-shaped geometry integrated in the channel. A fluid-tight connection is obtained by gluing the tubing with two-component epoxy resin adhesive to the chip holder. An alternative method that does not require the gluing of the tubing to the holder is depicted in Figure 3c. Here, the tubing is placed in a Labsmith one-piece fitting (T132-100 [30]), which clamps the tube when mounted in a metal 2-56 UNC nut, integrated in the holder. The other end of the tube is connected to a syringe by a Luer-lock connector.

A series of leakage tests was conducted to investigate the O-ring-based chip-to-world sealing. More than ten 3D-printed chip holders, each containing four to six fluidic connections, were each assembled up to fifty times. Fluidic tests were performed up to 100 kPa. In none of the O-rings did leakage occur. For three measurements, a pressure of 700 kPa was applied to a holder assembly by a Hamilton Gastight 1 mL syringe (Sigma-Aldrich), placed in a KD Scientific Legato 180 dual syringe pump (KD Scientific, Holliston, MA, USA). A fitting glass slide was placed to seal off the O-rings. To allow slow pressure increase, a gas plug was used in the syringe; the rest was filled with water. The applied pressure was measured by a LabSmith uPS0800-C360 pressure sensor set (LabSmith, Livermore, CA, USA) containing fluidic connection adapters and PEEK tubing. Within approximately four minutes at a flow rate of 0.1 mL/min, a pressure of 700 kPa was reached. Also in these tests, no leakage was detected. This demonstrates the reliability of the presented O-ring method.

3.3. Electric Chip Holder Connections

The electric connection between the bonding pads of the chip and 3D-printed holder can be realized by integrating spring probes within the holder. Because the probes are positioned at fixed, predefined locations, they are automatically aligned to the bonding pads during chip and holder assembly. In Figure 4, a 3D render of a chip holder with two electric connections is depicted. The recesses printed into the holder have a 100-μm wider diameter than the spring probes. They are fixed into the holder by soldering a wire at the end of the pin. The reliability of spring probe connections, a well-known solution in microelectronics, remains high, even after multiple re-assemblies of the chip and holder [32].

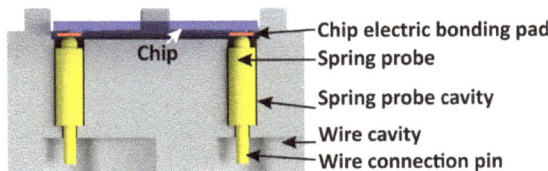

Figure 4. A method to electrically connect a microfluidic chip to the outer world by applying spring probes fitted into a 3D-printed holder.

3.4. 3D-Printed Microfluidic Chip Holders for Cell-Growth and Cell-Analysis Experiments

In this section, four dedicated 3D-printed chip holder designs for different cell growth and cell analysis experiments are presented. The holder aspects illustrated by these designs are the method of fluid supply connection (O-ring (#1), screwed (#2) or glued(#3,4)), the realization of fluid reservoirs on top of the chip (#1) and an assembly with multiple electric connections (#2,3). In addition, a holder that furthermore allows optics is presented (#4). All printed holder parts that are in contact with biological samples were coated with parylene-C. The first design, a 3D-printed chip holder for a mammalian cell growth chip, is depicted in Figure 5 [29]. The 18 mm × 13 mm × 1.4 mm chip is sandwiched between a top and a bottom holder fitted by M3-screws and nuts. The total system, chip and holder, has a dimension of 22 mm × 29 mm × 8 mm. Each top holder contains three fluid reservoirs of 20–40 μL, which are sealed on the microfluidic chip inlets by O-rings. Because of the open reservoirs, liquids can be supplied much more easily and more reliably with a pipette compared to pipetting directly into the chip. Furthermore, these reservoirs can be designed with a depth and spacing allowing a pipetting robot to handle the liquids. The optical window, the area located in the center of the chip, is easily accessible with a microscope objective, allowing optical investigation of biological samples. When measurements are conducted with mammalian cells or bacteria, it is recommended to use medical-grade O-rings. Standard O-rings are oiled and therefore might be toxic for the biological sample. We used O-rings made from E3609-70, which is a special Ethylene Propylene Diene Monomer rubber (EPDM), purchased from Parker Hannifin GmbH, Kaarst, Germany. According to the manufacturer, this material passed ISO 10993-5 and -10 testing, and it is compliant with USP Class VI and USP <87> , proving its biocompatibility and non-cytotoxicity.

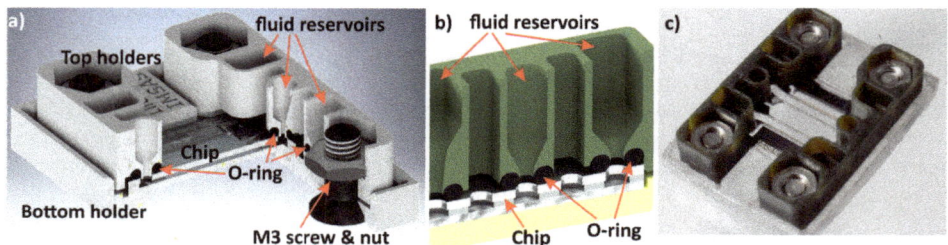

Figure 5. The assembly of a microfluidic chip and holder (22 mm × 29 mm × 8 mm). The chip is sandwiched between two top holder parts and a bottom holder. The three printed parts are fixed by M3 screws and nuts. Liquid reservoirs, integrated in the top holder, have a volume of 20–40 μL and are connected with O-rings to the chip inlets. (**a**) The schematic; (**b**) cut-out visualizing the fluid reservoirs and O-rings; and (**c**) a photo of the assembly.

The second design, a microfluidic device utilized for bacterial growth experiments, is depicted in Figure 6. The chip has integrated heating and temperature measurement elements. These require electrical connections from the chip to an external control circuitry. Four spring probes with a diameter of 1.6 mm and a length of 6.2 mm (Harwin Part Number P70-2300045R) were used to accomplish these electric connections. The fluidic chip-to-world connection consist of an O-ring placed between the chip and holder, an inner holder channel construction with a geometry fitting a 2-56 UNC nut and a conically-shaped structure allowing a Labsmith one-piece fitting tube connection.

The third design, depicted in Figure 7, is a microfluidic device used to position and analyze biological samples at predefined regions in the chip [48]. The chip holder has an optical window accessible with an inverted microscope, three fluidic and nine electric connections. The spring probes have a diameter of 0.68 mm, a length of 16.55 mm, and a travel distance of 2.65 mm (Multicomp Part Number P50-B-120-G).

In the fourth design, a measurement assembly fitting the microfluidic chip and external optical components are 3D-printed (Figure 8). This measurement setup is used for monitoring the oxygen consumption of mammalian cell cultures [49]. The setup is constructed from a 3D-printed chip holder (Figure 8b) and a 3D-printed optical alignment structure, fitting the chip holder and additional components, such as LEDs, a Raspberry Pi camera module and an optical filter at any desired position. The optical alignment structure was printed by a Form 1 3D printer (Formlabs, Somerville, MA, USA).

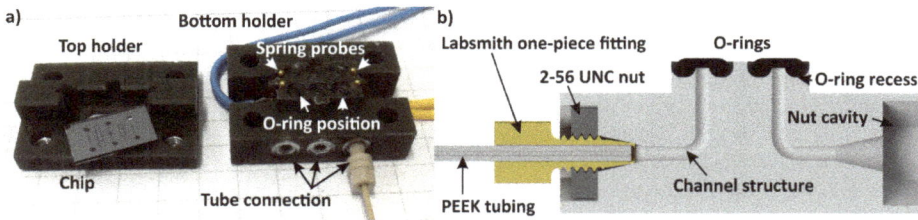

Figure 6. (**a**) A photo of a microfluidic chip holder (19 mm × 26.5 mm × 10 mm) with six fluidic and four electric connections. The fluidic outer connection is realized by placing 2-56 UNC metal nuts at the end of a channel structure. The chip (7 mm × 12 mm × 1 mm) has three measurement chambers that are deep reactive-ion etched in a silicon wafer. The chambers are closed by an anodically-bonded borosilicate glass wafer. (**b**) A 2D cut to visualize the fluidic inner channel structure and Labsmith one-piece fitting connection.

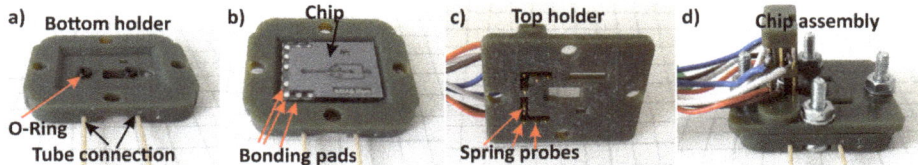

Figure 7. A microfluidic chip holder (21 mm × 30 mm × 8(21) mm) including three fluidic and nine electric connections. (**a**) The bottom holder part contains doughnut-shaped recesses where O-rings connecting the chip to the channel structure within the printed holder fit. (**b**) A glass-silicon-glass chip (13 mm × 17 mm × 1.4 mm) fitted in the bottom holder. (**c**) The top holder including spring probes to connect the bonding pads of the chip. (**d**) A photo of the assembled chip and holder.

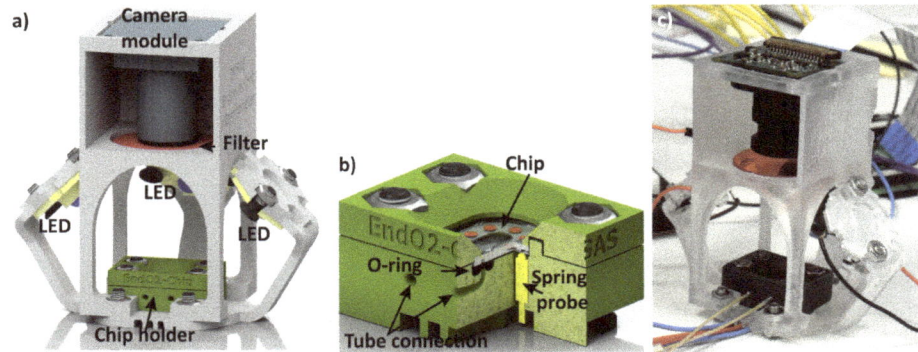

Figure 8. The measurement setup to investigate biological samples. The LEDs, optical filter and Raspberry Pi camera are utilized to conduct fluorescence experiments of the samples, located in the chip. (**a**) A 3D rendered image of the measurement setup; (**b**) the microfluidic chip holder (17 mm × 24 mm × 11.5 mm) containing fluidic and electric connections; (**c**) photo of the measurement setup (52 mm × 68 mm × 69 mm).

A 3D-printed part with a thickness of approximately 2.5 mm is strong enough to support a microfluidic chip. The holder design allows the easy exchange of the chip. None of the connections between the chip and holder are glued. Because of the 3D printing, the assembly including all fluidic, electric and optical components is much more compact compared to standard laboratory systems. Consequently, the devices can be easily transported out of the lab and be used even at remote locations.

The 3D printing of microfluidic chip holders brings high flexibility to research laboratories. When multiple parts are required, the fabrication can easily be up-scaled for small series production. One of the current drawbacks of 3D printing for mass production is the speed to fabricate devices. However, 3D printing technology is evolving quite fast, and the printing speed is a parameter that is increasing rapidly. Two interesting 3D printing technologies have recently been presented. TNO (The Hague, The Netherlands) has developed a microstereolithography device (Lepus) that constructs up to 1200 layers per hour (30 mm/h at a resolution of 25 μm per layer). The light synthesis technology (CLIP) from Carbon (USA) yields printing speeds of 100–1000 mm/h [50,51]. This reduces the printing time by a factor of a hundred. In the hearing aid industry [52], dentistry [53] and footwear industry [54], 3D printing has already evolved from a rapid prototyping technology into a manufacturing tool for mass production.

Other advantages of 3D printing that could revolutionize the manufacturing industry are on-demand direct fabrication of production parts stored in a digital library, customized product designs without additional cost, reduction of assembly work because structures of higher complexity can be constructed from a single printed part and the reduction of logistic hassle because the storage of production parts can be reduced [55].

4. Conclusions

The given examples show that 3D printing is a versatile tool to realize microfluidic chip holders. The flexibility of 3D printing allows the quick redesign of chip holders and adaptation for other chip geometries. Reliable chip and holder assembling is achieved by applying O-rings and spring probes to attach the fluid and electric connections. This solves the challenge of connecting fluidic and electrical parts of microfluidics to the outer world. When the 3D-printed structures are used in combination with biological samples, cytotoxic effects due to resin components can be prevented by coating 3D-printed structures with parylene-C.

The combination of silicon/glass microfluidic chips fabricated with highly-reliable clean-room technology and 3D-printed chip holders for the chip-to-world connection is a promising solution for applications where biocompatibility, optical transparency and accurate sample handling must be assured.

Acknowledgments: The authors would like to thank André Bödecker and Lukas Immoor for depositing parylene-C on the 3D-printed structures.

Author Contributions: The work presented in this paper was a collaboration of all authors. S.v.d.D., F.L. and F.B. designed and fabricated the devices. The biocompatibility tests, conducted by F.B., were designed by S.v.d.D., F.B. and M.J.V. The results were interpreted by S.v.d.D., F.L., F.B. and M.J.V. The manuscript was written by S.v.d.D. and discussed by all authors.

Conflicts of Interest: The authors declare no conflict of interest.

Abbreviations

The following abbreviations are used in this manuscript:

CLIP	Continuous Liquid Interface Production
DLP	Digital Light Processing
EPDM	Ethylene Propylene Diene Monomer
GMEM	Glasgow Minimum Essential Medium
MDCK	Madin-Darby Canine Kidney
PEEK	Polyether Ether Ketone
STL	Standard Tessellation Language

References

1. Vaezi, M.; Seitz, H.; Yang, S. A review on 3D micro-additive manufacturing technologies. *Int. J. Adv. Manuf. Technol.* **2013**, *67*, 1721–1754.
2. Waheed, S.; Cabot, J.M.; Macdonald, N.P.; Lewis, T.; Guijt, R.M.; Paull, B.; Breadmore, M.C. 3D printed microfluidic devices: Enablers and barriers. *Lab Chip* **2016**, *16*, 1993–2013.

3. He, Y.; Wu, Y.; Fu, J.Z.; Gao, Q.; Qiu, J.J. Developments of 3D printing microfluidics and applications in chemistry and biology: A review. *Electroanalysis* **2016**, *28*, 1658–1678.

4. Chen, C.; Mehl, B.T.; Munshi, A.S.; Townsend, A.D.; Spence, D.M.; Martin, R.S. 3D-printed microfluidic devices: Fabrication, advantages and limitations—A mini review. *Anal. Methods* **2016**, *8*, 6005–6012.

5. Mao, M.; He, J.; Li, X.; Zhang, B.; Lei, Q.; Liu, Y.; Li, D. The emerging frontiers and applications of high-resolution 3D printing. *Micromachines* **2017**, *8*, 113.

6. Xu, Y.; Wu, X.; Guo, X.; Kong, B.; Zhang, M.; Qian, X.; Mi, S.; Sun, W. The boom in 3D-printed sensor technology. *Sensors* **2017**, *17*, 1166.

7. Krujatz, F.; Lode, A.; Seidel, J.; Bley, T.; Gelinsky, M.; Steingroewer, J. Additive Biotech—Chances, challenges, and recent applications of additive manufacturing technologies in biotechnology. *New Biotechnol.* **2017**, *39*, 222–231.

8. Cumpston, B.H.; Ananthavel, S.P.; Barlow, S.; Dyer, D.L.; Ehrlich, J.E.; Erskine, L.L.; Heikal, A.A.; Kuebler, S.M.; Lee, I.Y.; McCord-Maughon, D.; et al. Two-photon polymerization initiators for three-dimensional optical data storage and microfabrication. *Nature* **1999**, *398*, 51–54.

9. Xing, J.F.; Zheng, M.L.; Duan, X.M. Two-photon polymerization microfabrication of hydrogels: An advanced 3D printing technology for tissue engineering and drug delivery. *Chem. Soc. Rev.* **2015**, *44*, 5031–5039.

10. Niemöller, A.; Jakes, P.; Kayser, S.; Lin, Y.; Lehnert, W.; Granwehr, J. 3D printed sample holder for in-operando EPR spectroscopy on high temperature polymer electrolyte fuel cells. *J. Magn. Reson.* **2016**, *269*, 157–161.

11. Hill, R.T.; Kozek, K.M.; Hucknall, A.; Smith, D.R.; Chilkoti, A. Nanoparticle–film plasmon ruler interrogated with transmission visible spectroscopy. *ACS Photonics* **2014**, *1*, 974–984.

12. Desmaison, A.; Lorenzo, C.; Rouquette, J.; Ducommun, B.; Lobjois, V. A versatile sample holder for single plane illumination microscopy. *J. Microsc.* **2013**, *251*, 128–132.

13. Friedrichs, A.; Busch, J.; van der Woerd, H.; Zielinski, O. SmartFluo: A method and affordable adapter to measure chlorophyll a fluorescence with smartphones. *Sensors* **2017**, *17*, 678.

14. Cevenini, L.; Calabretta, M.M.; Tarantino, G.; Michelini, E.; Roda, A. Smartphone-interfaced 3D printed toxicity biosensor integrating bioluminescent "sentinel cells". *Sens. Actuators B Chem.* **2016**, *225*, 249–257.

15. Yafia, M.; Ahmadi, A.; Hoorfar, M.; Najjaran, H. Ultra-portable smartphone controlled integrated digital microfluidic system in a 3D-printed modular assembly. *Micromachines* **2015**, *6*, 1289–1305.

16. Lücking, T.H.; Sambale, F.; Beutel, S.; Scheper, T. 3D-printed individual labware in biosciences by rapid prototyping: A proof of principle. *Eng. Life Sci.* **2015**, *15*, 51–56.

17. Shemelya, C.; Cedillos, F.; Aguilera, E.; Espalin, D.; Muse, D.; Wicker, R.; MacDonald, E. Encapsulated copper wire and copper mesh capacitive sensing for 3-D printing applications. *IEEE Sens. J.* **2015**, *15*, 1280–1286.

18. Plevniak, K.; Campbell, M.; Myers, T.; Hodges, A.; He, M. 3D printed auto-mixing chip enables rapid smartphone diagnosis of anemia. *Biomicrofluidics* **2016**, *10*, 054113.

19. Bhargava, K.; Ermagan, R.; Thompson, B.; Friedman, A.; Malmstadt, N. Modular, discrete micromixer elements fabricated by 3D printing. *Micromachines* **2017**, *8*, 137.

20. Gong, H.; Bickham, B.P.; Woolley, A.T.; Nordin, G.P. Custom 3D printer and resin for 18 μm × 20 μm microfluidic flow channels. *Lab Chip* **2017**, *17*, 2899–2909.

21. Au, A.K.; Bhattacharjee, N.; Horowitz, L.F.; Chang, T.C.; Folch, A. 3D-printed microfluidic automation. *Lab Chip* **2015**, *15*, 1934–41.

22. Gong, H.; Woolley, A.T.; Nordin, G.P. High density 3D printed microfluidic valves, pumps, and multiplexers. *Lab Chip* **2016**, *16*, 2450–8.

23. Van den Driesche, S.; Bunge, F.; Lucklum, F.; Vellekoop, M.J. 3D-Printing: An attractive tool to realize microfluidic chip holders. In Proceedings of the 3rd International Conference on MicroFluidic Handling Systems, Enschede, The Netherlands, 4–6 October 2017; pp. 94–97.

24. Oomen, P.; Mulder, J.; Verpoorte, E.; Oleschuk, R. Controlled, synchronized actuation of microdroplets by gravity in a superhydrophobic, 3D-printed device. *Anal. Chim. Acta* **2017**, *988*, 50–57.

25. Bhargava, K.C.; Thompson, B.; Malmstadt, N. Discrete elements for 3D microfluidics. *Proc. Natl. Acad. Sci. USA* **2014**, *111*, 15013–15018.

26. Paydar, O.; Paredes, C.; Hwang, Y.; Paz, J.; Shah, N.; Candler, R. Characterization of 3D-printed microfluidic chip interconnects with integrated O-rings. *Sens. Actuators A Phys.* **2014**, *205*, 199–203.

27. Alharbi, N.; Wismeijer, D.; Osman, R. Additive manufacturing techniques in prosthodontics: Where do we currently stand? A critical review. *Int. J. Prosthodont.* **2017**, *30*, 474–484.

28. Osman, R.B.; van der Veen, A.J.; Huiberts, D.; Wismeijer, D.; Alharbi, N. 3D-printing zirconia implants; a dream or a reality? An in-vitro study evaluating the dimensional accuracy, surface topography and mechanical properties of printed zirconia implant and discs. *J. Mech. Behav. Biomed. Mater.* **2017**, *75*, 521–528.

29. Bunge, F.; van den Driesche, S.; Vellekoop, M. Microfluidic platform for the long-term on-chip cultivation of mammalian cells for lab-on-a-chip applications. *Sensors* **2017**, *17*, 1603.

30. Labsmith One-Piece Fitting. Available online: http://products.labsmith.com/one-piece-fitting (accessed on 25 October 2017).

31. Buesink, W.; Heeren, H. Design Guideline for Microfluidic Side Connect. Available online: http://www.enablingmnt.co.uk/wp-content/uploads/2017/06/Design-Guidelines-for-Microfluidic-Side-Connect-version-1.0.pdf (accessed on 5 Februar 2018).

32. Bullema, J.E.; Heeren, H.; Buesink, W. Guidelines for Packaging of Microfluidics: Electrical Interconnections. Available online: http://mf-manufacturing.eu/wp-content/uploads/Guidelines-for-Electrical-Interconnections-to-microfluidic-devices-version-1.0.pdf (accessed on 5 Februar 2018).

33. Heeren, H.; Atkins, T.; Verplanck, N.; Peponnet, C.; Hewkin, P.; Blom, M.; Buesink, W.; Bullema, J.E.; Dekker, S. Design Guideline for Microfluidic Device and Component Interfaces (Part 1). Available online: http://www.enablingmnt.co.uk/wp-content/uploads/2016/05/Design-for-Microfluidic-Interfacing-White-Paper-part-1-version-2.pdf (accessed on 5 Februar 2018).

34. Heeren, H.; Atkins, T.; Verplanck, N.; Peponnet, C.; Hewkin, P.; Blom, M.; Buesink, W.; Bullema, J.E.; Dekker, S. Design Guideline for Microfluidic Device and Component Interfaces (Part 2). Available online: http://www.enablingmnt.co.uk/wp-content/uploads/2017/06/Design-for-Microfluidic-Interfacing-White-Paper-part-2-version-3.0.pdf (accessed on 5 Februar 2018).

35. Guckenberger, D.J.; de Groot, T.E.; Wan, A.M.D.; Beebe, D.J.; Young, E.W.K. Micromilling: A method for ultra-rapid prototyping of plastic microfluidic devices. *Lab Chip* **2015**, *15*, 2364–2378.

36. Bhattacharjee, N.; Urrios, A.; Kang, S.; Folch, A. The upcoming 3D-printing revolution in microfluidics. *Lab Chip* **2016**, *16*, 1720–1742.

37. Sun, C.; Fang, N.; Wu, D.; Zhang, X. Projection micro-stereolithography using digital micro-mirror dynamic mask. *Sens. Actuators A Phys.* **2005**, *121*, 113–120.

38. EnvisionTec HTM140 Resin. Available online: https://envisiontec.com/wp-content/uploads/2016/09/2018-HTM-140-V2-1.pdf (accessed on 25 October 2017).

39. EnvisionTec E-Shell 300 Resin. Available online: https://envisiontec.com/wp-content/uploads/2016/09/2018-E-Shell-300-Series.pdf (accessed on 26 January 2018).

40. Kip, S.N.; Strehler, E.E. Characterization of PMCA isoforms and their contribution to transcellular Ca^{2+} flux in MDCK cells. *Am. J. Physiol. Ren. Physiol.* **2003**, *284*, F122–F132.

41. Rehders, M.; Grosshäuser, B.B.; Smarandache, A.; Sadhukhan, A.; Mirastschijski, U.; Kempf, J.; Dünne, M.; Slenzka, K.; Brix, K. Effects of lunar and mars dust simulants on HaCaT keratinocytes and CHO-K1 fibroblasts. *Adv. Space Res.* **2011**, *47*, 1200–1213.

42. Parylene Coating. Available online: https://www.plasmaparylene.de/en/parylene-coating/parylene-rosenheim.php (accessed on 5 February 2018).

43. ISO 10993-6:2016, Biological Evaluation of Medical Devices—Part 6: Tests for Local Effects after Implantation. Available online: https://www.iso.org/standard/61089.html (accessed on 5 February 2018).

44. Chang, T.Y.; Yadav, V.G.; De Leo, S.; Mohedas, A.; Rajalingam, B.; Chen, C.L.; Selvarasah, S.; Dokmeci, M.R.; Khademhosseini, A. Cell and protein compatibility of parylene-C surfaces. *Langmuir* **2007**, *23*, 11718–11725.

45. Schmelzer, E.; Over, P.; Gridelli, B.; Gerlach, J.C. Response of primary human bone marrow mesenchymal stromal cells and dermal keratinocytes to thermal printer materials in vitro. *J. Med. Biol. Eng.* **2016**, *36*, 153–167.

46. Stratasys MED610 Resin. Available online: http://www.stratasys.com/de/materials/search/biocompatible (accessed on 29 January 2018).

47. Akhtar, M.; van den Driesche, S.; Bödecker, A.; Vellekoop, M.J. Long-term storage of droplets on a chip by Parylene AF4 coating of channels. *Sens. Actuators B Chem.* **2018**, *255*, 3576–3584.

48. Van den Driesche, S.; Bunge, F.; Tepner, S.; Kotitschke, M.; Vellekoop, M.J. Travelling-wave dielectrophoresis allowing flexible microchannel design for suspended cell handling. *Proc. SPIE* **2017**, *10247*, 102470H.

49. Bunge, F.; van den Driesche, S.; Waite, A.; Mirastschijski, U.; Vellekoop, M.J. μRespirometer to determine the oxygen consumption rate of mammalian cells in a microfluidic cell culture. In Proceedings of the 2017 IEEE 30th International Conference onMicro Electro Mechanical Systems (MEMS), Las Vegas, NV, USA, 22–26 January 2017; pp. 414–417.

50. TNO–Improved Stereolithography. Available online: https://www.tno.nl/en/focus-areas/industry/ roadmaps/flexible-free-form-products/additive-manufacturing/improved-micro-stereolithography/ (accessed on 5 February 2018).

51. Tumbleston, J.R.; Shirvanyants, D.; Ermoshkin, N.; JanuszIewicz, R.; Johnson, A.R.; Kelly, D.; Chen, K.; Pinschmidt, R.; Rolland, J.P.; Ermoshkin, A.; et al. Continuous liquid interface production of 3D objects. *Science* **2015**, *347*, 1349–1352.

52. Sandström, C.G. The non-disruptive emergence of an ecosystem for 3D Printing—Insights from the hearing aid industry's transition 1989–2008. *Technol. Forecast. Soc. Chang.* **2016**, *102*, 160–168.

53. Dawood, A.; Marti, B.M.; Sauret-Jackson, V.; Darwood, A. 3D printing in dentistry. *Br. Dent. J.* **2015**, *219*, 521–529.

54. Jones, D.; Larson, R. Footwear Assembly Method with 3D Printing. U.S. Patent 9,005,710, 14 April 2015.

55. Weller, C.; Kleer, R.; Piller, F.T. Economic implications of 3D printing: Market structure models in light of additive manufacturing revisited. *Int. J. Prod. Econ.* **2015**, *164*, 43–56.

micromachines

MDPI

Article

Microfluidics: A New Layer of Control for Extrusion-Based 3D Printing

Ludovic Serex *, Arnaud Bertsch and Philippe Renaud

EPFL STI IMT LMIS4, Station 17, CH-1015 Lausanne, Switzerland; arnaud.bertsch@epfl.ch (A.B.);
philippe.renaud@epfl.ch (P.R.)
* Correspondence: ludovic.serex@epfl.ch; Tel.: +41-21-69-36728

Received: 1 February 2018; Accepted: 15 February 2018; Published: 16 February 2018

Abstract: Advances in 3D printing have enabled the use of this technology in a growing number of fields, and have started to spark the interest of biologists. Having the particularity of being cell friendly and allowing multimaterial deposition, extrusion-based 3D printing has been shown to be the method of choice for bioprinting. However as biologically relevant constructs often need to be of high resolution and high complexity, new methods are needed, to provide an improved level of control on the deposited biomaterials. In this paper, we demonstrate how microfluidics can be used to add functions to extrusion 3D printers, which widens their field of application. Micromixers can be added to print heads to perform the last-second mixing of multiple components just before resin dispensing, which can be used for the deposition of new polymeric or composite materials, as well as for bioprinting new materials with tailored properties. The integration of micro-concentrators in the print heads allows a significant increase in cell concentration in bioprinting. The addition of rapid microfluidic switching as well as resolution increase through flow focusing are also demonstrated. Those elementary implementations of microfluidic functions for 3D printing pave the way for more complex applications enabling new prospects in 3D printing.

Keywords: micro-fluidic; additive manufacturing; 3D printing; bio-printing; lab on a tip

1. Introduction

Three-dimensional (3D) printing, also commonly referred to as additive manufacturing or rapid prototyping, is a set of techniques that consist in building 3D parts layer by layer. This fabrication principle dates from the early 1980s [1] and has seen a number of different implementations based on the use of multiple deposition techniques [2]. While photopolymers and thermoplastic polymers were initially used in 3D printing techniques, the choice of materials that can be used has been significantly widened, and includes metals [3,4], ceramics [5] and biomaterials [6–11]. Current research in this field includes the development of "smart materials" [12,13] that can evolve with time and bring additional functions to the fabricated objects. If 3D printing was first used for automotive and aerospace applications [14,15], many other application fields currently use these techniques, including medical application [16–19], tissue enginering [20–24], biosensors [25], microfabrication [26,27] or even construction [28] and the food industry [29]. With hobbyists now having access to 3D printing, it is likely that its field of applications will expand even more.

Among these 3D printing methods, stereolithography (SLA) and extrusion based system dominate the market. SLA is known for its very high resolution [30], but is subject to limitations directly related to the process itself, such as the limited biocompatibility of the materials that can be used (often linked to the use of photoinitiators) and the very challenging implementation of multi-material printing machines. On the other hand, extrusion-based processes are increasingly popular [31] as they are relatively cheap and easy to use. However, this printing technique suffers from its limited number of printable materials: only low melting temperature materials such as ABS for fused deposition modeling or fast crosslinking materials for bio-printing can be used. Additionally, in recent years, the need for smarter dispensing tools has emerged, in particular in the field of bioprinting to answer the need to print complex materials for cells [32–34]. As the field increasingly aims toward regenerative medicine [9,35–37]. A few implementations of such smart dispensing tools have been already presented in the literature, such as print heads made from needles used for manufacturing perfusable vascular constructs [38]. To further overcome this limitation, extrusion-based 3D printing can benefit from more complex microfluidic systems, which can implement a number of fluidic manipulation functions at the micro-scale. Microfluidics has seen major developments in recent years, and has contributed to the emergence of the concept of "Lab on a chip" by allowing the implementation of many fluidic functions such as micro-mixers [39–41], switching valve [42–44], flow focusing [45], particles focusing [46–48], in-channel detection [49] or particles and cell sorting [49–52] in compact, microfabricated devices.

Up until now, these functionalities have mainly been used on chip to perform various analysis. In this paper, we propose to exploit the potential of microfluidics to develop a "lab on a tip" that could perform various operations on the dispensing solution directly in the print head of the 3D printer. Using this principle, we demonstrate multiple smart printing heads that allow the use of new materials, enhance the print resolution, or allow the printing of composite parts or multi-material parts that were only possible using expensive 3D printing techniques.

2. Materials and Methods

2.1. Probe Fabrication

All the print head dispensing tips that are presented in this paper were manufacture in a similar way using specific microchannel designs depending on the targeted application. A 500 nm SiO_2 layer is first grown by wet oxidation on a single-side polished silicon wafer. Standard photolithography is performed to pattern the fluidic channels and the silicon oxide is etched to create a hard mask. In the case of the herringbone mixer presented in Section 3.3, which requires a two-level microchannel to be manufactured, an additional photoresist mask, silicon etching and resist striping is performed as described in the previous literature [39]. Then, the channels are etched to 300 μm in depth using deep reactive ion etching, and the top-layer oxide is etched away as well to expose the silicon for anodic bonding. Next, a borofloat® 33 glass wafer is mechanically drilled to create the probes inlets. The two wafers are cleaned with a piranha solution and bonded by anodic bonding. The wafer stack is then diced, revealing the dispensing outlet in a similar fashion as described in the previous literature [53].

2.2. 3D Printer Setup

The printer setup is similar to the one presented in previous work [53]. A DLT-180, Double Delta 3D printer has been purchased from He3D and adapted to fit the fabricated probes. The printing path and speed was either hard-coded in G-code or automatically generated using a slicer as is usually done for standard 3D printing.

3. Results

In this section we present four different probes that can perform specific tasks. Those probes demonstrate how microfluidics can be implemented to improve extrusion-based 3D printing in terms of printed materials, resolution or material switching.

3.1. Multi-Material

One limitation of extrusion-based 3D printers is that each print head can only print a single material. Printing an object made of multiple materials is usually achieved by using a 3D printing machine with multiple extrusion heads placed next to each other. By switching between print heads, different materials can then be printed [54]. In addition to requiring specific equipment, printing multi-material components with extrusion-based 3D printers is generally a very slow process, and a smooth transition between printed materials is not always guaranteed. A number of microfluidic systems allowing switching between different liquids have already been presented in the literature, including some that could be implemented in 3D printers [44]. Here, we will show a very simple system that is easy to implement and allows a fast switching between materials.

The print head we designed is composed of three micro-channels merging into one right before the ejection point, as presented in Figure 1A. Each of these three channels can be connected to a syringe containing a different material and actuated by a syringe pump. By choosing the sequence of actuation of the syringe pumps and synchronizing it with the geometry of the part being built, a seamless transition between multiple materials can be achieved during the object manufacturing process. Furthermore, as microfluidic benefits from very small dead volumes, it is possible to rapidly switch between materials. In our case, a complete transition between two materials can be performed in 500 ms, as presented in Figure 1B, where the transition between different colored liquids was recorded using a camera connected to a microscope focused at the tip of the print head we used.

The print head was then mounted on a 3D printer, and colored alginate solutions were printed on an agarose 4% and $CaCl_2$ 1% bed, inducing the alginate gelation by diffusion of the calcium ions into the alginate deposit. Figure 1C shows the smooth transition between colors obtained as the printing process goes.

By combining extrusion 3D printing with ultraviolet (UV) light irradiation of the printed layers, it is possible to print and crosslink photosensitive inks using the same print head based on merging microchannels. Thus, multimaterial components can be made by the extrusion method by selecting photosensitive resins having the desired properties, chosen from the large catalogue of materials developed for the stereolithography process. As an example, we used Formlabs resins RS-F2-GPCL-04 (which results in a transparent and rigid material once cured), RS-F2-GPWH-04 (which is a white resin) and RS-F2-GPBK-04 (which is a black resin). By switching between these resins, we were able to print multicolor parts, as shown in Figure 1D. Formlabs also developed a flexible resin (RS-F2-GPGR-02). By alternating rigid and flexible areas, we can create parts with varying Young's moduli. In this example, a two-hinge part was printed (Figure 1E).

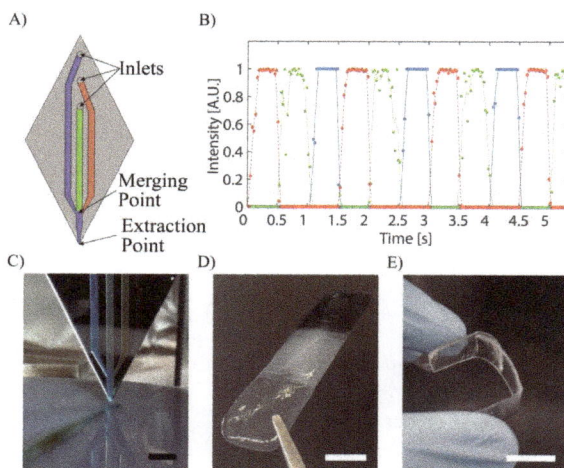

Figure 1. Multi-channel print head. (**A**) Schematic of the microfluidic channel design. (**B,C**) Switching of three colored liquids as recorded at the tip of the print head. (**D**) Clear, white and black part. (**E**) Two-hinges part printed with two different inks (one rigid and one flexible). All scale bars correspond to 1 cm.

Switching between two different materials can be performed in a cleaner and faster way that the one used in the very simple multi-channel print head presented in Figure 1, for example by using actuated virtual valves similar to the ones presented by Braschler et al. [42]; however, this method induces the waste of larger amounts of printing material, which can be a problem.

In the examples we presented, the materials were deposited alternately one after the other, using microfluidic switching, but multiplexing can also be implemented directly in the print head, allowing the alternation of materials both laterally and during printing. Microfluidic devices capable of performing such types of multiplexing have been demonstrated in Lab on Chips, using polydimethylsiloxane (PDMS) as structural material, as it allows a simple fabrication and actuation of valves, but its use in print heads for 3D printing may be limited, as the flexibility of PDMS would limit the extrusion pressure [55].

3.2. Enhanced Resolution

Microfluidic flows present a laminar behavior [45]. This property can be exploited to change the resolution of the printed material by focusing it while it is dispensed using sheath flows [25]. Hydrodynamic focusing only requires a very rudimentary microfluidic setup, the most common configuration being a 3-channel device, where the center flow stream is pinched between two side flow streams, resulting in a shrinking of the width of the center flow. Implementing hydrodynamic focusing in the print head of an extrusion-based 3D printer, its resolution can be not only be significantly increased, but it can be adjusted while printing by simply varying the ratio between the core and lateral flows. Figure 2A demonstrates the principle of sheath flow. Alginate filaments (in blue) are focused using de-ionized (DI) water with 4% $CaCl_2$. Depending on the ratio of the alginate core flow to the $CaCl_2$ solution sheath flow (Figure 2B), the obtained filaments can vary in diameter from 800 μm to below 200 μm, providing a fivefold resolution increase. Figure 2C shows the evolution of the size of the deposited filaments as a function of the ratio between sheath and core flows.

One issue of the use of sheath flows for increasing the printing resolution is the fact that the liquid used in the sheath flow for focusing is also "printed" alongside the actual material of interest. In our example, using a water-based solution, large quantities of water are ejected from the probe and may disrupt the printing. On Figure 2D, a printed filament with changing diameter is presented. The first section of the filament is small and soaked in the sheath flow that will ultimately dry. The filament then widens and the surplus of liquid produced by the sheath flow becomes almost nonexistent as it is reduced to a minimum. To solve the problem of residual fluid, Perfluoro(methyldecalin) (PFD) was used as sheath flow. Given that it is a highly volatile material, it will quickly evaporate, leaving only the desired focused filament. One limitation of this technique is that when the sheath flow becomes much larger than the core flow, flow instabilities occur, limiting the use of hydrodynamic focusing for enhancing the resolution of the dispensed material. In this case, there is always the possibility of reducing the dimensions of the microfabricated channels provided on the print head for improving the resolution. However, this implies the use of larger values of pressure to drive the various liquids in the microchannels. Furthermore, it is important to note that as liquids are being printed, the tuning of the printed filament through flux control of the core flow only without flow focusing can be achieved. With a constant printing speed, an increase of the flux will result in a larger printed filament (Figure 2E). This presents the advantage of not having residual liquid from the sheath flow, but the crosslinking must be performed from the probe either by UV light or, in this case, by printing alginate on a $CaCl_2$ bed, which is not suitable for high structures.

Adjusting the printer resolution on the fly opens new ways of controlling the printed part. It would thus be possible to choose the resolution depending on the importance of the part being printed while printing. This methods allows the printing of coarse filament for the filling of the printed part, while at the same time, for the wall, the filament can be refined to obtain fast and precise printing.

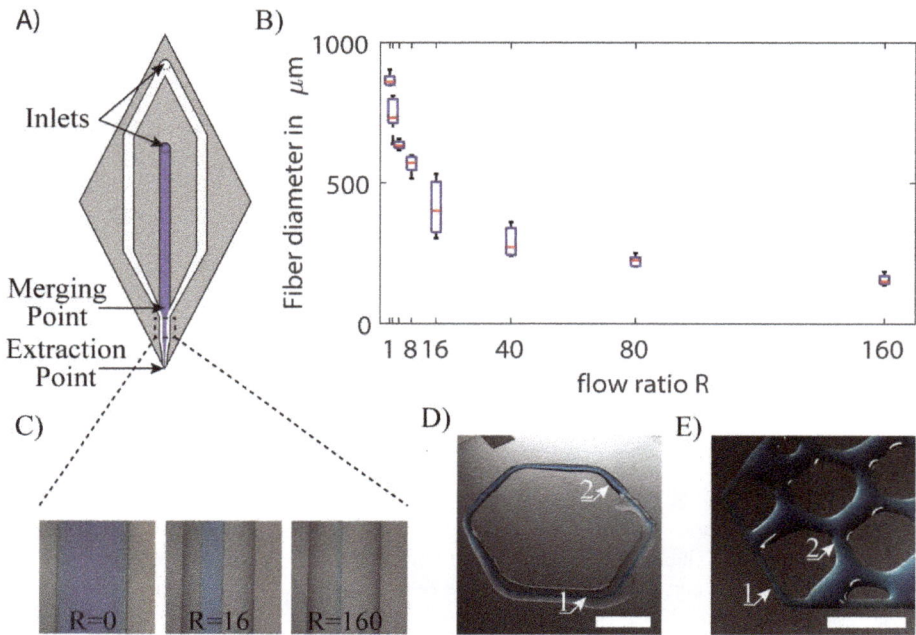

Figure 2. Print head based on flow focusing. (**A**) Principle of hydrodynamic flow focusing with the core flow in blue and the sheath flow in white. (**B**) Flow focusing at different sheath flow/core flow ratios (R). (**C**) Printed fiber diameter as a function of the flow ratio. (**D**) Printed filament going from small filament (1) to large filament (2) whilst the ratio R is changed. (**E**) First layer of a printed part with fine edge (1) and coarse filling (2). Scale bars = 5 mm.

3.3. New Materials

Laminar flows have the particularity that they do not mix easily, as turbulence does not occur in this flow regime. Mixing relies solely on the diffusion phenomenon, which can be enhanced by splitting and rearranging the flow using a mixer. Many micro-mixers [41] have been devised for microfluidic applications; of these static micro-mixers have the advantage of being passive structures that provide efficient mixing in laminar conditions and that can be integrated directly in the print head nozzles used for 3D printing.

In a previous work, we demonstrated the use of a meander-based static micro-mixer for performing the mixing of slow reacting materials used to 3D print carboxymethylcellulose-based cryogels [53]. A micro-fabricated dispensing print head allowing both the last-second mixing of cryogel precursors and the temperature control of the extruded material during printing was used. This allowed the 3D printing of multi-layer cryogels with on-demand local pore size change through the control in temperature of the dispensed solution. The seeding and culture of cells in specific areas of the obtained 3D cryogel structure was also demonstrated after its functionalization.

There are, however, static mixers much more efficient than meander-based micro-mixers. Herrigbone mixers [39] still remain to this day among the most efficient static micro-mixers, but their manufacturing using microfabrication technologies is slightly more complicated than that of meander-based mixers, as they are two-level structures and require a larger number of fabrication steps. By implementing herringbone micro-mixers at the tip of a dispensing print head, even viscous materials can be mixed in structures of only a few millimeters in length. This principle is demonstrated in Figure 3A, where colored glycerol streams (in blue and white) are pushed through two three-inlet devices, one being a simple straight microchannel, the other accommodating a herringbone micro-mixer. Mixing occurs within the first 5 mm of the herringbone structure, whereas there is no apparent mixing in the straight channel. Because different materials can be efficiently mixed, gradients can be created by variating their ratio. As an example, black and white Formlabs resins were printed going from fully black to fully white in a smooth gradient (Figure 3B). Therefore by adding the 3 primary colors to the mix, any color can be printed in a very simple manner.

Moreover, as mixing is performed, fast-reacting material can be brought in contact, just before they are deposited by 3D printing. As a demonstration, we performed the mixing of a two-component polymerizable resin based on polyethylene glycol 400 diacrylate, one component containing 3% N,N,N′,N′ Tetramethyl-ethylenediamine, the other containing 3% benzoyl peroxide. The benzoyl peroxide/amine system is known as an initiator for free-radical polymerization of acrylates [56] and the reaction is fast when all materials are brought in contact. By adequately choosing the amine component, the crosslinking speed can be adapted. Figure 3C shows a 3D printed filament made by mixing these components at the last second.

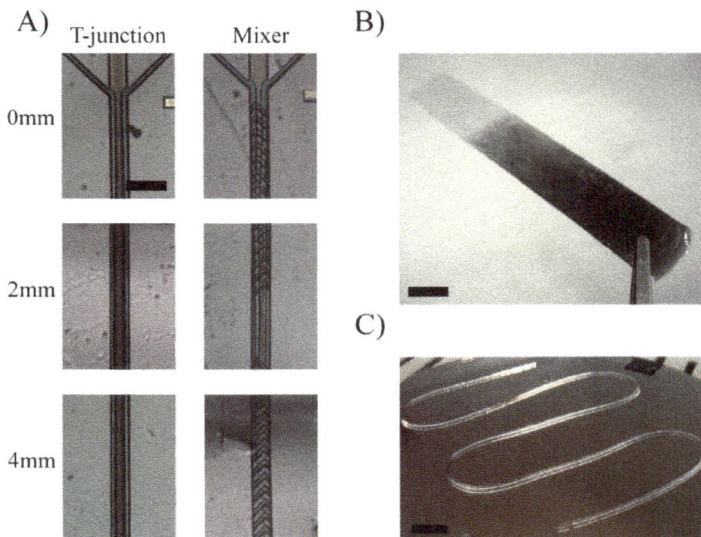

Figure 3. Herrigbone micro-mixer integrated at the tip of the probe. (**A**) Colored glycerol streams pushed through a simple T-junction and through a herringbone micro-mixer at 0 mm, 2 mm and 4 mm after the merging point. The mixer allows rapid and efficient mixing of viscous materials. Scale bar = 300 µm. (**B**) Mixing of white and black Formlabs resins at different ratios results in smooth gradients from one material to the other Scale bar = 1 cm. (**C**) Printed acrylate filament made using a two-component material. Scale bar = 1 cm.

The use of micro-mixers placed at the output of the print head of extrusion based 3D printers allows to mix reactive components right before dispensing them. This allows materials that are not usually used in 3D printing to be printed, such as polymer systems initiated by peroxides, as well as biomaterials such as hydrogels or cryogels. Additionally, more conventional materials can be mixed with a chosen ratio between them, allowing blends of materials to be created. This is of interest in the production of materials with gradients of properties in an object. It can simply be a gradient of colors if two resins of different color are used, but gradients of mechanical properties can also be fabricated, for example by mixing resins resulting in flexible and hard polymers or by mixing composite resins, charged with a filler material.

3.4. Concentration

In the previous paragraphs, print heads based on very simple microfluidic components were presented; however, microfabrication techniques can be used to manufacture much more complex structures. In this paragraph, we demonstrate a print head allowing the concentration of particles in solution, which can be used to increase the filler concentration of a composite resin just before it is printed or to concentrate cells in bioprinting. Figure 4A shows a schematic of the print head. A solution containing particles or cells enters the device by the inlet and flows in a straight channel equipped with a crossflow filter towards the outlet. This crossflow filter allows some of the liquid to be withdrawn between inlet and outlet simply by adequately imposing the flows both at the inlet and the waste outlet, thus concentrating the particles in solution. The extracted liquid is removed from the device through the waste outlet, while the concentrated solution is 3D printed. To separate micrometric-sized particles, the crossflow filter must feature small openings. This is made by a series of 50 μm tall, 2 μm wide pillars with a 4 μm pitch, as shown in Figure 4B. This structure is fabricated in a single etching step. Figure 4C–E demonstrates the concentration of a solution containing 8 μm beads.

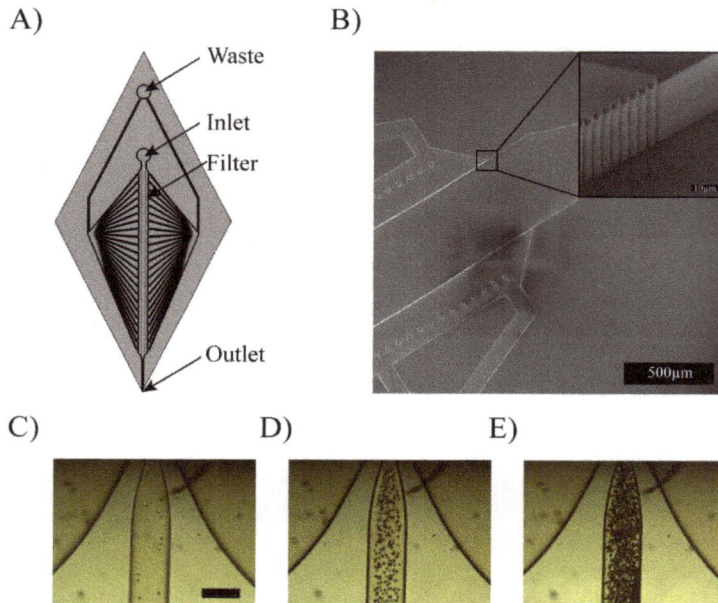

Figure 4. Crossflow filter for particle concentration. (**A**) Schematic of the print head. (**B**) Scanning electron micrograph of the crossflow filter. (**C–E**) Concentration of 8μm beads by adjusting the withdrawing factor. Scale bar = 250 μm.

Concentrating particles in solution before their being dispensed by 3D printing can find applications in various fields. It is known that the concentration of particles in composite materials is a major factor in their physical properties [57]. With the device we presented, the physical properties of the printed composites could be changed during printing. This is of particular interest when high concentrations of filler material need to be present in the printed material, making it very viscous. Concentrating the particles in a composite material at the last second when printing makes it possible to work with diluted (thus of low viscosity) solutions for the whole process and still produce composites of the desired composition, removing the excess liquid just before dispensation.

Concentrating cells in solution is of even bigger interest in the field of bio-printing, as intercellular communications rely heavily on intercellular distance [58], and thus on the cell concentration. However, working with high concentrations of cells in bioprinting is often inappropriate, as a large quantity of cells would be lost in pipes and other dead volumes, and the cells would be subjected to large shear forces because of the high viscosity of the solution. In general bioprinting techniques use solutions containing roughly 1 million cells per milliliter of solution, which results in cell-laden gels containing a cell concentration orders of magnitude lower than in native tissue [56].

By concentrating the cell right at the tip of the printer, it is possible to print tissue with biologically relevant cell concentration, without wasting large numbers of cells in dead volumes or subjecting them to high shear stress.

4. Discussion

Stereolithography and extrusion-based 3D printing are the most commonly used additive manufacturing techniques. Extrusion-based techniques are easy to use and are cost effective, but have seen almost no improvement in terms of process development since their creation. On the contrary, microfluidic techniques have seen major developments in terms of device design and fabrication, and these developments can now be applied to significantly improve the print heads of extrusion-based 3D printers.

In this paper, we explored the integration of four relatively simple microfluidic systems at the tip of the print head of extrusion-based 3D printers to create smart dispensing tools that can perform different functions right before printing. The integration of additional functions to print heads is of particular importance for bioprinting applications. We demonstrated fast switching between two or more materials to create multimaterial parts, but when applied to bioprinting, it can be used to switch between the different cell types needed in a cell co-culture. Enhancing the printing resolution using sheath flows can also be used for centering cells inside hydrogel filaments. Micro-mixers can be used to prepare hydrogels or cryogels from their components, to insert cells into the printing medium or to create gradients of composition or of cell types while printing. Concentration using crossflow filters makes it possible to reach cell concentrations close to native tissue without adverse effects during their manipulation.

Thanks to microfabrication, more complex functions can also be integrated in the print heads, taking advantage of all the knowledge available in the field of Lab on Chip devices, but such developments would probably have precise needs in terms of specific applications. For example, combining fluidic functions with in-channel spectroscopy could provide continuous feedback on the printed material quality, implementing dielectrophoresis electrodes could be used for filtering particles and cells and could make it possible to print only the viable cells, and detecting the passage of cells of interest while they flow through the microchannels and filtering out the excess of liquid medium could make it possible to print each cell at a precisely chosen location. Many other microfluidic functions can be integrated, such as fast switching valves, digital microfluidic, microfluidic gradient generators, cell concentrator or micro-heaters, each bringing the potential of using new materials and developing new applications for extrusion-based 3D printing.

The integration of customized microfluidic functions into the print heads of 3D printers has the potential to be a game changer in fields where additive manufacturing is used, and in particular for bioprinting. Compared to more conventional 3D printing, there is an additional level of complexity involved when making biological constructs out of different types of cells and scaffold materials, and the use of cells in bioprinting applications requires more care than the usual materials used in additive manufacturing techniques. Bioprinting is often described as the key development to go towards artificial organs and engineered biological tissues, but to obtain the level of complexity needed to form such complex cell constructs, major improvements in 3D printing techniques are needed, and one of the keys to achieving this is clearly the integration of microfluidic functions into print heads.

5. Conclusions

New opportunities generally arise when merging different technologies, and the present paper attempts to show the value of bringing microfluidics closer to extrusion-based 3D printing techniques. By applying some concepts developed over the years for Lab on Chip applications to create smart print heads, we tried to open the way and demonstrate improved dispensing functionalities. Four different microfluidic functions were implemented at the tip of print heads: Micro-mixers were used to mix reactive components before dispensing and to blend materials with varying ratios. Sheath flows were implemented to improve the print resolution. Fast switching was demonstrated to create multimaterial components and crossflow filtration was shown to concentrate particles or cells in the deposited material. By using microfluidic functions in the world of 3D printing, an additional layer of control can be added, enabling the fabrication of components that were impossible to print before. This methodology can be scaled, and has the potential to revolutionize the way we print, in particular in the field of bioprinting, where new challenges need to be faced in going towards the fabrication of engineered 3D biological tissues.

Acknowledgments: The authors of the paper would like to acknowledge the staff of the EPFL center of micronanotechnology for their support in the cleanrooms as well as Matthieu Rüegg for his help.

Author Contributions: Ludovic Serex conceived, designed and performed the experiments. Arnaud Bertsch designed the chemistry of the printed media. Ludovic Serex and Arnaud Bertsch wrote the paper. Research discussions were executed by all authors.

Conflicts of Interest: The authors confirm that there are no known conflict of interest associated with this publication and there has been no significant financial support for this work that could have influenced its outcome.

References

1. Kodama, H. Automatic method for fabricating a threedimensional plastic model with photo hardening polymer. *Rev. Sci. Instrum.* **1981**, *52*, 1770–1773. [CrossRef]
2. Gibson, I.; Rosen, D.; Stucker, B. Additive Manufacturing Technologies. In *Additive Manufacturing Technologies*, 2nd ed.; 3D Printing, Rapid Prototyping, and Direct Digital Manufacturing; Springer: Berlin, Germay, 2015. [CrossRef]
3. Todd, I. Metallurgy: Printing steels. *Nat. Mater.* **2017**, *17*, 13–14. [CrossRef] [PubMed]
4. Sames, W.J.; List, F.A.; Pannala, S.; Dehoff, R.R.; Babu, S.S. The metallurgy and processing science of metal additive manufacturing. *Int. Mater. Rev.* **2016**, *61*, 315–360. [CrossRef]
5. Eckel, Z.C.; Zhou, C.; Martin, J.H.; Jacobsen, A.J.; Carter, W.B.; Schaedler, T.A. Additive manufacturing of polymer-derived ceramics. *Science* **2016**, *351*, 58–62. [CrossRef] [PubMed]
6. Hockaday, L.A.; Kang, K.H.; Colangelo, N.W.; Cheung, P.Y.C.; Duan, B.; Malone, E.; Wu, J.; Girardi, L.N.; Bonassar, L.J.; Lipson, H.; et al. Rapid 3D printing of anatomically accurate and mechanically heterogeneous aortic valve hydrogel scaffolds. *Biofabrication* **2012**, *4*, 035005. [CrossRef] [PubMed]
7. Jang, J.; Park, J.Y.; Gao, G.; Cho, D.W. Biomaterials-based 3D cell printing for next-generation therapeutics and diagnostics. *Biomaterials* **2018**, *156*, 88–106. [CrossRef] [PubMed]
8. Murphy, S.V.; Atala, A. 3D bioprinting of tissues and organs. *Nat. Biotechnol.* **2014**, *32*, 773–785. [CrossRef] [PubMed]
9. Hospodiuk, M.; Dey, M.; Sosnoski, D.; Ozbolat, I.T. The bioink: A comprehensive review on bioprintable materials. *Biotechnol. Adv.* **2017**, *35*, 217–239. [CrossRef] [PubMed]
10. Kolesky, D.B.; Homan, K.A.; Skylar-Scott, M.A.; Lewis, J.A. Three-dimensional bioprinting of thick vascularized tissues. *Proc. Natl. Acad. Sci. USA* **2016**, *113*, 3179–3184. [CrossRef] [PubMed]
11. Pataky, K.; Braschler, T.; Negro, A.; Renaud, P.; Lutolf, M.P.; Brugger, J. Microdrop Printing of Hydrogel Bioinks into 3D Tissue-Like Geometries. *Adv. Mater.* **2012**, *24*, 391–396. [CrossRef] [PubMed]
12. Sydney Gladman, A.; Matsumoto, E.A.; Nuzzo, R.G.; Mahadevan, L.; Lewis, J.A. Biomimetic 4D printing. *Nat. Mater.* **2016**, *15*, 413–418. [CrossRef] [PubMed]
13. Khoo, Z.X.; Teoh, J.E.M.; Liu, Y.; Chua, C.K.; Yang, S.; An, J.; Leong, K.F.; Yeong, W.Y. 3D printing of smart materials: A review on recent progresses in 4D printing. *Virtual Phys. Prototyp.* **2015**, *10*, 103–122. [CrossRef]

14. Guo, N.; Leu, M.C. Additive manufacturing: Technology, applications and research needs. *Front. Mech. Eng.* **2013**, *8*, 215–243. [CrossRef]

15. Chu, M.Q.; Wang, L.; Ding, H.Y.; Sun, Z.G. Additive Manufacturing for Aerospace Application. *Appl. Mech. Mater.* **2015**, *798*, 457–461. [CrossRef]

16. Prasad, L.K.; Smyth, H. 3D Printing technologies for drug delivery: A review. *Drug Dev. Ind. Pharm.* **2016**, *42*, 1019–1031. [CrossRef] [PubMed]

17. Bose, S.; Ke, D.; Sahasrabudhe, H.; Bandyopadhyay, A. Additive Manufacturing of Biomaterials. *Prog. Mater. Sci.* **2017**, *93*, 45–111. [CrossRef]

18. Davia-Aracil, M.; Hinojo-Pérez, J.J.; Jimeno-Morenilla, A.; Mora-Mora, H. 3D printing of functional anatomical insoles. *Comput. Ind.* **2018**, *95*, 38–53. [CrossRef]

19. Rengier, F.; Mehndiratta, A.; Von Tengg-Kobligk, H.; Zechmann, C.M.; Unterhinninghofen, R.; Kauczor, H.U.; Giesel, F.L. 3D printing based on imaging data: Review of medical applications. *Int. J. Comput. Assist. Radiol. Surg.* **2010**, *5*, 335–341. [CrossRef] [PubMed]

20. Zhang, Y.S.; Yue, K.; Aleman, J.; Mollazadeh-Moghaddam, K.; Bakht, S.M.; Yang, J.; Jia, W.; Dell'Erba, V.; Assawes, P.; Shin, S.R.; et al. 3D Bioprinting for Tissue and Organ Fabrication. *Ann. Biomed. Eng.* **2017**, *45*, 148–163. [CrossRef] [PubMed]

21. Wu, Z.; Su, X.; Xu, Y.; Kong, B.; Sun, W.; Mi, S. Bioprinting three-dimensional cell-laden tissue constructs with controllable degradation. *Sci. Rep.* **2016**, *6*, 24474. [CrossRef] [PubMed]

22. Dvir-Ginzberg, M.; Gamlieli-Bonshtein, I.; Agbaria, R.; Cohen, S. Liver tissue engineering within alginate scaffolds: Effects of cell-seeding density on hepatocyte viability, morphology, and function. *Tissue Eng.* **2003**, *9*, 757–766. [CrossRef] [PubMed]

23. Zadpoor, A.A.; Malda, J. Additive Manufacturing of Biomaterials, Tissues, and Organs. *Ann. Biomed. Eng.* **2017**, *45*, 1–11. [CrossRef] [PubMed]

24. Roth, E.A.; Xu, T.; Das, M.; Gregory, C.; Hickman, J.J.; Boland, T. Inkjet printing for high-throughput cell patterning. *Biomaterials* **2004**, *25*, 3707–3715. [CrossRef] [PubMed]

25. Lind, J.U.; Busbee, T.A.; Valentine, A.D.; Pasqualini, F.S.; Yuan, H.; Yadid, M.; Park, S.J.; Kotikian, A.; Nesmith, A.P.; Campbell, P.H.; et al. Instrumented cardiac microphysiological devices via multimaterial three-dimensional printing. *Nat. Mater.* **2017**, *16*, 303–308. [CrossRef] [PubMed]

26. Vaezi, M.; Seitz, H.; Yang, S. A review on 3D micro-additive manufacturing technologies. *Int. J. Adv. Manuf. Technol.* **2013**, *67*, 1721–1754. [CrossRef]

27. Au, A.K.; Huynh, W.; Horowitz, L.F.; Folch, A. 3D-Printed Microfluidics. *Angew. Chem. Int. Ed.* **2016**, *55*, 3862–3881. [CrossRef] [PubMed]

28. Salet, T.A.M.; Bos, F.P.; Wolfs, R.J.M.; Ahmed, Z.Y. 3D concrete printing—A structural engineering perspective. In Proceedings of the 2017 fib Symposium—High Tech Concrete: Where Technology and Engineering Meet, Maastricht, The Netherlands, 12–14 June 2017.

29. Kouzani, A.Z.; Adams, S.; Oliver, R.; Nguwi, Y.Y.; Hemsley, B.; Balandin, S. 3D printing of a pavlova. In Proceedings of the Region 10 Conference (TENCON), Singapore, 22–25 November 2016; pp. 2281–2285. [CrossRef]

30. Bertsch, A.; Bernhard, P.; Vogt, C.; Renaud, P. Rapid prototyping of small size objects. *Rapid Prototyp. J.* **2000**, *6*, 259–266. [CrossRef]

31. Turner, B.N.; Strong, R.; Gold, S.A. A review of melt extrusion additive manufacturing processes: I. Process design and modeling. *Rapid Prototyp. J.* **2014**, *20*, 192–204. [CrossRef]

32. Liu, W.; Zhang, Y.S.; Heinrich, M.A.; De Ferrari, F.; Jang, H.L.; Bakht, S.M.; Alvarez, M.M.; Yang, J.; Li, Y.C.; Trujillo-de Santiago, G.; et al. Rapid Continuous Multimaterial Extrusion Bioprinting. *Adv. Mater.* **2017**, *29*, 1–8. [CrossRef] [PubMed]

33. Hardin, J.O.; Ober, T.J.; Valentine, A.D.; Lewis, J.A. Microfluidic printheads for multimaterial 3D printing of viscoelastic inks. *Adv. Mater.* **2015**, *27*, 3279–3284. [CrossRef] [PubMed]

34. Nie, M.; Mistry, P.; Yang, J.; Takeuchi, S. Microfluidic enabled rapid bioprinting of hydrogel μfiber based porous constructs. In Proceedings of the 2017 IEEE 30th International Conference on Micro Electro Mechanical Systems (MEMS), Las Vegas, NV, USA, 22–26 January 2017; pp. 589–591. [CrossRef]

35. Tan, X.P.; Tan, Y.J.; Chow, C.S.L.; Tor, S.B.; Yeong, W.Y. Metallic powder-bed based 3D printing of cellular scaffolds for orthopaedic implants: A state-of-the-art review on manufacturing, topological design, mechanical properties and biocompatibility. *Mater. Sci. Eng. C* **2017**, *76*, 1328–1343. [CrossRef] [PubMed]

36. Yu, Y.; Moncal, K.K.; Li, J.; Peng, W.; Rivero, I.; Martin, J.A.; Ozbolat, I.T. Three-dimensional bioprinting using self-Assembling scalable scaffold-free 'tissue strands' as a new bioink. *Sci. Rep.* **2016**, *6*, 1–11. [CrossRef] [PubMed]

37. Duan, B. State-of-the-Art Review of 3D Bioprinting for Cardiovascular Tissue Engineering. *Ann. Biomed. Eng.* **2017**, *45*, 195–209. [CrossRef] [PubMed]

38. Jia, W.; Gungor-Ozkerim, P.S.; Zhang, Y.S.; Yue, K.; Zhu, K.; Liu, W.; Pi, Q.; Byambaa, B.; Dokmeci, M.R.; Shin, S.R.; et al. Direct 3D bioprinting of perfusable vascular constructs using a blend bioink. *Biomaterials* **2016**, *106*, 58–68. [CrossRef] [PubMed]

39. Stroock, A.D.; Dertinger, S.K.; Ajdari, A.; Mezić, I.; Stone, H.A.; Whitesides, G.M. Chaotic mixer for microchannels. *Science* **2002**, *295*, 647–651. [CrossRef] [PubMed]

40. Liu, Y.Z.; Kim, B.J.; Sung, H.J. Two-fluid mixing in a microchannel. *Int. J. Heat Fluid Flow* **2004**, *25*, 986–995. [CrossRef]

41. Lee, C.Y.; Chang, C.L.; Wang, Y.N.; Fu, L.M. Microfluidic Mixing: A Review. *Int. J. Mol. Sci.* **2011**, *12*, 3263–3287. [CrossRef] [PubMed]

42. Braschler, T.; Theytaz, J.; Zvitov-Marabi, R.; Van Lintel, H.; Loche, G.; Kunze, A.; Demierre, N.; Tornay, R.; Schlund, M.; Renaud, P. A virtual valve for smooth contamination-free flow switching. *Lab Chip* **2007**, *7*, 1111–1113. [CrossRef] [PubMed]

43. Liepsch, D.; Versorgungstechnik, F.; Germany, W.; Karlsruhe, U.; Vlachos, N.S. Measurement and calculations of laminar in A ninety degree bifurcation. *J. Biomech.* **1982**, *15*, 473–485. [CrossRef]

44. Devaraju, N.S.G.K.; Unger, M.A. Pressure driven digital logic in PDMS based microfluidic devices fabricated by multilayer soft lithography. *Lab Chip* **2012**, *12*, 4809–4815. [CrossRef] [PubMed]

45. Knight, J.B.; Vishwanath, A.; Brody, J.P.; Austin, R.H. Hydrodynamic focusing on a silicon chip: Mixing nanoliters in microseconds. *Phys. Rev. Lett.* **1998**, *80*, 3863–3866. [CrossRef]

46. Martel, J.M.; Toner, M. Inertial focusing in microfluidics. *Annu. Rev. Biomed. Eng.* **2014**, *16*, 371–396. [CrossRef] [PubMed]

47. Xuan, X.; Zhu, J.; Church, C. Particle focusing in microfluidic devices. *Microfluid. Nanofluid.* **2010**, *9*, 1–16. [CrossRef]

48. Aoki, R.; Yamada, M.; Yasuda, M.; Seki, M. In-channel focusing of flowing microparticles utilizing hydrodynamic filtration_Supplementary material. *Microfluid. Nanofluid.* **2009**, *6*, 571–576. [CrossRef]

49. Gascoyne, P.R.C.; Vykoukal, J. Review Particle separation by dielectrophoresis. *Electrophoresis* **2002**, *23*, 1973–1983. [CrossRef]

50. Huang, L.R.; Cox, E.C.; Austin, R.H.; Sturm, J.C. Lateral Displacement. 2004; *304*, 987–991.

51. Valero, A.; Braschler, T.; Demierre, N.; Renaud, P. A miniaturized continuous dielectrophoretic cell sorter and its applications. *Biomicrofluidics* **2010**, *4*, 022807. [CrossRef] [PubMed]

52. Yamada, M.; Seki, M. Hydrodynamic filtration for on-chip particle concentration and classification utilizing microfluidics. *Lab Chip* **2005**, *5*, 1233–1239. [CrossRef] [PubMed]

53. Serex, L.; Braschler, T.; Filippova, A.; Rochat, A.; Béduer, A.; Bertsch, A.; Renaud, P. Pore Size Manipulation in 3D Printed Cryogels Enables Selective Cell Seeding. *Adv. Mater. Technol.* **2018**, 1700340. [CrossRef]

54. Colosi, C.; Costantini, M.; Barbetta, A.; Dentini, M. Microfluidic bioprinting of heterogeneous 3d tissue constructs. *Methods Mol. Biol.* **2017**, *1612*, 369–380. [PubMed]

55. Constantinescu, V.N. *Laminar Viscous Flow*; Springer: New York, NY, USA, 1995. [CrossRef]

56. Sideridou, I.D.; Achilias, D.S.; Karava, O. Reactivity of benzoyl peroxide/amine system as an initiator for the free radical polymerization of dental and orthopaedic dimethacrylate monomers: Effect of the amine and monomer chemical structure. *Macromolecules* **2006**, *39*, 2072–2080. [CrossRef]

57. Alsson, B.E.O.P. Effective intercellular communication distances are determined by the relative time constants for cyto/chemokine secretion and diffusion. *Engineering* **1997**, *94*, 12258–12262.

58. Bianconi, E.; Piovesan, A.; Facchin, F.; Beraudi, A.; Casadei, R.; Frabetti, F.; Vitale, L.; Pelleri, M.C.; Tassani, S.; Piva, F.; et al. An estimation of the number of cells in the human body. *Ann. Hum. Biol.* **2013**, *4460*, 463–471. [CrossRef] [PubMed]

micromachines

MDPI

Review

3D-Printed Chips: Compatibility of Additive Manufacturing Photopolymeric Substrata with Biological Applications

Megan Carve [1,†] **and Donald Wlodkowic** [1,2,*,†]

1   School of Science, RMIT University, Melbourne, VIC 3083, Australia; s3437621@student.rmit.edu.au
2   Centre for Additive Manufacturing, School of Engineering, RMIT University,
    Melbourne, VIC 3083, Australia
*   Correspondence: donald.wlodkowic@rmit.edu.au; Tel.: +61-3-9925-7157
†   These authors contributed equally to this work.

Received: 27 January 2018; Accepted: 19 February 2018; Published: 23 February 2018

Abstract: Additive manufacturing (AM) is ideal for building adaptable, structurally complex, three-dimensional, monolithic lab-on-chip (LOC) devices from only a computer design file. Consequently, it has potential to advance micro- to milllifluidic LOC design, prototyping, and production and further its application in areas of biomedical and biological research. However, its application in these areas has been hampered due to material biocompatibility concerns. In this review, we summarise commonly used AM techniques: vat polymerisation and material jetting. We discuss factors influencing material biocompatibility as well as methods to mitigate material toxicity and thus promote its application in these research fields.

Keywords: lab-on-a-chip; bioassay; toxicity; additive manufacturing; polymers; 3D printing

---

1. Introduction

Additive manufacturing (AM), colloquially known as three-dimensional (3D) printing, is an automated computer-assisted design (CAD) and fabrication method developed in the 1980's [1]. Via AM, structurally complex monolithic devices can be built from a range of materials, such as liquid-polymer or powder grains, with various material, mechanical, and physical properties. AM is a versatile and agile technology platform suited to the rapid development of prototypes and the fabrication of adaptable high-value products. Expansion of innovative desktop AM systems, exhibiting fast build-speeds and improved printing resolutions, coupled with decreasing prices of AM technologies, is enabling their rapid adoption in small-scale industries and laboratories. By 2020, the global AM industry revenue is predicted to exceed $21 billion [2]. It's growing popularity is due to the advantages it has over traditional 'subtractive' or 'formative' manufacturing techniques, which rely on economies of scale and are comparatively wasteful [3].

Demand for inexpensive prototyping and fabrication of lab-on-a-chip (LOC) devices has stimulated substantial interest in alternative fabrication methods. In this regard, AM has attracted interest within the LOC community, which aims to reduce the cost and complexity of prototyping and developing bespoken devices [4]. AM circumvents cumbersome processes inherent to traditional manufacturing methods (e.g., photolithography), such as complex multi-step fabrication techniques and the use of expensive clean room facilities. Specifically, the vat polymerisation AM methods, stereolithography (SLA), multi-photon polymerisation, and digital light processing (DLP), provide elegant fabrication alternatives.

Vat polymerization AM can produce optically transparent structurally complex monolithic devices with functional elements such as integrated fluidic interconnects, valves, pumps, and lenses [5]. In this method, a solid object is built in a layer-by-layer fashion by the selective exposure of a photosensitive polymer resin to a focused laser beam or light projection (Figure 1) [6]. Presently, most consumer-grade 3D printers build objects by a process known as fused deposition modelling (FDM). This involves melting and depositing a filament of plastic material such as acrylonitrile butadiene styrene (ABS) and poly(lactic acid) acid (PLA) in a layer-by-layer fashion. Decreasing costs of vat polymerisation machines means they are no longer only affordable to large-scale industries but are accessible to the individual consumer.

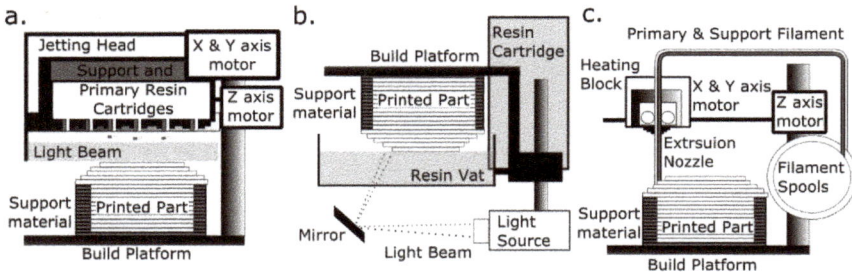

Figure 1. Examples of additive manufacturing (AM) systems that use polymer materials: (**a**) Material jetting, where a photo-sensitive photopolymer is deposited in droplets that are then polymerized by a passing light source. The process is repeated so that the solid object is built up layer by layer on the build tray. (**b**) Vat polymerization, where a photosensitive polymer contained in a vat is polymerized in a layer-by-layer fashion either by a light-beam (stereolithography and multi-photon polymerization) or by light-projection (digital light processing). (**c**) Material extrusion, where the polymer filament is softened as it passes through a heating block and extruded in a layer-by-layer fashion.

A critical limitation of AM systems for LOC applications, however, is the printing resolution. In general, AM systems are capable of producing a minimum printable feature size 70–300 μm and a z-axis layer height of >20 μm [1,7]. While current generation AM systems appear less suited to the fabrication of true microfluidic systems, they are suited to the fabrication of milllifluidic or mesofluidic devices, the latter is characterised by a channel's cross-dimensions ranging from 250 μm to 2000 μm and high aspect ratios. Design and structural constraints make them particularly difficult to fabricate using conventional photolithography techniques, thus demanding non-standard fabrication approaches ideally suited to AM [8–10]. Emerging applications in biomedical research and ecological toxicity testing include in-situ analysis of embryos and small metazoan organisms [11–15].

Despite the perceived advantages, reports on the toxicity of AM photosensitive polymeric materials have curtailed their application in AM-LOC technologies aimed at in vitro bioassays and biotests [13,16–19]. Resin compounds such as photoinitiators (PIs), monomers, short-chain polymers, and auxiliary compounds may leach from final printed parts in aqueous media and consequently may compromise part biocompatibility. Furthermore, attempting to eliminate these potentially toxic compounds by following standard manufacturer-recommended part-cleaning procedures might not adequately mitigate toxicity. This review focuses on the perceived toxicity of AM photosensitive polymeric materials as well as current approaches to alleviate toxicity, with the aim of facilitating applications in biomedical and biological research.

2. Overview of Additive Manufacturing (AM) Fabrication Technologies

AM technologies are classified according to the initial condition of the build material and the physical principles underlying the solidification process (Figure 1) [1]. In 2017, photo-polymerisation AM, which encompasses material jetting (MJ) and vat polymerisation (SLA and DLP), accounted for the majority of materials used in the global AM market [20].

The vat polymerization process produces a solidified 3D object by localised irradiation and polymerisation of a light sensitive resin (contained in a vat) by a spatially controlled light source [1]. The light source used in SLA is a spot laser that irradiates the resin as it scans in an x–y direction of each plane, whereas the light source used in DLP is a projector and the whole x–y plane is irradiated in a single flash. For both, the final object is built in a layer-by-layer fashion as the build platform moves through the resin along the z-axis, whereas multi-photon, also known as two-photon, polymerisation (MPP), produces a solid object by focusing a femtosecond laser beam into the resin vat. The laser moves in any direction within the resin, so it is not a layer-by-layer technique [21,22].

In contrast, MJ is a process similar to inkjet printing and involves jetting droplets of photosensitive polymer resin onto the build platform (Figure 1). Resin is solidified by a passing light source, and the process is repeated in a layer-by-layer fashion, as the object is built from the bottom up. Vat polymerization technology, more than MJ, can accelerate micro/milllifluidic device development since it is capable of achieving finer resolutions. It has been used to build LOC and micro-devices from optically transparent materials [5,23–25].

SLA and DLP has been used in LOC design and manufacture. The authors of Au et al. [26] built a peristaltic pump and cell perfusion chamber with integrated fluidic valves, and the authors of Comina et al. [27] built a monolithic LOC with integrated check-valves. In addition, LOC devices have been fabricated via replica moulding in poly(dimethylsiloxane) PDMS from moulds [28,29]. SLA and DLP are applied in dental implant design and production, surgical planning, the building of anatomical prosthetic devices, and tissue engineering [5,30,31].

MPP is suited to the design of microfluidic devices, as it can build structures down to 200 nm from materials classified as biocompatible [32]. It has been used to build micro-needle arrays for vaccine and drug delivery [33] and develop complex suspended micro-channel resonators for LOC biomechanical sensing applications [34] as well as integrative optofluidic refractometers with microtubes of various diameters and wall thicknesses [35].

3. Photopolymerization and Stereolithographic Resins

Polymerization, by chemical reactions, connects unsaturated monomer molecules to propagate a polymer-chain network [36,37]. To create a 3D solid object, a photo-sensitive stereolithographic polymer resin is solidified by the successive vitrification of the photopolymer by a light-initiated chemical reaction that leads to the cross-linking of polymers by produced reactive species (free radicals, cations or anions) [38]. The photo-polymerisation process is consistent with 'chain-growth' polymerization in that it requires initiation, propagation, chain transfer, and termination [36,37,39].

Resin components include a photo-sensitive polymer and a photoinitiator system along with other additives and fillers, such as inhibitors, plasticizers, light stabilizers, and pigments (e.g., shown in Tables 1–3). Auxiliary compounds may improve polymerisation (in the industry termed 'curing') efficiency and enhance mechanical, physical, and aesthetic properties of the final printed part. Ultimately, final material properties depend on the post-cleaning and post-curing process.

In radical polymerization, photolytic energy is absorbed by the PI and the produced reactive species is a free radical, a molecular fragment having one unpaired electrons (Figure 2) [36]. The propagating site of reactivity is a carbon radical [40]. Free radicals are transferred from monomer to monomer during chain growth. Radical polymerization terminates when two radicals, either the forming polymer chains or PI fragments, combine.

In cationic polymerization, photolytic energy is absorbed by the PI to release free radicals and a Lewis or a Brönsted acid [40,41]. The propagating site of reactivity is a carbocation. Charge is transferred from monomer to monomer during chain growth. Termination occurs via unimolecular rearrangement within the counterion, i.e., where an anionic fragment of the counterion combines with the propagating chain end, inactivating the growing chain, and reducing the concentration of the initiator complex [42].

Two polymerisation steps are required to produce the final object built using vat polymerisation AM [31]. The first-step occurs in the vat of the machine and is by the light emitted by the machine. This step produces a partially cured object, referred to as 'green'. While the object at this stage has its final shape and form, a second curing step is need to optimise its mechanical properties. Typically, the 'green' object is first rinsed in a solvent to remove polymerised resin. The second curing step occurs in a purpose-built 'post-curing' oven that emits a (stroboscopic) light at a wavelength that should be matched to the resin type. The object typically must be systematically re-orientated during this step, so that exposure of each surface of the geometry to the light source is maximised.

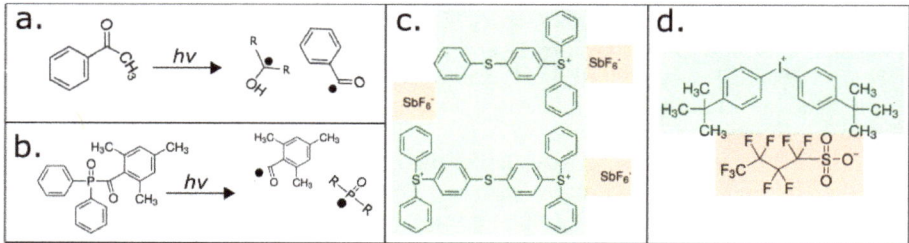

Figure 2. The two photoinitiator (PI) mechanisms are radical (**a,b**) and cationic (**c,d**). In the radical system, absorption of light (*hv*) produces a free radical by homolytic cleavage, propagating polymerization, for example, as in (a) hydroxyacetophenone and (b) phosphine-oxide, where R represents a methyl group in (a) and a phenyl ring in (b). In the cationic PI system, in PIs such as (c) triarylsulfonium hexafluoroantimonate salts and (d) bis(4-tert-butylphenyl)iodonium perfluoro-1-butanesulfonate, light is absorbed causing heterolytic and homolytic cleavage that forms a cationic portion (labeled in green) and anionic portion (labeled in pink). The generated molecules are reactive with monomers, forming an acid and free radicals, propagating polymerization (open source image adapted from [36]).

3.1. Photoinitiator Systems

A PI compound (examples are shown in Table 1) acts as catalyst for photopolymerisation by converting absorbed light energy into chemical energy, free radicals, and/or cations [43]. To enhance the efficiency and depth of polymerization, stereolithographic resins will typically contain multiple 'PIs' and a thermal initiator compound. A thermal initiator, such as benzoyl peroxide or Lauroyl peroxide, ideally does not affect the resin during printing, but operates during 'post-curing' to aid in interior part polymerization. Radical PI systems are classified by their mode of operation as either a Norrish Type I or Norrish Type II. The Norrish Type I PI molecule undergoes homolytic cleavage to produce free radicals upon the absorption of light [38], whereas the Norrish Type II PI is a bimolecular system, where absorption of light, typically ultra-violet (UV)-C, causes a light absorbing molecule (or sensitizer) to interact with a second molecule (a co-initiator) to generate free radicals [44].

A cationic PI system consists of a cationic and anionic pair, and each component has an explicit role in the polymerization mechanism [37]. The cationic portion of the PI molecule absorbs light to produce an excited electron state, whereas the anionic portion of the PI molecule absorbs light to produce an acid [41]. The excited electron state causes the PI molecule to undergo homolytic cleavage to yield several free radicals and heterolytic cleavage to produce another cationic species and an acid. Typically, cationic PIs are either sulfonium or iodonium salts (e.g., 4-octyloxy-phenyl-phenyl iodonium hexafluoroantimonate). Cationic photopolymerization is used for monomers, such as epoxides and vinyl ethers, which cannot be polymerized by free radical mechanisms [45].

For photopolymerization to proceed efficiently, the absorption wavelength of the PI must overlap with the emission spectrum of the light source and there must be minimal competing absorption by the other resin components [46]. Optimal absorption of each PI occurs at a specific wavelength range, for example, phosphine-oxide compounds will typically have an absorption maximum between 360 nm and 400 nm, and camphorquinone has an absorption maximum at a wavelength of 468 nm [30]. Catatonic PIs absorb in the wavelength range between 220 nm and 280 nm, with the different molecular configurations around the sulphur or iodine atom determining the specific absorption maximum for each PI [39].

Table 1. Summary of toxicity data available for photoinitiators used in stereolithography resins.

Compound	*w/w*	Available	Toxicological Information
Phosphine oxide compounds [1] (Type II)	0.1–5%	FORMlabs e.g., Dental and E-Shell series	Fertility impairing effect [47], acute and chronic toxic for aquatic organisms [48], toxic effect on mouse NIH 3T3 cells [38]. Not readily biodegradable by OECD criteria [48,49]. LD_{50} Oral rat > 5000 mg/kg (OECD) [48] LC_{50} (48 h) *Oryzias latipes*—6.53 mg/L (JIS K 0102-71) [49,50] EC_{50} (48 h) *Daphnia magna*—3.53 mg/L (OECD 202) [48,50] EC_{50} (72 h) *Pseudokirchneriella subcapitata*—1.56 mg/L (OECD 201) [49]
Hydroxy-acetophenone (Type II)			Readily biodegradable (OECD 301B) [51] LD_{50} Oral Rat—2.240 mg/kg [51] LC_{50} (96 h) *Salmo gairdneri*—25 mg/L [51] EC_{50} (48 h) *Daphnia magna*—50 mg/L [51]
Benzophenone compounds [2] (Type II)	<10%	UV-cured inks	Causes liver hypertrophy and kidney adenoma in rats [52] EC_{50} (24 h) *Daphnia magna*—0.28 mg/L [53] LC_{50} (96 h) *Pimephales promelas*—14.2 mg/L [53] BP-3 and BP-4: LC_{50} (48 h) *Daphnia magna*—1.09 and 47.47 mg/L LC_{50} (96 h) *Brachydanio rerio*—3.89 and 633.00 mg/L
Camphorquinone		Dental resins	EC_{50} mouse fibroblasts—235 μM [30]
1-hydroxy cyclo hexyl phenyl ketone		FORMlabs Irgacure 184	LC_{50} (96 h) *Danio rerio*—24 mg/L [54] EC_{50} (48 h) *Daphnia magna*—59.3 mg/L (OECD 202) [54] EC_{50} (72 h) *Desmodesmus subspicatus*—14.4 mg/L (OECD 201) [54]
Triarylsulfonium salt (Cationic)[3]	1–10%	3D Systems	EC_{50} (24 h) *Daphnia magna*—4.4 mg/L [55] EC_{50} (48 h) *Daphnia magna*—0.68 mg/L [55]

[1] Including diphenyl(2,4,6-trimethylbenzoyl)phosphine oxide (TPO) and bis acyl phosphine oxide (BAPO). [2] Including benzophenone-3 (BP-3), benzophenone-4 (BP-4) [56]. [3] 50% propylene carbonate and 50% mixed triarylsulfonium salts, i.e., an antimonate mixture.

3.2. Photopolymer Matrix/Systems

Photopolymers are mono-, di- and trifunctional monomers, acting as cross linkers, and hyper-branched oligomers that are multifunctional, acting as both cross linkers and diluents [57]. Classical photopolymer systems contain a single type of photopolymer (e.g., acrylic, epoxide, vinyl-ether, or a thiol-ene), whereas commercially available resins have combinations of photopolymers to enhance mechanical properties of the final fabricated object and to overcome the limitations of single polymer systems (e.g., low strength and high anisotropy). Examples of photopolymers are shown in Table 2.

Most free-radical curing resins employ the unsaturated functionality of acrylate and methacrylates; however, allyl- and vinyl-based formulations are also available [1]. An acrylate oligomer backbone can be modified to be of a different polymer class to achieve the material properties desired in the final object. For example, material properties are customized when an acrylate oligomer is combined with highly flexible polybutadienes, rigid and chemically resistant bisphenol A epoxies, and flexible yet tough polyurethanes [58]. Polyester- and acrylic-based acrylate oligomers are also commercially available.

The biocompatibility of candidate photopolymeric materials for the AM of biomedical devices has been carefully scrutinised. Organically modified ceramic (ORMOCER) composite resins, for example, have been used with MPP and SLA to build devices including scaffolds for tissue engineering, microneedles for drug delivery, and dental implants [33,59,60]. ORMOCER-based photopolymer resins contain silicon alkoxides, organically modified silicon alkoxides, several metal alkoxides and in/organic monomers [61]. Typically, they operate either as methacrylate alkoxysilane systems, initiated via radical polymerization, or epoxysilane systems initiated via cationic polymerization. The biocompatibility of ORMOCER copolymer resins increases as the content of inorganic co-polymers decreases, such that leaching of toxic residues is minimised [62]. For example, a polymerized object made from an ORMOCER-based resin may contain up to 50% unreacted methacrylate monomer. These are liable to leach in an aqueous medium and cause toxicity.

Polyethylene glycols (PEGs) are polymers of ethylene oxide [63]. PEGs are non-toxic, except when they are administered at exceedingly high doses (for humans, >10 mg/(kg·day^{-1})). Material properties of PEGs can be fine-tuned by modifying the molecular weight (MW) of the PEG backbone and by cross-linking them with acrylate groups. Polyethylene glycol diacrylate (PEG-DA) contains double-bond acrylate groups at each end of the PEG chain.

PEG-DA is a relatively biocompatible material [64,65]. While PEG-DA is used in biomedical applications, toxicity of PEG-DA can arise from unreacted monomers and PIs. Consequently, assessment on biocompatibility precedes each application [64]. The authors of Urrios et al. [66], for example, built transparent microfluidic devices in PEG-DA (MW 250) using SLA. To mitigate the toxicity of unreacted PEG-DA monomers and PI, the surfaces were subjected to an additional curing step in a UV bath to allow residual compounds to leach. The authors of Traore and Behkam [67] combined PEG-DA hydrogels with photolithography to produce LOC devices for use in cellular assays. To mitigate toxicity, gels were soaked in PBS overnight to remove excess PI and unreacted monomers.

Direct AM of PDMS is still in its early stages of development. The authors of Femmer et al. [68] used AM to create the first membrane made from PDMS using DLP. They fabricated features in the range of 150 and 300 μm from a custom PDMS-based resin. However, the optical clarity of the material was reduced because PI concentration and biocompatibility assays were not undertaken. PDMS materials have been applied, to a degree, in the development of microfluidic LOC using AM. The authors of He et al. [69] produced an LOC using FDM, sugar, and PDMS. Micro-channels were created by extruding melted sugar on a PDMS-based layer. PDMS was cast over the sugar layer and the sugar was removed with water. The authors of Comina et al. [29] used AM master moulds to cast PDMS into LOC. In addition, PDMS has been used to coat AM LOC to improve biocompatibility.

Table 2. Summary of toxicity data available for photopolymers used in stereolithography resins.

Compound	w/w	Available	Toxicological Information
Acrylate monomers, Acrylate and Urethane acrylate oligomers	5–60%	FORMlabs Autodesk Envisiontec 3D Systems	Toxic or harmful to various species of fish, algae and water microorganisms [49]. Potential mutagens and a reproductive and developmental toxicant. LD_{50} Oral rat >5000 mg/kg [49] LC_{50} (96 h) *Brachydanio rerio*—10.1 mg/L (OECD 203) [70] LC_{50} (96 h) *Cyprinus carpio*—1.2 mg/L (OECD 203) [71] LC_{50} (96 h) *Pimephales promelas*—34.7 mg/L (OECD 203) [70]
Methyl methacrylate monomers [1], and oligomers	5–90%	FORMlabs Envisiontec Dental resin	Assessment of repeated dose toxicity indicates potential to affect the liver and kidneys as indicated in animal studies [72]. Potential mutagen, and a reproductive and developmental toxicant, aquatic toxicant, and genotoxic in mammalian cell culture [73]. LC_{50} (96 h) *Salmo gairdneri*—3.4 mg/L (OECD 203) [73] LC_{50} (96 h) *Cyprinodon variegatus*—1.1 mg/L (OECD 203) [73] EC_{50} (48 h) *Daphnia magna*—2.6 mg/L (OECD 202) [73] EC_{50} (72 h) *Selenastrum capricornutum*—3.55 mg/L (OECD 201) [73] EC_{50} (96 h) *Mysidopsis bahia*—1.6 mg/L (OPP 72-3) [73] LC_{50} (96 h) *Lepomis macrochirus*—283 mg/L * [74] LC_{50} (96 h) *Oncorhynchus mykiss*—5.2 mg/L * [75] EC_{50} (48 h) *Daphnia magna*—8.74 mg/L * [75] EC_{50} (72 h) *Pseudokirchneriella subcapitata*—5.2 mg/L * [75] LD_{50} Oral rat—7900 mg/kg * [74]
Tripropylene Glycol diacrylate		3D Systems	LD_{50} Oral rat—6800 mg/kg (OECD 401) [76] LC_{50} (96 h) *Leuciscus idus* >4.6–10 mg/L [76] EC_{50} (48 h) *Daphnia magna*—89 mg/L [76] EC_{50} (72 h) *Scenedesmus subspicatus*—65.9 mg/L [76]

Table 2. *Cont.*

Compound	w/w	Available	Toxicological Information
Hydroxyethyl Methacrylate		Dental resins	EC_{50} (48 h) *Daphnia magna*—380 mg/L (OECD 202) [77] EC_{50} (72 h) *Selenastrum capricornutum*—836 mg/L (OECD 201) [77]
3,4-Epoxy cyclohexylmethyl 3,4-epoxy-cyclohexane carboxylate	25–60%	3D Systems	EC_{50} (48 h) *Daphnia magna*—40 mg/L [78] LC_{50} (96 h) *Oncorhynchus mykiss*—24 mg/L [78] LC_{50} Oral rats—5000 mg/kg [78]
1,6-bis(2,3-epoxy propoxy) hexane	15–30%	3D Systems	Not easily biodegradable (according to OECD-criteria) [79] EC_{50} (48 h) *Daphnia magna*—47 mg/L [79] LC_{50} (96 h) *Leuciscus idus*—30 mg/L [79] LD_{50} Oral rats—2190 mg/Kg [79]
Bisphenol A-diglycidyl dimethacrylate (Bis-GMA)		Dental resins	EC_{50} mouse fibroblasts—9.35 µM [30]
Tetraacrylate [2,3]	30–60%	Autodesk Evisiontec FORMlabs	LC_{50} (96 h) *Cyprinus carp*—1.2 mg/L [2] [80] LC_{50} (96 h) *Danio rerio*—7.9 mg/L [3] [80]

[1] Degrades to methacrylic acid: LD_{50} Oral rat—1320 mg/kg, LC_{50} (96 h) *Oncorhynchus mykiss*—85 mg/L [81].
[2] Alkox. pentaerythritol tetraacrylate. [3] Di(trimethylolpropane)tetra-acrylate.

Additives and fillers are auxiliary compounds included for the purpose of enhancing and customizing qualities of the printed part. For example, vertical print resolution and transparency is improved, and solubility is reduced, by inclusion of 'light-blocker' compounds, such as carbon black and Naphthol-based pigments, and UV blockers, such as 2,2'-(2,5-thiophene diyl) bis(5-tert-butyl benzoxazol) Liu and He [82]. All acrylate-based resins require auxiliary compounds to extend resin shelf life. Inhibitors, which are usually a quinone-based compound such as butylated hydroxytoluene, hydroquinone, hydroquinone monomethyl ether, and benzoquinone, prevent spontaneous polymerization during storage [83].

In general, commercially available resins utilize a wide range of auxiliary compounds that serve various purposes such as tuning material, mechanical, physical, and aesthetic propertis. Several examples of auxillary compounds are shown in Table 3. Further example compounds include silica, titanium dioxide, zinc, iron oxides, silver, silicon nitrides, calcium, lead, cerium, tin, zirconium, strontium, barium, borosilicate glasses, kaolin, quartz, talc, rubber impact modifiers, thermoplastic and cross-linked polyurethanes, polyethylene, polypropylene, polycarbonates, and poly-epoxides [83–89].

4. Compatibility of AM Substrata with Biological Applications

Biocompatibility of polymeric substrata used to fabricate LOC technologies is critical for any biological and biomedical applications. Bioassays on living cells and tissues are conducted in aqueous media and leaching of any chemicals from 3D printed plastics after contact with water can directly affect test specimens. Until recently, critical biocompatibility issues and potential hazard risk implications of widespread usage of 3D printed polymers have received only marginal attention [13,17–19,56,90]. Data on toxicity, bioaccumulative potential, persistence, and degradability of photopolymer materials is largely lacking [4]. This is in part due to a lack of available information on resin composition, which constrains independent evaluation of environmental and human health risk.

Typically, commercial resin formulas are proprietary. Only known hazardous materials are declared in material product safety data sheets (SDS). Compounds considered by the manufacturer as posing no significant risk are omitted. Open source and patent documents showed limited insight on the potential formulas and compounds used in resins. This indicates that formulas are complex, potentially containing greater than 20 compounds [82–88]. Consequently, it is difficult to predict which compounds may leach and to estimate and measure the toxicity of individual and mixtures of compounds. Comprehensive biological and chemical testing is a consequence necessary to evaluate biocompatibility.

Generally, substrata used for MJ and vat photopolymerization AM can have greater toxicity than those used in material extrusion AM [17,91]. Typically, photopolymerization AM is reliant on acrylate- and methacrylate-based compounds, which have up to a 90% weight-to-weight ratio (*w/w*), and phosphine-oxide-based PI systems, which are less than 5% *w/w* (Tables 1 and 2). These compounds are known to be acutely and chronically toxic to aquatic test organisms, including fish, algae, and water microorganisms. In mammals, chronic exposure to these compounds has sublethal effects that include impaired reproductive, liver, and kidney function and abnormal growth and development. Despite this, studies investigating AM polymer toxicity have primarily focused on the emission hazards of fumes resulting from material extrusion techniques, and photopolymerization materials have been largely ignored [12,13,18,92,93].

The two principal mechanisms by which compounds can be released from a polymerized object are (1) degradation or erosion and (2) extraction by solvent or aqueous medium, the latter often referred to as a leaching of toxic residues [91]. Erosion, which may be caused by photo, thermal, mechanical, or chemical factors, and the solubility of the resin-matrix also influence the release of compounds. For example, UDMA-based resins (urethane dimethacrylate, 1,6-bis(methacryloyloxy-2-methoxycarbonyl amino)-trimethylhexane) are less water-soluble than materials containing bisphenol-A-glycidyl methacrylates and are thus more susceptible to erosion [91]. Salivary esterases can erode the surfaces of resin-based dental materials, releasing methacrylic substances that subsequently enter the intestine via swallowing, diffuse into the circulatory system, and remain in the body until metabolized [94]. Unpolymerized co-monomers, triethylene glycol dimethacrylate (TEGDMA) and 2-hydroxyethyl methacrylate (HEMA), released from polymerized parts used as dental implants, have been detected in the human circulatory system [95].

Table 3. Summary of toxicity data available for auxiliary compounds used in stereolithography resins.

Compound	*w/w*%	Available	Toxicological Information
Butylated hydroxytoluene		Dental resins	Toxic or harmful to various species of fish, algae, and water microorganisms [96] LD_{50} Oral rat >6000 mg/kg (OECD 401) [96] LC_{50} (48 h) *Oryzias latipes*—5.3 mg/L [96] EC_{50} (48 h) *Daphnia magna*—0.48 mg/L (OECD 202) [96] EC_{50} (24 h) Protozoa—1.7 mg/L [96]
Sebacate compounds [1]	<5%	FORMlabs e.g., Dental Envisiontec	Toxic to aquatic life with long lasting effects [49], not readily biodegradable (OECD 301B) [49,97] LD_{50} Oral rat—3230 mg/kg (OECD 423) [49,97] LC_{50} (96 h) *Lepomis macrochirus*—0.97 mg/L (OECD 203) [49,97] LC_{50} (96 h) *Oncorhynchus mykiss*—7.9 mg/L (OECD 203) [49] LC_{50} (96 h) *Brachydanio rerio*—0.9 mg/L (OECD 203) [49] LC_{50} (48 h) *Daphnia magna*—8.58 mg/L (OECD 202) [97] EC_{50} (72 h) *Pseudokirchneriella subcapitata*—1.1 mg/L (OECD 201) [97] EC_{50} (72 h) *Desmodesmus subspicatus*—1.68 mg/L (OECD 201) [49]
Methylthiophenol compounds [2]		Autodesk	LC_{50} (96 h) *Danio rerio*—9 mg/L [80] EC_{50} (72 h) *Pediastrum boryanum*—1.7 mg/L [80] EC_{50} (24 h) *Daphnia magna*—15 mg/L [80]
Hydroquinone		Dental resins	Evidence of mutagenicity in mammal studies, toxic to aquatic life; absorption, in sufficient concentrations, leads to cyanosis [98] LC_{50} (96 h) *Oncorhynchus mykiss*—0.04 mg/L [98] EC_{50} (48 h) *Daphnia magna* 0.13 mg/L [98] EC_{50} (72 h) *Pseudokirchneriella subcapitata*—0.34 mg/L [98] LD_{50} Oral rat—367.3 mg/kg [98]

[1] Bis(2,2,6,6-tetramethyl-4-piperidyl) sebacate, Pentamethyl-piperidyl sebacate.
[2] 2-methyl-1-(4-methylthiophenol)-2-Morpholinopropan-1-one.

Leaching of toxic residues to aqueous media is significantly more important for implementations of AM in bioanalytical LOC technologies. Recent reports have shown that leaching of compounds from plastic parts is relatively high in the first 24 h [30]. The rate of compound leaching is influenced partly by geometry, characteristics of resin components, and polymerization extent [36,99]. Several studies have isolated compounds such as residual monomers, additives, and photoinitiators by means of extraction with aqueous media including distilled water, natural or artificial saliva, Ringer's solution, and organic diluents such as methanol, ethanol, and acetone. We have recently tested a random sample of photopolymer leachate from several SLA systems using gas chromatography–mass spectrometry (GC-MS) analysis. Our qualitative data based on different retention times in GC confirmed the presence of a photoinitiator 1-hydroxycyclohexyl phenyl ketone (1-HCHPK) and a substance closely related but not identical to methacrylate monomer (Figure 3).

Interestingly, 1-HCHPK has recently been reported as toxicant leaching from polyethylene ampoules used for intravenous injections [100]. Such pilot GC-MS analysis highlighted its limited detection capability for substances that are polar and non-volatile. Indeed, subsequent toxicity profiling of identified pure compounds suggested that they were responsible for only half of the cumulative toxicity effect. This warrants additional analysis to enable quantitative and conclusive identification of compounds that leach out of the 3D-printed plastic parts. Combined, these data signify possible biological risks associated with photoinitiators use in different plastic materials.

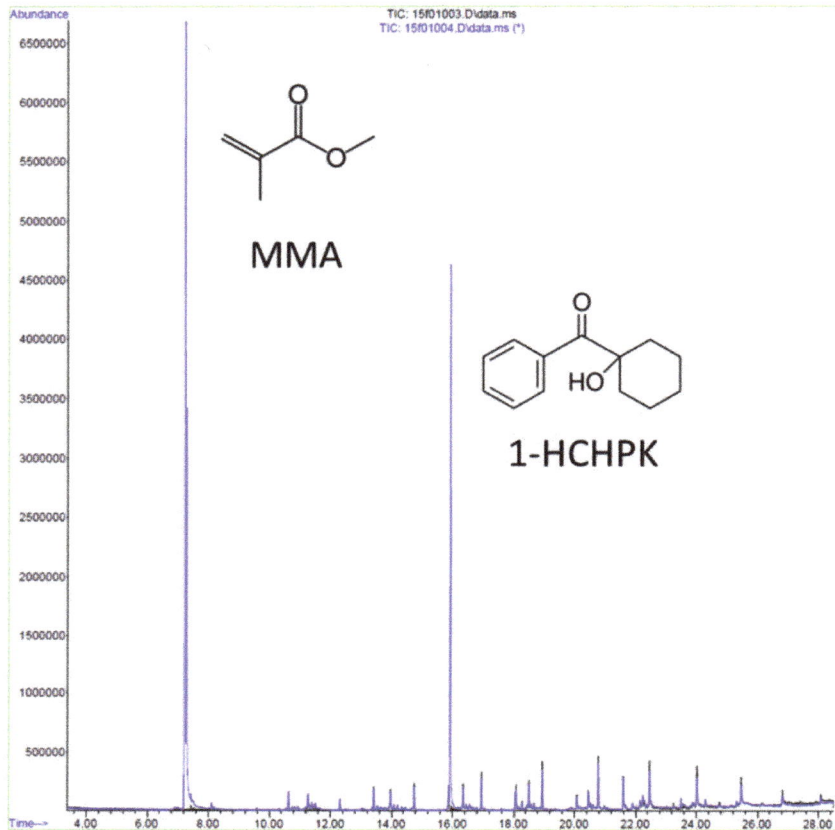

Figure 3. Qualitative analysis of Form Labs photopolymer leachate using gas chromatography-mass spectrometer (GC-MS) indicating the presence of photoinitiator 1-hydroxycyclohexyl phenyl ketone (1-HCHPK) and a substance closely related but not identical to methacrylate monomer.

Incomplete polymerization of photo-reactive resins amid MJ, SLA, and DLP processes can lead to greater levels of uncured and highly toxic substratum in the manufactured object, which can potentially increase the leaching rate [31]. Even under optimal conditions, the conversion of monomers to polymers is usually incomplete, and even the most efficient systems achieve approximately 55–60% of complete polymerization [101]. Early termination of polymerization may further result from a higher-than-optimal PI concentration in the resin, which over-produces reactive species, or the presence of oxygen [102]. The above issues can be further intensified by part geometry. The latter affects polymerization since initiation is due to 'line of sight', making shadowed areas more difficult to cure, particularly at the 'post-curing' stage. Since the PI, compounds are not typically indicated by the resin manufacturer, mismatch of the absorption characteristics of the resin to the emission characteristics of the chosen curing method, is a potential problem that may result in incomplete polymerization. For example, most UV photoinitiators, in particular those for cationic polymerization, exhibit very weak or no absorption at 365 nm, 400 nm, and 465 nm, which makes the MPM lamps and LEDs inefficient light sources for curing [45].

When considering AM systems such as SLA, MJ and DLP for any biological applications it is important to consider that PIs, auxiliary compounds, monomers, and short chain polymers, as well as their metabolites, may not be exhausted during or after polymerization. Furthermore, they are not entirely bound to or within the printed object [103,104]. For example, small photoinitiators of 200–250 Da used in food packaging, such as 4-methyl benzophenone (4-MBP) and isopropylthioxanthone, were found to migrate from the packaging into the food, raising significant food safety concerns [105]. The authors of Short et al. [90] found that the PI antimony, a toxic heavy metal, leached from AM objects over a 24 h period. In addition, several studies on human tissues have shown that PI metabolites were implicated in material toxicity [106].

Key metabolites of PIs are free radicals, also known as reactive oxygen species (ROS), such as peroxides and peroxy radicals [40,44]. These are implicated in the damage of DNA and proteins by oxidative stress mechanisms [107]. The pathophysiology of ageing and various age-related diseases, including chronic inflammatory of the gastrointestinal tract, diseases associated with cartilage, and other neurology disorders, have been linked to oxidative damage [108].

Recent clinical investigations of biocompatibility of AM-fabricated polymeric dental implants have shown a correlation between leachate and irritation of the oral mucosa [30]. Compounds such as MMA, formaldehyde (a degradation of a copolymer formed from oxygen and methacrylate during polymerization), and dibutyl phthalate (a plasticizer) have been detected in saliva. The authors of Schweikl et al. [109] demonstrated that methacrylate monomers, such as HEMA, leach from dental materials and induce cell apoptosis as a response to the associated DNA damage. Epoxides are highly reactive molecules that are also implicated in DNA damage, apoptosis, and carcinogenic and mutagenic effects [94]. The leached PI camphorquinone has moderate cytotoxic effects as shown in human submandibular-duct cells [110]. It has been associated with cytotoxicity and correlated with a significant increase in intracellular ROS in human pulp fibroblasts [111].

A commonly used endpoint for the determination of adverse systemic effects is acute oral toxicity in rats (Tables 1–3). This quantifies the single-dose required to kill 50% of test animals (LD_{50}). Since photopolymerised resin-based materials release compounds in relatively small amounts, acute oral toxicity in rats is less relevant to the assessment of the biocompatibility of these materials compared with tests documenting sublethal endpoints [91]. The latter determines the effective concentrations, which induce a response in 50% of the test animals (EC_{50}), sometimes determined along with the concentration required to kill 50% of the test organisms (LC_{50}). These values allow comparisons between relative biocompatibility of materials. For example, the authors of Geurtsen [111] investigated cytotoxic effects of 35 single compounds used in stereolithographic resins, in permanent 3T3 cells, and in three primary human oral fibroblast cultures. They showed that EC_{50} values varied significantly among compounds, and that the PI 2,6-di-t-butyl-4-methylphenol, the auxiliary compound 2-hydroxy-4-methoxy benzophenone, and the PI diphenyliodonium chloride had relatively elevated cytotoxic effects.

Several recent studies have investigated the biocompatibility of parts made by vat polymerization and MJ (Table 4). These studies utilized the fish embryo toxicity (FET) test, which is an established and sensitive, phenotype-based physiological analysis of developing zebrafish (*Danio rerio*) embryos. It is a relatively non-biased approach and is used extensively in the process of elucidating organ-specific toxicity and environmental adaptations at the organ, tissue, and systems level [112]. The embryonic developmental stage is considered one of the most sensitive to environmental perturbations and can be readily applied to assess the impacts of any potential toxic effects of chemicals or solid phases.

Despite SLA and MJ parts being post-processed according to manufacturer specifications, they appear to leach compounds. The authors of Macdonald et al. [13] investigated the biocompatibility of commercial resins—VisiJet Crystal EX200, VisiJet S300, Watershed 11122 XC, ABSplus P-430, and Fototec SLA-7150-Clear—using FET and found that these resins may leach toxic compounds (Table 4). Leachate from several caused malformations during embryonic development (teratogenicity) of zebrafish embryos. The authors of Oskui et al. [17] tested the biocompatibility of a Form Clear (Form Labs, Inc., Somerville, MA, USA) resin and found that exposure to leachate lowered survival rates of zebrafish embryos and elevated rates of malformations (yolk sac edema, heart edema, embryo length deformation, spine flexures, a lack of melanophore development, and a lack of swim bladders). In a similar study, the authors of Alifui-Segbaya et al. [16] showed that 48 h of exposure to a VisiJet (3D Systems Inc., Rock Hill, SC, USA) crystal-printed part leachate had lethal, sublethal, and teratogenic effects on zebrafish embryos.

Table 4. Summary of recent assessments on biocompatibility of parts printed with commercially available SLA and MJ printing polymers.

Resin	Organism	Toxicological Information
VisiJet Crystal	Algae [1]	At 24 h ~70% growth inhibition [18].
	Flea [2]	At 24 h 100% mortality [18]
	Rotifer [3]	At 24 h 100% mortality [18]
	Zebrafish [4]	Stunted growth, missing eyes, reduced pigmentation and yolk sac, abnormal shapes and also appear darker [13]. Greater than 90% mortality observed at 48 h [16] and 100% mortality observed at at 72 h [13].
Watershed 11122XC	Algae [1]	At 24 h >90% growth inhibition [18]
	Rotifer [3]	At 24 h ~ 100% mortality [18]
	Flea [2,5]	At 24 h ~ 100% mortality [18]
Fototec 7150 Clear	Algae [1]	At 24 h >90% growth inhibition [18]
	Rotifer [3]	At 24 h ~ 100% mortality [18]
	Flea [2,5]	At 24 h ~ 100% mortality [18]
Form Clear	Algae [1]	At 24 h ~60% growth inhibition [18]
	Rotifer [3]	At 24 h ~100% mortality [18]
	Flea [2,5]	At 24 h ~ 100% mortality [18]
	Zebrafish [4]	At 72 h higher rate of mortality, malformations (yolk sac edema, heart edema, embryo length deformation, spine flexures, lack of melanophore development, and a lack of swim bladders) [17].
VisiJet Clear	Zebrafish [4]	At 48 h >90% mortality of embryos [16].
	Algae [1]	At 24 h >90% growth inhibition [18]
	Rotifer [3]	At 24 h ~ 100% mortality [18]
	Flea [2,5]	At 24 h ~ 100% mortality [18]
MED610/620	Zebrafish [4]	>50% lethality [16].

[1] Freshwater algae (*Pseudokirchneriella subcapitata*)—OECD 201 Growth Inhibition Test. [2] Freshwater water flea *Daphnia sp.*—OECD 202 Acute Immobilization Test. [3] Freshwater rotifer *Brachionus calycifloru*—ASTM E1440-91 Acute Toxicity Test. [4] Zebrafish *Danio rerio* embryo—OECD 236 Fish Embryo Acute Toxicity (FET) Assay. [5] Freshwater water flea *Ceriodaphnia dubia*—USEPA Acute Toxicity Test.

Zhu et al. [18] expanded upon the above studies and performed a battery of cell-based and whole organism bioassays to assess the biocompatibility of several polymers including Watershed 11122 XC, Dreve Fototec 7150 Clear, VisiJet Crystal, Form Clear, and VisiJet SL Clear (Table 4). This work for the first time provided an in-depth and multispecies view of potential biological implications of fabricating devices using AM technologies. The published results demonstrated that leachate from polymerized parts were toxic to vertebrates and several invertebrate model organisms. All of the zebra fish larvae exposed to MJ and SLA leachates developed complete paralysis within 5 min of exposure, indicating the effect leachates may have on the central nervous system of zebrafish larvae.

These results identified VisiJet Crystal polymer as toxic despite its being classified as a substance with favourable biocompatibility, as evidenced by United States Pharmacopeia (USP) Class VI certification [13,18,113]. However, its USP Class VI certification is dependent on specific post-curing and cleaning steps [113]. These recent studies, though, have demonstrated that compounds, which are potentially toxic, remain and can leach. Furthermore, standard and specialised post-processing steps may not be adequate for removing toxicity, which still may occur in ranges affecting LOC bioassays.

While advances in AM of microfluidic and biomedical devices empower a rapidly growing number of applications, the above studies suggest that considerable caution must be exercised to mitigate biocompatibility concerns. The AM of biomicrofluidic LOC devices may not be suitable for in vitro bioassays applications. Ongoing research is aimed at solving these biocompatibility concerns by, for example, the development of AM systems capable of using biologically compatible substrata such as PDMS and PEG-DA [114].

5. Methods for Mitigating Toxicity of Polymeric Resins

AM is undoubtedly an elegant approach to the fabrication of monolithic functional LOC devices. Its ability to produced LOC with integrated fluidic interconnects and functional elements such as valves has been proven. In addition, several different approaches aimed at ensuring biocompatibility of photopolymer resins have been explored. Beyond customizing resin formulas to select for less toxic compounds, carefully managed post-cleaning and post-curing processes, or coating of surfaces with various biocompatible compounds, can increase biocompatibility.

While cross-linked polymer matrix may not leach, unreacted monomers, short-chain polymers, additives and PI residues are prone to leaching when the part is in aqueous media (Tables 1–3). For this reason, dentures and orthodontic devices, made using proprietary resins, are routinely stored in water for up to 24 h (depending on the type of resin) to allow uncured compounds to be released, thus reducing unwanted side effects [30]. For microfluidic LOC and biomedical devices made using proprietary photopolymer resins, coating with PDMS, a nontoxic and transparent elastomer, has also been used to overcome the limitation of unknown surface chemistry and toxicity [114,115].

Biocompatibility increases when polymerizing and post-curing light sources, and emission regimes, are matched to meet the absorption requirements of the resin compounds [46]. To optimize polymerization, the absorption maximum of PIs present in the photopolymer resin should overlap with the wavelength emitted by the polymerizing light source [45,116]. Commonly used light sources for polymerization are light-emitting diodes (LEDs), medium pressure mercury (MPM) lamps, and the UV-light-emitting diodes (UV LEDs). LED curing units emit at two wavelength peaks (400 nm and 465 nm), MPM lamps emit at one wavelength peak (365 nm) and UV LED can emit at four wavelength peaks (320 nm, 345 nm, 365 nm, and 390 nm). The total energy (light intensity *times* exposure time) irradiating the surface is also essential to polymerization [117]. Thus, it is necessary that the chosen polymerizing light source emits a sufficient intensity, and at wavelengths matching the absorption range of the PIs present in the SL resin.

Differences in the structure of printed parts will influence the effectiveness of post-curing treatments. To maximize exposure of the printed parts surface to the polymerizing light source, commercial curing ovens often have a rotating inner table and several floodlights that project from the inner walls and ceiling. However, LOCs have enclosed inner channels that are likely to be shaded from polymerizing light. The degree of which will depend on the penetration depth of the light source through the part, the thickness of the part, and part geometry. Shading will result in incomplete curing and less than optimal mechanical properties. For this reason, post-curing alone may not effectively remove toxicity.

Photopolymer resins classed as biocompatible require thorough post-cleaning [113]. For example, a standard cleaning procedure for a part printed in VisiJet clear consists of soaking the part in isopropyl alcohol (IPA), brushing lightly if needed, rinsing in IPA, and air-drying, followed by 10–30 min of post-curing. Cleaning procedures required for a USP Class VI certification involve soaking the part for 20 min in IPA, followed by scrubbing, and afterward four repeat cycles of soaking the part for 5 min in IPA, again followed by scrubbing. Each repeat soak is to be conducted in fresh IPA. Subsequently excess solvent is removed from the surface of the part using clean compressed air, and the part is air-dried for a minimum of 6 h, during which the part must be flipped periodically to facilitate equal drying. After this, the part must be cured for 1 h per side in a ProJet Curing Unit.

To reduce leaching, Macdonald et al. [13] experimented with coating SLA printed parts with wax and found that the coating was only effective at delaying the onset of toxicity (by ~40 h). The authors of van den Driesche et al. [118] removed the toxicity of parts printed by DLP in resins, E-Shell 300 and HTM140, by coating them in parylene-C, which is an ultra-thin biocompatible polymer coating. Coating parts with biocompatible hydrogels (e.g., PEG-DA) or PDMS may also be an effective strategy to remove toxicity of easily accessible photopolymeric resins.

In another attempt to mitigate toxicity of photopolymeric resins by minimizing leaching in situ, the authors of Popov and Evseev [119] experimented with post-processing SLA-printed anatomical implants with supercritical carbon dioxide and found that, without treatment, the implant caused severe inflammation, whereas, with treatment, the implant was biocompatible [119].

Commercially available resins vary in how effective post-curing and cleaning techniques are at reducing toxicity. The authors of Macdonald et al. [13] treated parts by washing them in 99% ethanol and found that this increased biocompatibility of some printed parts, such as Fototec 7150 polymers. However, VisiJet Crystal still showed toxicity after treatment and 100% mortality of zebrafish embryos at 72 h was observed. The authors of Oskui et al. [17] reduced toxicity of the Form Clear SLA polymer by post-curing each side for 30 min. However, while survival and hatching rates of embryos increased compared to those exposed to non-treated parts, the majority of larvae still had elevated rates of malformations (e.g., yolk sac and heart edema and slower swim bladder development).

To improve the safety of photosensitive polymers, alternative compounds can replace those that exhibit higher toxicity. PIs that are less toxic and less likely to leach are being developed. For example, PIs derived from grafting or condensing low-molecular-weight PIs to linear, dendritic, or hyper-branched polymers exhibit reduced leaching compared to their corresponding low-molecular-weight analogues [120]. The authors of Nguyen et al. [106] used riboflavin (Vitamin B2) as an alternative PI, and this had the effect of significantly increasing the biocompatibility of 3D-printed objects, compared to those made with other commercial PIs such as Irgacure 2959 and Irgacure 369. Acrylates, which are cytotoxic, can be replaced by less toxic methacrylates and thiol-ene systems (Table 2). In general, cationic photopolymerizable systems are characterized by decreased toxicity and are, to a degree, a viable alternative to acrylate-radical photopolymerization systems Sangermano [40]. Compounds such as methacryloxypropyl trimethoxysilane-zirconium propoxide copolymer have demonstrated biocompatibility and have been used to build microneedles [121]. As previously discussed, materials such as PDMS, ORMOCER, and PEG-DA are considered relatively non-toxic.

6. Outlook

The AM industry is projected to grow by 17.8% by 2025 due to the expansion of AM applications, the adoption of industry standards, the improvements in material quality, and its technological capabilities [2]. Currently, SLA and DLP AM occupy ~32% of the market share, and FDM and SLS,

~36% and ~33%, respectively [20]. Photopolymerization AM (namely, SLA, MJ, and DLP) systems represent by far the largest and most attractive market segment. SLA, DLP, and MPP are AM methods that are highly relevant to LOC design and manufacture as they offer higher resolution than alternatives and can utilize optically transparent materials.

While advances in AM of micro-milllifluidic devices is enabling a rapidly growing number of applications, the studies discussed in this work suggest that caution is required when devices are fabricated for biological applications. Further improvement in resin compositions by, for example, improving the depth of curing and selecting environmentally benign compounds will mitigate toxicity and simultaneously improve print resolution. The treatment of parts with organic solvents and dedicated surface modifications has been shown to minimize toxicity in some cases. These methods can be considered as a practical, albeit limited, stopgap solution to improve biocompatibility. However, due to limited knowledge about proprietary polymers and their interactions with biological specimens, such methods lack in-depth studies evaluating their effectiveness. Moreover, solvent extraction methods during post-processing steps are inherently variable and may not remove the potential for long-term effects resulting in chronic toxicity.

To solve biocompatibility issues, ongoing research efforts aim to develop AM systems proficient at using biologically compatible substrata such as ORMOCER PDMS and PEG-DA [114,122–124]. Despite some recent progress, these techniques are still at early stages of development and not readily available on the market.

This review highlights that, despite the obvious advantages of AM, the leaching and toxicity mechanisms of AM photopolymers need to be clarified so that innovative AM technologies can be used to advance the LOC field. In this context, we also have to become aware of the larger picture associated with AM technologies. We expect that, with their increasingly use, many relevant questions about waste disposal and about environmental and human health effects will soon emerge. Relatively little is known about long-term impacts of AM materials on human health as well as wellbeing of the environment. New research efforts exploring the above and related issues are necessary [125].

Acknowledgments: All sources of funding of the study should be disclosed. Please clearly indicate grants that you have received in support of your research work. Clearly state if you received funds for covering the costs to publish in open access.

Conflicts of Interest: The authors declare no conflict of interest.

Abbreviations

The following abbreviations are used in this manuscript:

MDPI	Multidisciplinary Digital Publishing Institute
AM	Additive manufacturing
3D	Three-dimensional
CAD	Computer-assisted design
LOC	Lab-on-a-Chip
ABS	Acrylonitrile butadiene styrene
PLA	Polylactic acid
SLA	Stereolithography
MJ	Material jetting
DLP	Digital light processing
2PP	Multiphoton polymerization
PI	Photoinitiator
UV	Ultra-violet
TEGDMA	Triethylene glycol dimethacrylate
HEMA	2-hydroxyethyl methacrylate
w/w	weight-to-weight
GC-MS	Gas Chromatography-Mass Spectrometry
1-HCHPK	1-hydroxycyclohexyl phenyl ketone

LED	Light-emitting diodes
MPM	Medium pressure mercury
USP	United States Pharmacopeia
PDMS	Poly(dimethylsiloxane)
PEG-DA	Poly(ethylene glycol) diacrylate
OECD	Organisation for Economic Co-operation and Development
FET	Fish embryo toxicity
EC	Effective concentration
LC	Lethal concentration
LD	Lethal dose
TPO	Diphenyl(2,4,6-trimethylbenzoyl)phosphine oxide
BP-3	Benzophenone-3
BP-4	Benzophenone-4
BAPO	Bis Acyl Phosphine oxide
MMA	Methyl methacrylate
UDMA	Urethane dimethacrylate
ROS	Reactive oxygen species
MW	Molecular weight
IPA	Isopropyl alcohol

References

1. Ligon, S.C.; Liska, R.; Stampfl, J.; Gurr, M.; Mulhaupt, R. Polymers for 3D Printing and Customized Additive Manufacturing. *Chem. Rev.* **2017**, *117*, 10212–10290.

2. Transparency Market Research (TMR). 3D Printing Market (Use—Commercial and Personal; Technology—PolyJet, Fused Deposition Modeling (FDM), Selective Laser Sintering (SLS), and Stereolithography (SLA); By Application—Consumer Products and Electronics, Automotive, Medical, Industrial, Aerospace, Military & Defense, Architecture, and Education)—Global Industry Analysis, Size, Share, Growth, Trends and Forecast 2017–2025. Available online: https://www.transparencymarketresearch.com/3d-printing-industry.html (accessed on 23 Feburary 2018).

3. Petrick, I.J.; Simpson, T.W. 3D Printing Disrupts Manufacturing: How Economies of One Create New Rules of Competition. *Res. Technol. Manag.* **2013**, *56*, 12–16.

4. Waheed, S.; Cabot, J.M.; Macdonald, N.P.; Lewis, T.; Guijt, R.M.; Paull, B.; Breadmore, M.C. 3D printed microfluidic devices: Enablers and barriers. *Lab Chip* **2016**, *16*, 1993–2013.

5. Campana, O.; Wlodkowic, D. The undiscovered country: Ecotoxicology meets microfluidics. *Sens. Actuators B Chem.* **2018**, *257*, 692–704.

6. Lee, J.Y.; An, J.; Chua, C.K. Fundamentals and applications of 3D printing for novel materials. *Appl. Mater. Today* **2017**, *7*, 120–133.

7. Bhushan, B.; Caspers, M. An overview of additive manufacturing (3D printing) for microfabrication. *Microsyst. Technol.* **2017**, *23*, 1117–1124.

8. Akagi, J.; Khoshmanesh, K.; Evans, B.; Hall, C.J.; Crosier, K.E.; Cooper, J.M.; Crosier, P.S.; Wlodkowic, D. Miniaturized embryo array for automated trapping, immobilization and microperfusion of zebrafish embryos. *PLoS ONE* **2012**, *7*, e36630.

9. Huang, Y.; Reyes Aldasoro, C.C.; Persoone, G.; Wlodkowic, D. Integrated microfluidic technology for sub-lethal and behavioral marine ecotoxicity biotests. In Proceedings of the SPIE—The International Society for Optical Engineering, Barcelona, Spain, 5–6 May 2015.

10. Huang, Y.; Persoone, G.; Nugegoda, D.; Wlodkowic, D. Enabling sub-lethal behavioral ecotoxicity biotests using microfluidic Lab-on-a-Chip technology. *Sens. Actuators B Chem.* **2016**, *226*, 289–298.

11. Lee, W.; Kwon, D.; Choi, W.; Jung, G.Y.; Au, A.K.; Folch, A.; Jeon, S. 3D-Printed Microfluidic Device for the Detection of Pathogenic Bacteria Using Size-based Separation in Helical Channel with Trapezoid Cross-Section. *Sci. Rep.* **2015**, *5*, 1–7.

12. Zhu, F.; Skommer, J.; Macdonald, N.P.; Friedrich, T.; Kaslin, J.; Wlodkowic, D. Three-dimensional printed millifluidic devices for zebrafish embryo tests. *Biomicrofluidics* **2015**, *9*, doi:10.1063/1.4927379.

13. Macdonald, N.P.; Zhu, F.; Hall, C.J.; Reboud, J.; Crosier, P.S.; Patton, E.E.; Wlodkowic, D.; Cooper, J.M. Lab on a Chip Assessment of biocompatibility of 3D printed photopolymers using zebrafish embryo toxicity. *Lab Chip* **2016**, *16*, 291–297.

14. Fuad, N.M.; Kaslin, J.; Wlodkowic, D. Lab-on-a-Chip imaging micro-echocardiography (iμEC) for rapid assessment of cardiovascular activity in zebrafish larvae. *Sens. Actuators B Chem.* **2018**, *256*, 1131–1141.

15. Wlodkowic, D.; Khoshmanesh, K.; Akagi, J.; Williams, D.E.; Cooper, J.M. Wormometry-on-a-chip: Innovative technologies for in situ analysis of small multicellular organisms. *Cytom. Part A* **2011**, *79A*, 799–813.

16. Alifui-Segbaya, F.; Varma, S.; Lieschke, G.J.; George, R. Biocompatibility of Photopolymers in 3D Printing. *3D Print. Addit. Manuf.* **2017**, *4*, 185–191.

17. Oskui, S.M.; Diamante, G.; Liao, C.; Shi, W.; Gan, J.; Schlenk, D.; Grover, W.H. Assessing and Reducing the Toxicity of 3D-Printed Parts. *Environ. Sci. Technol. Lett.* **2016**, *3*, 1–6.

18. Zhu, F.; Friedrich, T.; Nugegoda, D.; Kaslin, J. Assessment of the biocompatibility of three-dimensional-printed polymers using multispecies toxicity tests. *Biomicrofluidics* **2015**, *9*, doi:10.1063/1.4939031.

19. Zhu, F.; Skommer, J.; Friedrich, T.; Kaslin, J.; Wlodkowic, D. 3D printed polymers toxicity profiling: A caution for biodevice applications. *Proc. SPIE* **2015**, *9668*, doi:10.1117/12.2202392.

20. Columbus, L. The State of 3D Printing, 2017.

21. Fourkas, J.T. Fundamentals of Two-Photon Fabrication. In *Three-Dimensional Microfabrication Using Two-Photon Polymerization*; William Andrew: Norwich, NY, USA, 2016; Chapter 1.3, pp. 45–61.

22. Obata, K.; El-Tamer, A.; Koch, L.; Hinze, U.; Chichkov, B.N. High-aspect 3D two-photon polymerization structuring with widened objective working range (WOW-2PP). *Light Sci. Appl.* **2013**, *2*, 8–11.

23. Horváth, B.; Ormos, P. Nearly Aberration-Free Multiphoton Polymerization into Thick Photoresist Layers. *Micromachines* **2017**, *8*, 219.

24. Sochol, R.D.; Sweet, E.; Glick, C.C.; Venkatesh, S.; Avetisyan, A.; Ekman, K.F.; Raulinaitis, A.; Tsai, A.; Wienkers, A.; Korner, K.; et al. 3D printed microfluidic circuitry via multijet-based additive manufacturing. *Lab Chip* **2016**, *16*, 668–678.

25. Yazdi, A.A.; Popma, A.; Wong, W.; Nguyen, T.; Pan, Y.; Xu, J. 3D printing: An emerging tool for novel microfluidics and lab-on-a-chip applications. *Microfluid. Nanofluid.* **2016**, *20*, 1–18.

26. Au, A.K.; Lee, W.; Folch, A. Mail-Order Microfluidics: Evaluation of Stereolithography for the Production of Microfluidic Devices. *Lab Chip* **2014**, *14*, 1294–1301.

27. Comina, G.; Suska, A.; Filippini, D. 3D printed unibody lab-on-a-chip: Features survey and check-valves integration. *Micromachines* **2015**, *6*, 437–451.

28. Suzuki, H.; Mitsuno, K.; Shiroguchi, K.; Tsugane, M.; Okano, T.; Dohi, T.; Tsuji, T. One-step micromolding of complex 3D microchambers for single-cell analysis. *Lab Chip* **2017**, *17*, 647–652.

29. Comina, G.; Suska, A.; Filippini, D. PDMS lab-on-a-chip fabrication using 3D printed templates. *Lab Chip* **2014**, *14*, 424–430. [arXiv:cs/9605103].

30. Schmalz, G.; Arenholt-Bindslev, D. *Biocompatibility of Dental Materials*; Springer: Berlin/Heidelberg, Germany, 2009; pp. 1–379.

31. Melchels, F.P.W.; Feijen, J.; Grijpma, D.W. Biomaterials A review on stereolithography and its applications in biomedical engineering. *Biomaterials* **2010**, *31*, 6121–6130.

32. Nanoscribe GmbH. *Nanoscribe*; Nanoscribe GmbH.: Eggenstein-Leopoldshafen, Germany.

33. Gittard, S.D.; Ovsianikov, A.; Chichkov, B.N.; Doraiswamy, A.; Narayan, R.J. Two Photon Polymerization of Microneedles for Transdermal Drug Delivery. *Expert Opin. Drug Deliv.* **2010**, *7*, 513–533.

34. Accoto, C.; Qualtieri, A.; Pisanello, F.; Ricciardi, C.; Pirri, C.F.; Vittorio, M.D.; Rizzi, F. Two-Photon Polymerization Lithography and Laser Doppler Vibrometry of a SU-8-Based Suspended Microchannel Resonator. *J. Microelectromech. Syst.* **2015**, *24*, 1038–1042.

35. Li, Y.; Fang, Y.; Wang, J.; Wang, L.; Tang, S.; Jiang, C.; Zheng, L.; Mei, Y. Integrative optofluidic microcavity with tubular channels and coupled waveguides via two-photon polymerization. *Lab Chip* **2016**, *16*, 4406–4414.

36. Fouassier, J.P.; Lalevée, J. Photochemical Production of Interpenetrating Polymer Networks; Simultaneous Initiation of Radical and Cationic Polymerization Reactions. *Polymers* **2014**, *6*, 2588–2610.

37. Ophardt, C.; Reusch, W. *Organic Chemistry*; 2015, Available online: https://chem.libretexts.org/Core/Organic_Chemistry (accessed on 23 Feburary 2018).

38. Bail, R.; Patel, A.; Yang, H.; Rogers, C.M.; Rose, F.R.A.J.; Segal, J.I.; Ratchev, S.M. The effect of a type I photoinitiator on cure kinetics and cell toxicity in projection-microstereolithography. *Procedia-Soc. Behav. Sci.* **2013**, *5*, 222–225.

39. Gotro, J. *Cationic Photopolymerization*; 2016, Available online: https://polymerinnovationblog.com/uv-curing-part-five-cationic-photopolymerization/ (accessed on 23 Feburary 2018).

40. Sangermano, M. Advances in cationic photopolymerization. *Pure Appl. Chem.* **2012**, *84*, 2065–2133.

41. Crivello, J.V. The discovery and development of onium salt cationic photoinitiators. *J. Polym. Sci. Part A Polym. Chem.* **1999**, *37*, 4241–4254.

42. Subramanian, M.N. *Basics of Polymer Chemistry*; River Publishers: London, UK, 2017.

43. Wang, X.; Jiang, M.; Zhou, Z.; Gou, J.; Hui, D. 3D printing of polymer matrix composites: A review and prospective. *Compos. Part B Eng.* **2017**, *110*, 442–458.

44. Lago, M.A.; de Quirós, A.R.-B.; Sendón, R.; Nieto, M.T.; Paseiro, P.; Lago, M.A. Food Additives & Contaminants: Part A Photoinitiators: A food safety review. *Food Addit. Contam. Part A* **2015**, *32*, 779–798.

45. Nowak, D.; Ortyl, J.; Kamińska-Borek, I.; Kukuła, K.; Topa, M.; Popielarz, R. Photopolymerization of hybrid monomers: Part I: Comparison of the performance of selected photoinitiators in cationic and free-radical polymerization of hybrid monomers. *Polym. Test.* **2017**, *64*, 313–320.

46. Czech, Z.; Klementowska, P.; Drzycimska, A. Choosing the right initiator. *Eur. Coat. J.* **2007**, *2*, 26.

47. Clear Photoreactive Resin for Formlabs 3D Printers. Available online: https://formlabs.com/media/upload/Clear__Resin_SDS_EU.pdf (accessed on 21 February 2018).

48. Sigma-Aldrich. *SAFETY DATA SHEET: Diphenyl(2,4,6-trimethylbenzoyl)phosphine Oxide*, version 5.4; Sigma-Aldrich: St. Louis, MO, USA, 2016.

49. Dental LT Clear. Available online: https://formlabs.com/media/upload/DentalLTClear-SDS-EN.pdf (accessed on 21 February 2018).

50. SAFETY DATA SHEET: NextDent Ortho IBT, ID:M-NOIBT-2015-01-UK. Available online: https://nextdent.com/wp-content/uploads/2016/02/UK_SDS-NextDent-Ortho-IBT-V01-2015.pdf (accessed on 21 February 2018).

51. SIGMA-ALDRICH. *SAFETY DATA SHEET: 4-Hydroxyacetophenone*, version 3.5; Sigma-Aldrich: St. Louis, MO, USA, 2016.

52. Anadón, A.; Bell, D.; Binderup, M.L.; Bursch, W.; Castle, L.; Crebelli, R.; Engel, K.H.; Franz, R.; Gontard, N.; Haertlé, T.; et al. Toxicological Evalatuion of Benzophenone. *Eur. Food Saf. Auth.* **2009**, *2009*, doi:10.2903/j.efsa.2009.1104.

53. SIGMA-ALDRICH. *SAFETY DATA SHEET: Benzophenone*, version 3.12; Sigma-Aldrich: St. Louis, MO, USA, 2017.

54. SIGMA-ALDRICH. *SAFETY DATA SHEET: 1-Hydroxycyclohexyl Phenyl Ketone*, version 5.2; Sigma-Aldrich: St. Louis, MO, USA, 2014.

55. 3D Systems Inc. *Safety Data Sheet: VisiJet SL Clear, ID: 24672-S12-04-A*; 3D Systems Inc.: Rock Hill, SC, USA, 2016.

56. Du, Y.; Wang, W.Q.; Pei, Z.T.; Ahmad, F.; Xu, R.R. Acute Toxicity and Ecological Risk Assessment of in Ultraviolet (UV)-Filters. *Int. J. Environ. Res. Public Health* **2017**, *14*, 414.

57. Zhang, J.; Xiao, P.; Dietlin, C.; Campolo, D.; Dumur, F.; Gigmes, D.; Morlet-savary, F.; Fouassier, J.P.; Lalevée, J. Cationic Photoinitiators for Near UV and Visible LEDs: A Particular Insight into One-Component Systems. *Macromol. Chem. Phys.* **2016**, *217*, 1214–1227.

58. Maurya, S.D.; Kurmvanshi, S.K.; Mohanty, S.; Nayak, S.K. A Review on Acrylate-Terminated Urethane Oligomers and Polymers: Synthesis and Applications. *Polym.-Plast. Technol. Eng.* **2017**, *57*, 625–656.

59. Gittard, S.D.; Narayan, R.J. Laser direct writing of micro- and nano-scale medical devices. *Expert Rev. Med. Dev.* **2010**, *7*, 343–356.

60. Ovsianikov, A.; Chichkov, B.; Mente, P.; Monteiro-Riviere, N.A.; Doraiswamy, A.; Narayan, R.J. Two photon polymerization of polymer-ceramic hybrid materials for transdermal drug delivery. *Int. J. Appl. Ceram. Technol.* **2007**, *4*, 22–29.

61. Haas, K.H.; Wolter, H. Synthesis, properties and applications of inorganic-organic copolymers (ORMOCER®s). *Curr. Opin. Solid State Mater. Sci.* **1999**, *4*, 571–580.

62. Comparative evaluation of residual monomer content and polymerization shrinkage of a packable composite and an ormocer. *J. Conserv. Dent.* **2012**, *15*, 161, doi:10.4103/0972-0707.94592

63. Webster, R.; Elliott, V.; Park, B.K.; Walker, D.; Hankin, M.; Taupin, P. PEG and PEG conjugates toxicity: Towards an understanding of the toxicity of PEG and its relevance to PEGylated biologicals. In *PEGylated Protein Drugs: Basic Science and Clinical Applications*; Veronese, F.M., Ed.; Birkhäuser: Basel, Switzerland, 2009; pp. 127–146.

64. McAvoy, K.; Jones, D.; Thakur, R. Synthesis and Characterisation of Photocrosslinked poly(ethylene glycol) diacrylate Implants for Sustained Ocular Drug Delivery. *Pharm. Res.* **2018**, *35*, doi:10.1007/s11095-017-2298-9.

65. Hahn, M.S.; Taite, L.J.; Moon, J.J.; Rowland, M.C.; Ruffino, K.A.; West, J.L. Photolithographic patterning of polyethylene glycol hydrogels. *Biomaterials* **2006**, *27*, 2519–2524.

66. Urrios, A.; Parra-Cabrera, C.; Bhattacharjee, N.; Gonzalez-Suarez, A.M.; Rigat-Brugarolas, L.G.; Nallapatti, U.; Samitier, J.; DeForest, C.A.; Posas, F.; Garcia-Cordero, J.L.; et al. 3D-printing of transparent bio-microfluidic devices in PEG-DA. *Lab Chip* **2016**, *16*, 2287–2294.

67. Traore, M.A.; Behkam, B. A PEG-DA microfluidic device for chemotaxis studies. *J. Micromech. Microeng.* **2013**, *23*, doi:10.1088/0960-1317/23/8/085014.

68. Femmer, T.; Kuehne, A.J.C.; Wessling, M. Print your own membrane: Direct rapid prototyping of polydimethylsiloxane. *Lab Chip* **2014**, *14*, 2610.

69. He, Y.; Qiu, J.; Fu, J.; Zhang, J.; Ren, Y.; Liu, A. Printing 3D microfluidic chips with a 3D sugar printer. *Microfluid. Nanofluid.* **2015**, *19*, 447–456.

70. Safety Data Sheet: E-shell 600 Clear. Available online: https://envisiontec.com/wp-content/uploads/2016/09/MSDS-EShell-600.pdf (accessed on 21 February 2018).

71. Autodesk-Inc. *SAFETY DATA SHEET: Autodesk Resin: PR57-K-v.2 Black*; Autodesk-Inc.: San Rafael, CA, USA, 2016.

72. Dormer, W.; Gomes, R.; Meek, M. *Concise International Chemical Assessment Document: Methyl methacrylate*; Technical Report; World Health Organization: Geneva, Switzerland, 1998.

73. BASF CORPORATION. *Safety Data Sheet: METHYL ACRYLATE*, version: 3; BASF CORPORATION: Ludwigshafen, Germany, 2016.

74. SIGMA-ALDRICH. *SAFETY DATA SHEET: Methyl Methacrylate*, version 5.4; Sigma-Aldrich: St. Louis, MO, USA, 2015.

75. SIGMA-ALDRICH. *SAFETY DATA SHEET: Methyl Acrylate*, version 3.9; Sigma-Aldrich: St. Louis, MO, USA, 2015.

76. Allnex-USA. *SAFETY DATA SHEET: Tripropyleneglycol Diacrylate*, version 0021157; Allnex-USA: Alpharetta, GA, USA, 2016.

77. SIGMA-ALDRICH. *Safety Data Sheet: 2-Hydroxyethyl Methacrylate*, version 4.9; Sigma-Aldrich: St. Louis, MO, USA, 2015.

78. Safety Data Sheet: Accura 40. Available online: http://infocenter.3dsystems.com/materials/sites/default/files/sds-files/production/sla/Accura_40/24022-s12-02-asds_ghsenglishaccura_40.pdf (accessed on 21 February 2018).

79. PAGEL Spezial-Beton GmbH & Co. KG. *Safety Data Sheet: EH130 A*, version 2.21; PAGEL Spezial-Beton GmbH & Co. KG: Essen, Germany, 2016.

80. Autodesk-Inc. *SAFETY DATA SHEET: Autodesk PR57 CMYKW Resins*; Autodesk-Inc.: San Rafael, CA, USA, 2016.

81. SIGMA-ALDRICH. *SAFETY DATA SHEET: Methacrylic Acid*, version 4.7; Sigma-Aldrich: St. Louis, MO, USA, 2014.

82. Liu, H.; He, C. Additive use in Photopolymer Resin for 3D Printing to Enhance the Appearance of Printed Parts. U.S. Patent US9574039B1, 21 February 2017.

83. Sun, B.J.; Kennedy, C.R.; Sundar, V.; Lichkus, A.M. Three-Dimensional Fabricating Material Systems for Producing Dental Products. U.S. Patent US20140131908A1, 15 May 2014.

84. Murphy, E.J.; Ansel, R.E.; Krajewski, J.J. Method of forming a three-dimensional object by stereolithography and composition therefore, U.S. Patent US4942001A, 17 July 1990.

85. Ramos, M.J.; Harrison, J.P.; Coats, A.L.; Hay, J.S.; Harrison, J.P.; Hay, J.S.; Ramos, M.J. Stereolithography Resins and Methods. U.S. Patent US7211368B2, 1 May 2007.

86. Steinkraus, W.J.; Woods, J.; Rooney, J.M.; Jacobine, A.F.; Glaser, D.M. Thiolene Compositions on Based Bicyclic 'Ene Compounds. U.S. Patent US4808638A, 28 February 1989.

87. Sun, F. Methods for Making Dental Restorations Using Two-Phase Light Curing Materials. U.S. Patent US7939575B2, 10 May 2011.

88. Vanmaele, L.; Daems, E.; De Voeght, F.; Van Thillo, E. 3D-Inkjet Printing Methods. U.S. Patent US8142860B2, 27 March 2012.

89. EnvisionTEC. *Safety Data Sheet: Photopolymer ABS Tough Series (Includes ABS Tough, ABS Tough M, ABS 3SP Tough)*; EnvisionTEC: Gladbeck, Germany, 2015.

90. Short, D.B.; Volk, D.; Badger, P.D.; Melzer, J.; Salerno, P.; Sirinterlikci, A. 3D Printing (Rapid Prototyping) Photopolymers: An Emerging Source of Antimony to the Environment. *3D Print. Addit. Manuf.* **2014**, *1*, 24–33.

91. Geurtsen, W. Biocompatibility of resin-modified filling materials. *Crit. Rev. Oral Biol. Med.* **2000**, *11*, 333–355.

92. Azimi, P.; Zhao, D.; Pouzet, C.; Crain, N.E.; Stephens, B. Emissions of Ultrafine Particles and Volatile Organic Compounds from Commercially Available Desktop Three-Dimensional Printers with Multiple Filaments. *Environ. Sci. Technol.* **2016**, *50*, 1260–1268.

93. Zhu, F.; Macdonald, N.P.; Cooper, J.M.; Wlodkowic, D. Additive manufacturing of lab-on-a-chip devices: Promises and challenges. *SPIE—Int. Soc. Opt. Eng.* **2013**, *8923*, doi:10.1117/12.2033400.

94. Yang, Y.; He, X.; Shi, J.; Hickel, R.; Reichl, F.X.; Högg, C. Effects of antioxidants on DNA double-strand breaks in human gingival fibroblasts exposed to dental resin co-monomer epoxy metabolites. *Dent. Mater.* **2017**, *33*, 418–426.

95. De Souza Costa, C.A.; Do Nascimento, A.B.L.; Teixeira, H.M. Response of human pulps following acid conditioning and application of a bonding agent in deep cavities. *Dent. Mater.* **2002**, *18*, 543–551.

96. SIGMA-ALDRICH. *SAFETY DATA SHEET: Butylated Hydroxytoluene*, version 5.2.; Sigma-Aldrich: St. Louis, MO, USA, 2014.

97. SIGMA-ALDRICH. *SAFETY DATA SHEET: 1-Bis(2,2,6,6-tetramethyl-4-piperidyl) Sebacate*, version 4.4; Sigma-Aldrich: St. Louis, MO, USA, 2015.

98. SIGMA-ALDRICH. *SAFETY DATA SHEET: Hydroquinone*, version 4.2; Sigma-Aldrich: St. Louis, MO, USA, 2017.

99. Cebe, M.A.; Cebe, F.; Cengiz, M.F.; Cetin, A.R.; Arpag, O.F.; Ozturk, B. Elution of monomer from different bulk fill dental composite resins. *Dent. Mater.* **2015**, *31*, e141–e149.

100. Yamaji, K.; Kawasaki, Y.; Yoshitome, K.; Matsunaga, H.; Sendo, T. Quantitation and Human Monocyte Cytotoxicity of the Polymerization Agent 1-Hydroxycyclohexyl Phenyl Ketone (Irgacure 184) from Three Brands of Aqueous Injection Solution. *Biol. Pharm. Bull.* **2012**, *35*, 1821–1825.

101. Ferracane, J.L.; Condon, J.R. Post-cure heat treatments for composites: Properties and fractography. *Dent. Mater.* **1992**, *8*, 290–295.

102. Ogunyinka, A.; Palin, W.M.; Shortall, A.C.; Marquis, P.M. Photoinitiation chemistry affects light transmission and degree of conversion of curing experimental dental resin composites. *Dent. Mater.* **2007**, *23*, 807–813.

103. Aparicio, J.L.; Elizalde, M. Migration of photoinitiators in food packaging: A review. *Packag. Technol. Sci.* **2015**, *28*, 181–203.

104. Schmalz, G.; Galler, K.M. Biocompatibility of biomaterials – Lessons learned and considerations for the design of novel materials. *Dent. Mater.* **2017**, *33*, 382–393.

105. Anderson, W.A.; Castle, L. Benzophenone in cartonboard packaging materials and the factors that influence its migration into food. *Food Addit. Contam.* **2003**, *20*, 607–618.

106. Nguyen, A.K.; Gittard, S.D.; Koroleva, A.; Schlie, S.; Gaidukeviciute, A.; Chichkov, B.N.; Narayan, R.J. Two-photon polymerization of polyethylene glycol diacrylate scaffolds with riboflavin and triethanolamine used as a water-soluble photoinitiator. *Regen. Med.* **2013**, *8*, 725–738.

107. Stohs, S.J. The role of free radicals in toxicity and disease. *J. Basic Clin. Physiol. Pharmacol.* **1995**, *6*, 205–228.

108. Finkel, T.; Holbrook, N.J. Oxidants, oxidative stress and the biology of ageing. *Nature* **2000**, *408*, 239–247.

109. Schweikl, H.; Petzel, C.; Bolay, C.; Hiller, K.A.; Buchalla, W.; Krifka, S. 2-Hydroxyethyl methacrylate-induced apoptosis through the ATM- and p53-dependent intrinsic mitochondrial pathway. *Biomaterials* **2014**, *35*, 2890–2904.

110. Atsumi, T.; Murata, J.; Kamiyanagi, I.; Fujisawa, S.; Ueha, T. Cytotoxicity of photosensitizers camphorquinone and 9-fluorenone with visible light irradiation on a human submandibular-duct cell line in vitro. *Arch. Oral Biol.* **1998**, *43*, 73–81.

111. Geurtsen, W. Substances released from dental resin composites and glass ionomer cements. *Eur. J. Oral Sci.* **1998**, *106*, 687–695.
112. Scholz, S.; Fischer, S.; Gündel, U.; Küster, E.; Luckenbach, T.; Voelker, D. The zebrafish embryo model in environmental risk assessment—Applications beyond acute toxicity testing. *Environ. Sci. Pollut. Res.* **2008**, *15*, 394–404.
113. 3D Systems Inc. *VisiJet SL Clear cleaning procedure for USP Class VI*; Library Technology Reports; 3D Systems Inc.: Rock Hill, SC, USA, 2013; pp. 1–2.
114. Hinton, T.J.; Hudson, A.; Pusch, K.; Lee, A.; Feinberg, A.W. 3D Printing PDMS Elastomer in a Hydrophilic Support Bath via Freeform Reversible Embedding. *ACS Biomater. Sci. Eng.* **2016**, *2*, 1781–1786.
115. Gross, B.C.; Anderson, K.B.; Meisel, J.E.; McNitt, M.I.; Spence, D.M. Polymer Coatings in 3D-Printed Fluidic Device Channels for Improved Cellular Adherence Prior to Electrical Lysis Bethany. *Anal. Chem.* **2015**, *87*, 6335–6341.
116. Abedin, F.; Ye, Q.; Camarda, K.; Spencer, P. Impact of light intensity on the polymerization kinetics and network structure of model hydrophobic and hydrophilic methacrylate based dental adhesive resin. *J. Biomed. Mater. Res. B Appl. Biomater.* **2016**, *104*, 139–148.
117. Emami, N.; Söderholm, K.J.M.; Berglund, L.A. Effect of light power density variations on bulk curing properties of dental composites. *J. Dent.* **2003**, *31*, 189–196.
118. van den Driesche, S.; Lucklum, F.; Bunge, F.; Vellekoop, M. 3D Printing Solutions for Microfluidic Chip-To-World Connections. *Micromachines* **2018**, *9*, 71.
119. Popov, V.K.; Evseev, A.V. Laser stereolithography and supercritical fluid processing for custom-designed implant fabrication. *J. Mater. Sci. Mater. Med.* **2004**, *15*, 123–128.
120. Han, Y.; Wang, F.; Lim, C.; Chi, H.; Chen, D.; Wan, F. High-Performance Nano-Photoinitiators with Improved Safety for 3D Printing. *Appl. Mater. Interfaces* **2017**, *9*, 32418–32423.
121. Versace, D.L.; Soppera, O.; Lalevée, J.; Croutxé-Barghorn, C. Influence of zirconium propoxide on the radical induced photopolymerisation of hybrid sol–gel materials. *New J. Chem.* **2008**, *32*, 2270–2278.
122. Lee, D.S.; Kim, S.J.; Sohn, J.H.; Kim, I.G.; Kim, S.W.; Sohn, D.W.; Kim, J.H.; Choi, B. Biocompatibility of a PDMS-coated micro-device: Bladder volume monitoring sensor. *Chin. J. Polym. Sci. (Engl. Ed.)* **2012**, *30*, 242–249.
123. Peterson, S.L.; McDonald, A.; Gourley, P.L.; Sasaki, D.Y. Poly(dimethylsiloxane) thin films as biocompatible coatings for microfluidic devices: Cell culture and flow studies with glial cells. *J. Biomed. Mater. Res—Part A* **2005**, *72*, 10–18.
124. xia Zheng, G.; jie Li, Y.; lin Qi, L.; ming Liu, X.; Wang, H.; ping Yu, S.; hua Wang, Y. Marine phytoplankton motility sensor integrated into a microfluidic chip for high-throughput pollutant toxicity assessment. *Mar. Pollut. Bull.* **2014**, *84*, 147–154.
125. Kellens, K.; Baumers, M.; Gutowski, T.G.; Flanagan, W.; Lifset, R.; Duflou, J.R. Environmental Dimensions of Additive Manufacturing: Mapping Application Domains and Their Environmental Implications. *J. Ind. Ecol.* **2017**, *21*, S49–S68.

micromachines

MDPI

Article

Highly Fluorinated Methacrylates for Optical 3D Printing of Microfluidic Devices

Frederik Kotz [†], Patrick Risch [†], Dorothea Helmer and Bastian E. Rapp *

Institute of Microstructure Technology, Karlsruhe Institute of Technology, Hermann-von-Helmholtz-Platz 1, 76344 Eggenstein-Leopoldshafen, Germany; Frederik.Kotz@kit.edu (F.K.); Patrick.Risch@kit.edu (P.R.); Dorothea.Helmer@kit.edu (D.H.)
* Correspondence: Bastian.Rapp@kit.edu; Tel.: +49-721-608-28981
† Contributed equally to this work.

Received: 31 January 2018; Accepted: 7 March 2018; Published: 8 March 2018

Abstract: Highly fluorinated perfluoropolyether (PFPE) methacrylates are of great interest for transparent and chemically resistant microfluidic chips. However, so far only a few examples of material formulations for three-dimensional (3D) printing of these polymers have been demonstrated. In this paper we show that microfluidic chips can be printed using these highly fluorinated polymers by 3D stereolithography printing. We developed photocurable resin formulations that can be printed in commercial benchtop stereolithography printers. We demonstrate that the developed formulations can be printed with minimal cross-sectional area of 600 μm for monolithic embedded microfluidic channels and 200 μm for open structures. The printed and polymerized PFPE methacrylates show a good transmittance above 70% at wavelengths between 520–900 nm and a high chemical resistance when being exposed to organic solvents. Microfluidic mixers were printed to demonstrate the great variability of different designs that can be printed using stereolithography.

Keywords: 3D printing; perfluoropolyether; additive manufacturing; microfluidics; stereolithography

1. Introduction

Three-dimensional (3D) printing is of high interest for microfluidics due to the multitude of channel systems that need to be tested in the course of lab-on-a-chip device development. With 3D printing, prototypes and devices can be produced in a fast and effective manner. While many different techniques exist for 3D printing of microfluidic devices [1–3], for example, fused deposition modeling (FDM) [4], inkjet printing [5], two-photon printing [6], or stereolithography. While two-photon printing has been used to fabricate optical submicron structures [7,8], stereolithography remains the method of choice for most laboratories to fabricate microfluidic chips as it combines affordable machinery with high resolution [9]. Microfluidics started out with chips made from silicon and glass, but today polymers (thermosets, elastomers) are the preferred material class because polymer processing and structuring is considerably easier when compared to glass and silicon. One of the main perspectives of microfluidics is automation and miniaturization in biological as well as in chemical devices. While alternative materials, such as glass [10,11], are extremely chemically resistant, polymers are still the preferred choice for rapid fabrication as they do not require specialized fabrication equipment [12,13]. However, most polymers that are used for printing and fabrication of microfluidic chips are not stable when exposed to even low concentrations of chemicals, especially organic solvents. This is why there has been increasing interest in recent years in developing resistant materials, which could pave the way to simple fabrication techniques of solvent-resistant microfluidic chips. Concerning durability and chemical resistance, fluoropolymers stand out specifically. Fluorination affords the lowest known surface energies: the difluoromethylene ($-CF_2-$) group, the difluoromethyl group ($-CF_2H$), and the trifluoromethyl group ($-CF_3$) possess surface energies of ~18 mN/m, ~15 mN/m, and ~6 mN/m, respectively [14]. For comparison, methyl groups (-CH$_3$) have surface energies of ~23 mN/m [14]. In addition, the carbon/fluorine bond is the shortest bond in organic chemistry which makes fluorinated polymers outstandingly stable also in direct contact with organic solvents. There is a high interest in using fluoropolymers for microfluidic chips, and many applications have been reported. Commercial fluorinated thermoplasts such as Viton and Dyneon can be structured by hot embossing [15,16], or selectively bonded to parts of the chip that require chemical stability [17]. Commercially available perfluoropolyether (PFPE) polymers, like SIFEL, have been used for chip fabrication via, e.g., casting [18] or spin-coating [19]. Another strategy is to employ chemical vapor deposition techniques to deposit fluoropolymer films on microfluidic devices [20]. Photolithographic direct structuring of highly fluorinated polymers has been first reported using PFPE acrylates [21], which are now widely used for the fabrication of microfluidic devices [22,23]. Despite the interesting features of fluoropolymers, few attempts have been reported to structure fluoropolymers via 3D printing or for the fabrication of microfluidic devices. Recently, a method to fabricate pillar arrays (100 μm diameter, 400 μm height, 50 μm layer thickness, 150–200 μm spacing) by printing a custom-synthesized PFPE tetraacrylate was reported [24]. Here, we present a simple method for 3D printing of PFPE dimethacrylates for the fabrication of microfluidic devices. To the best of our knowledge, this is the first technique for 3D printing of microfluidic chips in highly fluorinated polymers.

2. Materials and Methods

Materials: All of the chemicals were used as received and were not purified any further. PFPE Fluorolink MD700 was purchased from Acota, Shrewsbury, UK. Acetone, 2-propanol, methanol (MeOH), dichloromethane (DCM), dimethylformamide (DMF), tetrahydrofuran (THF), toluene, and *n*-heptane were purchased from Merck, Darmstadt, Germany. Diphenyl(2,4,6-trimethylbenzoyl) phosphine oxide (TPO), phenylbis(2,4,6-trimethylbenzoyl)phosphine oxide (PPO) was purchased from Sigma Aldrich, Taufkirchen, Germany. Tinuvin 384-2 (T384-2) and Tinuvin 326 (T326) were kindly provided by BASF, Ludwigshafen, Germany. Sudan Orange G (SOG) was purchased from Sigma Aldrich. Elastosil RT 601 A/B was purchased from Wacker, Munich, Germany. Formlabs Tough was purchased from Formlabs, Somerville, MA, USA. The microfluidic channels were filled with a black, blue, and yellow printing ink that was purchased from ESM online, Hirschberg, Germany.

Stereolithography: The benchtop stereolithography printer Asiga Pico 2 was used for printing, postcuring of the printed parts was done with the ultraviolet radiation lamp Asiga Flash DR-301C (both Asiga, Alexandria, Australia). Single layer thickness was 50 μm and an overcuring of 125 μm was adjusted to avoid delamination of the individual layers during the printing process. Single layer thickness of 50 μm was chosen for giving good printing results at adequate printing speeds. The overcure of 125 μm was necessary to prevent the parts from delaminating during printing. Overcuring does not cause the channel structures to be blocked because the Asiga Slicing Software compensates for the overcuring when void or free hanging structures are printed. Compensation is achieved by the so-called z-compensation, which adds the overcure thickness to the size of the channel void (see Supplementary Figure S1 for details). The spectrum of the blue LED was measured using a BTC112E Fibre Coupled TE Cooled Linear CCD array spectrometer (B&W Tek, Newark, DE, USA). UV-Vis measurements were performed on an Evolution 201 UV-Vis spectrophotometer (Thermo Fisher, Waltham, MA, USA). Quartz Suprasil high precision cells were purchased from Hellma, Müllheim, Germany. HybriWell chambers were purchased from Grace Bio-Labs, Bend, OR, USA. Layer thicknesses were measured using a MT 60 M length gauge (Heidenhain, Traunreut, Germany). A Stemi 508 microscope (Zeiss, Oberkochen, Germany) with an AxioCam ERc 5s was used for microscopy.

LED spectrum measurements: The spectrum of the blue LED of the Asiga Pico 2 was measured using a BTC112E spectrometer. A background spectrum of the surrounding area was recorded, then the printer LED light was switched on and the spectrometer was held above the optical setup, which was close to the printing focus of the (removed) build tray. Spectra were recorded with the BWSpec software (version 4.03_23_C, B&W Tek, Newark, DE, USA). Spectra have been normalized to the maximum value.

UV-Vis measurements: Initiators were weighed into microcentrifuge tubes, dissolved in acetone and diluted to a concentration of 12.5 mg/mL. The solutions were transferred to a quartz glass Suprasil high precision cell and measured in an UV-Vis spectrophotometer against a blank measurement of pure acetone. Absorbers were weighed in microcentrifuge tubes, dissolved in acetone, and were diluted to a concentration of 3.76 mg/mL. The SOG solution was further diluted to a concentration of 0.12 mg/mL.

Preparation of printing mixtures: Stock solutions of the initiator TPO or PPO and acetone were prepared in a vortex mixer of type Reax Top (Heidolph, Schwabach, Germany). The required amount of absorber SOG, T326, or T384-2 was added in a glass vial and the initiator/acetone stock solution was added up to their respective concentrations. After mixing with the vortex mixer, the PFPE Fluorolink MD700 was added to this solution up to the final concentrations listed in Table 1. After agitating the glass vial, entrapped air bubbles in the solution where removed by using an ultrasonic bath Sonorex Da 300 (Bandelin electronic, Berlin, Germany) for 2 min at 25 °C.

Table 1. Composition of printing formulations mix 1–5.

	Mix 1	Mix 2	Mix 3	Mix 4	Mix 5
Initiator (mg/mL) [1]	6.25 TPO	9 PPO	6.25 TPO	6.25 TPO	9 PPO
Absorber (mg/mL) [1]	1.88 T326	10 T384-2	0.235 SOG	-	0.5 T326

[1] Final concentration in MD700.

Preparation of poly(dimethylsiloxane) (PDMS): Elastosil RT 601 A and B were mixed in a ratio of 9:1 wt %. Entrapped air bubbles were removed using a desiccator and a vacuum pump. The material was polymerized at a temperature of 70 °C for 30 min.

Determination of optimal printing parameters: To determine the layer thicknesses in dependence of the exposure time under the LED of the printer, the printing mixtures were pipetted into a rectangular cutout of a PDMS form to a height of 2 mm and were placed above a circular light spot of the printer with a diameter of approximately 3 mm and were exposed to the light for time periods between 4 s and 20 s. The polymer layer thicknesses were measured at three different spots using a length gauge. The results were plotted and analyzed for the slope of the resulting plots. The light intensity was set to 8.8 mW·cm^{-2}.

Solvent compatibility: Solvent compatibility of the printed PFPE material was determined using 8 mm × 8 mm × 3 mm printed blocks of mix 3 and 5. The blocks were weighted three times and consecutively immersed in eight solvents (water, MeOH, DCM, DMF, THF, toluene, acetone, and *n*-heptane—approximately 5 mL each) for 24 h. The blocks were retrieved from the solvents, the solvent on the outside was quickly dabbed away and the blocks were immediately placed on the balance and the initial weight was recorded. The block was then re-immersed in the solvent and after weighting of the next eight blocks, the block was measured again. In this way, the weight was determined a total of three times.

Contact angle measurement: Contact angles were measured with an OCA15 Pro (Data Physics, San Jose, CA, USA). Static contact angles were measured using 5 µL water droplets. The surface energy was calculated using the Owens, Wendt, Rabel, and Kaelble (OWRK) method. 5 µL of diiodomethan and water were used as the nonpolar and polar testing liquid, respectively.

3. Results and Discussion

Several printing mixtures were used to produce PFPE microfluidic devices. To determine the optimum mixture, spectra of initiators and absorbers were recorded, mixtures were prepared and the optimum printing parameters were tested. Microfluidic devices were printed and tested for their performance. The solvent compatibility and the spectra of the printed materials were characterized.

3.1. UV-Vis Spectra of Initiators and Absorbers

In order to obtain embedded microchannels with a height of less than 1 mm, a printing formulation requires an effective initiator and a significant amount of absorber, thus ensuring that the light does not penetrate too deep into the material which would lead to undesired polymerization in layers below the lowermost layer facing the light source. The absorber must cover the whole spectral range of the light source. The UV LED of the Asiga Pico 2 emits light at 370–430 nm (see Figure 1). Two phosphine based initiators, PPO and TPO were tested for their absorption properties. At the relevant concentrations for 3D printing, PPO effectively absorbs light between 325–465 nm, while TPO shows a less broad absorption spectrum between 325 nm–415 nm (see Figure 1a). Both initiators absorb within the spectrum of the UV LED of the Asiga Pico 2 printer and were suitable for the printing process. Three absorbers, T326, T384-2, and SOG were tested for their absorption properties (Figure 1b). The absorption spectrum of T326 is well suited for printing with the UV light source, but the powder could only be mixed at a significant amount with MD700 if dissolved in acetone first (mix 1). However, this renders the printing formulation unstable for storage due to the volatility of acetone. T384-2 could also only be mixed with MD700 if being dissolved in acetone first and additionally didn't cover the whole range of the UV LED. SOG is a very effective absorber at wavelengths between 325 nm and 560 nm. In addition, it was readily dissolvable in MD700 and the formulation was stable for storage.

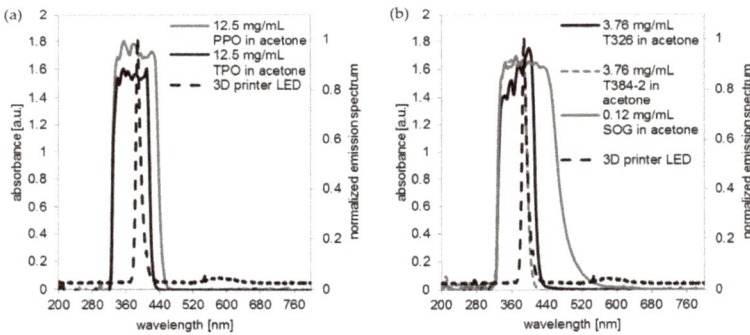

Figure 1. Absorption spectra of initiators and absorbers tested in comparison to the spectrum emitted by the UV LED of the Asiga Pico 2 printer (three-dimensional (3D) printer LED): (**a**) Absorption spectrum of the initiators phenylbis(2,4,6-trimethylbenzoyl)phosphine oxide (PPO) and Diphenyl(2,4,6-trimethylbenzoyl)phosphine oxide (TPO). TPO (325–415 nm) shows a less broad absorption spectrum than PPO (325–465 nm); (**b**) Absorption spectra of the absorbers tested: Tinuvin 326 (T326), Tinuvin 384-2 (T384-2), and Sudan Orange G (SOG).

3.2. Optimum Printing Parameters

The aim of this work was to print microfluidic channels with a layer thickness of 50 μm. To prevent the layers from delaminating during the printing process, an additional overcuring of 125 μm was required. The printing parameters for the different mixtures were optimized by analyzing the measured polymerization thickness in dependence of the exposure time (see Figure 2). The optimum printing curve should have a small slope (when plotted on a lin-log semi-logarithmic graph) to achieve most accurate printing results, but should allow for the printing of layers with a thickness of 175 μm (50 + 125 μm) within reasonable time [25]. The SOG absorber shows an optimum slope for a concentration of 0.235 mg SOG per mL of MD700, which affords a total layer thickness of 175 μm after an exposure time of 11.250 s (see Figure 2a). The T326 formulations (Figure 2b) show this optimum slope for 1.88 mg T326 per mL of MD700 (9.5 s for 175 μm layer thickness) and the T384-2 formulations at 10 mg T384-2 per mL of MD700 (6.5 s for 175 μm thickness, Figure 2c). As discussed above, these high amounts of absorbers (both T326 and T384-2) were not readily soluble in the final formulations and thus required acetone as a solvent, thereby impairing storage stability due to the volatility of acetone. Additionally, these mixtures led to a decrease of the optical transparency of the printed parts (see characterization in Section 3.5). Therefore, SOG was chosen as suitable absorber due to the fact that low concentrations are sufficient to obtain adequate light absorption and high transparency of the final parts. In the course of this work, mix 3 with an SOG concentration of 0.235 mg per mL of MD700 was used for printing the embedded microfluidic channels.

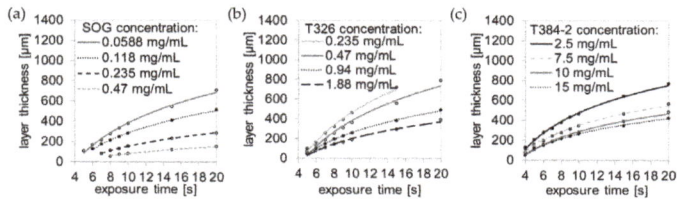

Figure 2. Layer thickness of printing formulations as a function of the exposure time. The polymerization curve flattens with increasing absorber concentration (final absorber concentration in MD700). Layer thickness as a function of the exposure time for different SOG (**a**), T326 (**b**), and T384-2 (**c**) concentrations.

3.3. Printing Embedded Microfluidic Chips

Two exemplary microfluidic gradient generators were printed with the formulation using SOG as absorber (mix 3, see Table 1). The chips were printed with a channel width and height of 800 μm and 600 μm, respectively, with a single layer thickness of 50 μm (Figure 3a–d). After printing, the microfluidic chips were washed out with 2-propanol in an ultrasonic bath for 15 min at a temperature of 45 °C. Afterwards, the printed parts were postcured for 15 min. The microfluidic chips were filled with black ink to highlight the 3D channel networks.

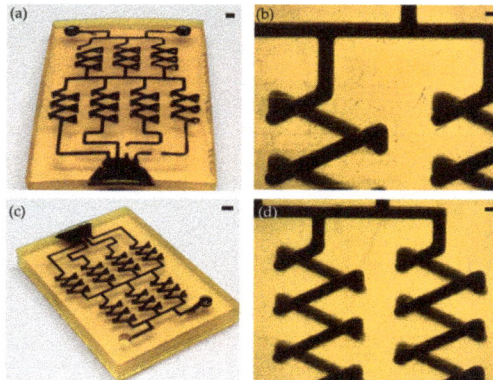

Figure 3. 3D printed perfluoropolyether (PFPE) microfluidic chips: (**a**) Front view of a gradient generator with a channel width and height of 800 μm, filled with black printing ink (scale bar: 2 mm); (**b**) Photomicrograph of the 800 μm gradient generator (scale bar: 500 μm); (**c**) Isometric view of the 600 μm gradient generator, filled with black printing ink (scale bar: 2 mm); and, (**d**) Photomicrograph of the 600 μm gradient generator (scale bar: 500 μm).

In Figure 4, the microfluidic gradient generator with a channel width and height of 800 μm was filled with a blue printing ink at the upper entrance and a yellow printing ink at the lower entrance to demonstrate the gradient generation along the mixer cascade. See Supplementary Information in Figure S2 for image analysis of the RGB red channel pixel brightness evaluation.

Figure 4. Gradient generation in a 3D printed PFPE microfluidic chip with a channel height and width of 800 μm. The upper entrance is filled with a yellow ink, the lower entrance is filled with a blue ink (scale bar: 2 mm).

3.4. Characterization of Printed Microfluidic Channels

We measured the channel height and width of the printed channel to characterize the homogeneity of the printed microfluidic chips. The channel height and width were measured on printed channels with a length of 24 mm at five different spots (see Figure 5a). The highest reduction of the channel height was measured in the middle of the channels at 12 mm from the inlet. For the 1000 μm and the 800 μm channel, a reduction of 20.3% and 21.5% of the theoretical channel height was measured at 12 mm from the inlet. The highest reduction of the channel height was measured for the 600 μm channel, with a reduction of 57.2%. The cross sections of the 800 μm and the 600 μm channel at the inlet and for the middle spot are shown in Figure 5b–e.

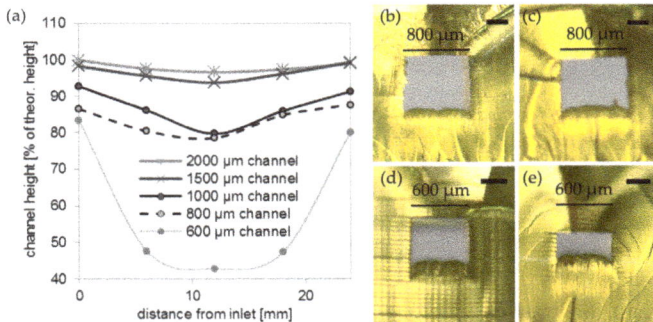

Figure 5. Characterization of the channel height of the printed PFPE methacrylates: (**a**) Height of a printed rectangular PFPE methacrylate channel for different spots from the inlet to the outlet. Channel heights decrease with distance from the inlet/outlet; (**b**) Lateral view of the 800 μm channel at the inlet, channel width of 800 μm, height of 692 μm (scale bar: 250 μm); (**c**) Cross section of the 800 μm channel in the middle, channel width of 800 μm, height of 628 μm (scale bar: 250 μm); (**d**) Lateral view of the 600 μm channel at the inlet, channel width of 600 μm, height of 500 μm (scale bar: 250 μm); and, (**e**) Cross section of the 600 μm channel in the middle, channel width of 600 μm, height of 257 μm (scale bar: 250 μm).

The minimum channel dimensions of 600 μm that could be printed with the developed MD700 formulations are comparable to minimum channel sizes of around 500 μm, which have been described in literature for printing microfluidic chips using benchtop stereolithography printers [26].

The minimum channel dimensions in this work were mainly limited by the high viscosity (850 mPas, according to the manufacturer's information) of the commercially available PFPE methacrylate MD700. These highly viscous resins are difficult to fully wash out of the printed microfluidic chips especially for long channels. Developing resin formulations with lower viscosities by using fluorinated acrylates with lower molecular weight and lower viscosity could help in solving this problem. Adding an appropriate solvent, like 1-butanol, to the resin formulation [27] could be another strategy to reduce the viscosity of the resin formulation with the additional advantage that the shrinkage of the material during solvent evaporation could further reduce the minimum channel dimensions.

The formulation for printing open-channel structures or replication masters (mix 4) was identical to mix 3 for printing embedded microfluidic chips with the sole exception that no absorber was required. Rectangular open channels with a height and width from 500 μm to 200 μm were printed. As can be seen in Figure 6a, the channels show a high uniformity of the width along the channel length. The width was measured at three different spots: The 500 μm channel possesses a width of 499 ± 14 μm, the 400 μm channel a width of 413 ± 14 μm, the 300 μm channel a width of 300 ± 13 μm, and the 200 μm channel a width of 205 ± 22 μm. Figure 6b shows the cross section of the printed open channels and demonstrates the high accuracy of the printed channel height. The 500 μm channel possess a height of 491 ± 6 μm, the 400 μm channel a height of 403 ± 8 μm, the 300 μm channel a height of 307 ± 6 μm, and the 200 μm channel a height of 209 ± 7 μm. Figure 6c,d shows an exemplary printed PFPE methacrylate structure with a 500 μm chess board structure.

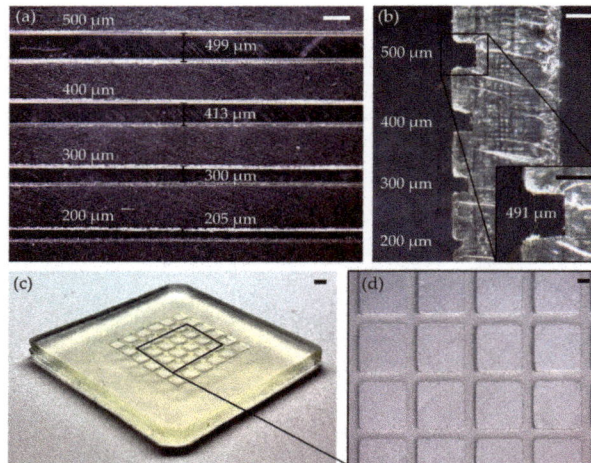

Figure 6. PFPE-printed open channels with a height and width of 500 μm, 400 μm, 300 μm, 200 μm and an exemplary printed PFPE chessboard structure: (**a**) The top view of the open channels shows a high conformity of the channel width (scale bar: 500 μm); (**b**) The lateral view shows the accordance of the channel heights with the set values. The inset shows the cross section of the 500 μm channel with a channel height of approximately 491 μm (scale bars: 500 μm); (**c**) Exemplary printed PFPE methacrylate chessboard structure (scale bar: 2 mm); and, (**d**) Microscope image of the chessboard from (**c**) (scale bar: 500 μm).

3.5. Characterization of Printed PFPE Methacrylates

Solvent compatibility of printed PFPE methacrylates of mix 3 and 5 were tested and compared to replicated PDMS components and the printed commercial stereolithography resin Formlabs Tough. Therefore, blocks ($8 \times 8 \times 3$ mm^3) were immersed in the following solvents: water, MeOH, DCM, DMF, THF, toluene, acetone, and *n*-heptane. As can be seen in Figure 7, the printed PFPE methacrylates show a higher solvent compatibility than the printed commercial resin and the replicated PDMS blocks. The printed PFPE methacryates show a maximum swelling of around 14% in THF and 11% in DCM. In comparison, PDMS showed a swelling of 146% in THF and 175% in DCM. The printed Formlabs Tough parts showed the highest swelling of around 78% in DCM and 76% in DMF. However, the printed Formlabs Tough parts that were immersed in DCM, DMF, THF, and acetone were strongly damaged by the solvents (see Figure 7b,c). We further tested if the used absorbers and initiators affect the chemical resistance of the printed PFPE methacrylates. For this we compared blocks made from mix 5 (initiator: PPO, absorber: T326) and from mix 3 (initiator: TPO, absorber: SOG). As can be seen from Figure 7, the printed polymers from both mixtures show a similar swelling, suggesting that the absorbers and initiators do not have major impacts on the swelling of the printed PFPE parts. We also determined the surface energies of the printed PFPE methacrylates, which were measured at 19.5 ± 0.5 mN/m. Given this relatively low value, the printed PFPE methacrylates showed a high water contact angle of $107 \pm 0.8°$, which is comparable to the values described in literature [21]. Although the contact angle for water on MD700 is relatively high, at 107° the capillary pressure (calculated using Laplace-Young equation [28]) is around -1.4 mbar for a printed channel with a circular cross-section of 600 μm. Therefore, the channels do not oppose microfluidic flow in a significant manner.

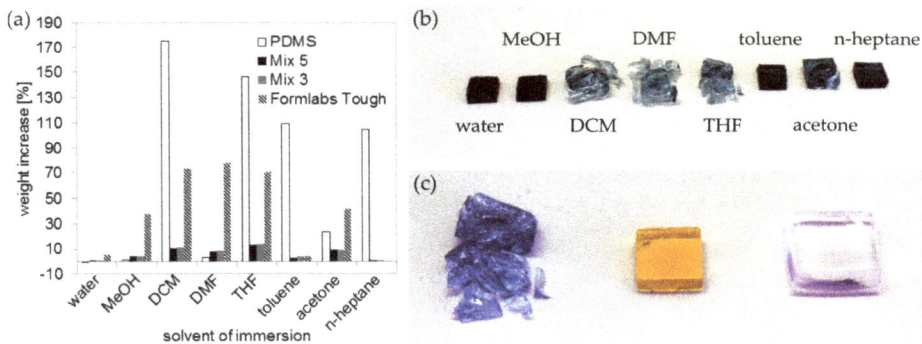

Figure 7. Solvent compatibility of printed PFPE methacrylates compared to poly(dimethylsiloxane) (PDMS) and the printed commercial stereolithography resin Formlabs Tough: (**a**) Rectangular $8 \times 8 \times 3$ mm^3 blocks were immersed in eight different solvents and the weight increase was measured after 24 h. The PFPE blocks show a lower weight increase than the PDMS blocks and the Formlabs Tough resin, which were immersed in the solvents dichloromethane (DCM), tetrahydrofuran (THF), toluene, acetone, and *n*-heptane. Mix 5 (initiator: PPO, absorber: T326) and mix 3 (initiator: TPO, absorber: SOG) show a similar swelling, which demonstrates that the absorbers and initiators have no major impact on the swelling of the printed PFPE parts; (**b**) Polymerized Formlabs Tough resin after immersion in the respective solvents for 24 h. The parts were strongly damaged by DCM, dimethylformamide (DMF), THF, and acetone; (**c**) Photograph of a Formlabs Tough block (left) and a PFPE methacrylate block (mix 3, middle) and PDMS (right) after 24 h immersion in THF.

We further characterized the transmittance of the printed PFPE methacrylates using UV-Vis spectroscopy. The transmittance of the printed PFPE methacrylates was further compared to polymerized PDMS parts. The transmission spectra of polymerized PFPE sheets with a thickness of 250 μm are shown in Figure 8a. Mix 1 with the absorber T326 shows a transparency above 30% between 400–900 nm, mix 2 (T384-2) a transparency above 35% between 400–900 nm. Mix 3 (SOG) shows a transmittance of over 70% between 520–900 nm. Mix 4 without any absorber shows the highest transparency above 75% between 350–900 nm. Mix 1 and mix 3 show a peak at about 330 nm and 280 nm in the UV-range, which is most likely an artefact from the photoinitiator TPO. Figure 8b shows two blocks printed from mix 1 and 2 and a microfluidic mixer that is printed from mix 3 (from left to right, thickness: 3.5 mm) placed on top of the logo of our lab (NeptunLab, Eggenstein-Leopoldshafen, Germany) showing the varying transparencies of the printed PFPE parts. As mentioned above, the high concentration of T326 and T384-2 required for printing microfluidic channels below 1 mm is not readily soluble in the formulation and required a prior dissolving in acetone. Consequently, a decrease of the transparency of the printed PFPE methacrylates was observed. SOG, which possesses a high light absorption at low concentrations, is readily soluble in the mixture and the printed parts have a higher optical clarity.

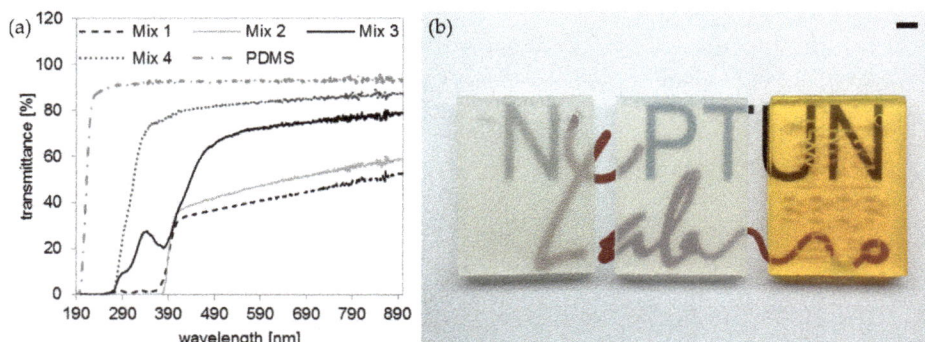

Figure 8. UV-Vis spectroscopy of PDMS and the printed PFPE methacrylates from mix 1, 2, 3 and 4 show different transparencies between 190–900 nm: (**a**) Transmission spectra of polymerized PDMS and PFPE methacrylates from mixtures 1, 2, 3 and 4 (thickness: 250 µm). Mix 1 shows a transparency of about 30% between 400–900 nm, mix 2 a transparency above 35% between 400–900 nm and mix 3 has a transmission above 70% between 520–900 nm. Mix 4 has a transparency of over 75% between 350–900 nm. Peaks at 330 and 280 nm are measured for the formulations prepared with TPO as photoinitiator. For comparison the UV-Vis spectrum of PDMS is shown; (**b**) Three microfluidic chips printed from mixtures 1, 2, and 3 (from left to right) show different transparencies as a function of the used absorber (scale bar: 4 mm).

4. Conclusions

In this work, we have demonstrated a new method to fabricate microfluidic chips in highly fluorinated PFPE methacrylates using stereolithography 3D printing. We have developed material formulations that can be printed in commercial stereolithography printers. We have shown that by adjusting the amount of initiator and absorber transparent microfluidic chips can be printed. The printed chips show high chemical resistance in the tested organic solvents. We believe that printing highly fluorinated PFPE methacrylates will find numerous applications from chemical resistant microfluidic valves to chemistry-on-chip applications.

Supplementary Materials: The following materials are available online at http://www.mdpi.com/2072-666X/9/3/115/s1, Figure S1: Overcuring and z-compensation of the Asiga Pico 2 printer: (a) Initial CAD file; (b) Slicing of the CAD file without the overcuring effect. The sliced CAD file has the same dimension like the initial CAD file; (c) Overcuring (l+o) is needed to prevent the slices from delamination during the printing process; (d) During the slicing process the overcuring is taken into account by the so called z-compensation. The shown sliced channel structure is higher than the initial CAD file; (e) The difference is polymerized by the overcuring during the printing process. The printed channel has the height of the initial CAD file. Figure S2: Determination of the color intensity in a PFPE-printed microfluidic channel (800 µm channel height and width) at different mixing positions: (a) The color intensity ((red+green+blue)/3) at different measurement spots shows the gradient generation in the microfluidic chip; (b) Greyscale image of the microfluidic chip showing the position of the measurement spots.

Acknowledgments: This work was funded by the German Ministry of Education and Research (BMBF), funding code 03X5527 "Fluoropor".

Author Contributions: F.K. printing experiments and formulations, P.R. printing experiments and formulations, D.H. UV-Vis measurements and writing of the paper, B.E.R. concept idea and writing of the paper.

Conflicts of Interest: The authors declare no conflict of interest.

References

1. Waheed, S.; Cabot, J.M.; Macdonald, N.P.; Lewis, T.; Guijt, R.M.; Paull, B.; Breadmore, M.C. 3D printed microfluidic devices: Enablers and barriers. *Lab Chip* **2016**, *16*, 1993–2013. [CrossRef] [PubMed]

2. Bhattacharjee, N.; Urrios, A.; Kanga, S.; Folch, A. The upcoming 3D-printing revolution in microfluidics. *Lab Chip* **2016**, *16*, 1720–1742. [CrossRef] [PubMed]

3. Au, A.K.; Huynh, W.; Horowitz, L.F.; Folch, A. 3D-printed microfluidics. *Angew. Chem. Int. Ed.* **2016**, *55*, 3862–3881. [CrossRef] [PubMed]

4. Tsuda, S.; Jaffery, H.; Doran, D.; Hezwani, M.; Robbins, P.J.; Yoshida, M.; Cronin, L. Customizable 3D printed 'plug and play' millifluidic devices for programmable fluidics. *PLoS ONE* **2015**, *10*, e0141640. [CrossRef] [PubMed]

5. Su, W.J.; Cook, B.S.; Fang, Y.N.; Tentzeris, M.M. Fully inkjet-printed microfluidics: A solution to low-cost rapid three-dimensional microfluidics fabrication with numerous electrical and sensing applications. *Sci. Rep.* **2016**, *6*. [CrossRef] [PubMed]

6. Ho, C.M.B.; Ng, S.H.; Li, K.H.H.; Yoon, Y.-J. 3D printed microfluidics for biological applications. *Lab Chip* **2015**, *15*, 3627–3637. [CrossRef] [PubMed]

7. Malinauskas, M.; Žukauskas, A.; Hasegawa, S.; Hayasaki, Y.; Mizeikis, V.; Buividas, R.; Juodkazis, S. Ultrafast laser processing of materials: From science to industry. *Light Sci. Appl.* **2016**, *5*, e16133. [CrossRef]

8. Jonušauskas, L.; Gailevičius, D.; Mikoliūnaitė, L.; Sakalauskas, D.; Šakirzanovas, S.; Juodkazis, S.; Malinauskas, M. Optically clear and resilient free-form μ-optics 3D-printed via ultrafast laser lithography. *Materials* **2017**, *10*, 12. [CrossRef] [PubMed]

9. Gong, H.; Bickham, B.P.; Woolley, A.T.; Nordin, G.P. Custom 3D printer and resin for 18 μm × 20 μm microfluidic flow channels. *Lab Chip* **2017**, *17*, 2899–2909. [CrossRef] [PubMed]

10. Kotz, F.; Plewa, K.; Bauer, W.; Schneider, N.; Keller, N.; Nargang, T.; Helmer, D.; Sachsenheimer, K.; Schafer, M.; Worgull, M.; et al. Liquid glass: A facile soft replication method for structuring glass. *Adv. Mater.* **2016**, *28*, 4646–4650. [CrossRef] [PubMed]

11. Kotz, F.; Arnold, K.; Bauer, W.; Schild, D.; Keller, N.; Sachsenheimer, K.; Nargang, T.M.; Richter, C.; Helmer, D.; Rapp, B.E. Three-dimensional printing of transparent fused silica glass. *Nature* **2017**, *544*, 337–339. [CrossRef] [PubMed]

12. De Mello, A. Focus: Plastic fantastic? *Lab Chip* **2002**, *2*, 31N–36N. [CrossRef] [PubMed]

13. Kotz, F.; Arnold, K.; Wagner, S.; Bauer, W.; Keller, N.; Nargang, T.M.; Helmer, D.; Rapp, B.E. Liquid PMMA: A high resolution polymethylmethacrylate negative photoresist as enabling material for direct printing of microfluidic chips. *Adv. Eng. Mater.* **2017**. [CrossRef]

14. Jarvis, N.L.; Zisman, W.A. *Surface Chemistry of Fluorochemicals*; Report of NRL Progress; U.S. Naval Research Laboratory (NRL): Washington, DC, USA, 1965.

15. Begolo, S.; Colas, G.; Viovy, J.L.; Malaquin, L. New family of fluorinated polymer chips for droplet and organic solvent microfluidics. *Lab Chip* **2011**, *11*, 508–512. [CrossRef] [PubMed]

16. Sharma, G.; Klintberg, L.; Hjort, K. Viton-based fluoroelastomer microfluidics. *J. Micromech. Microeng.* **2011**, *21*. [CrossRef]

17. Ogilvie, I.R.G.; Sieben, V.J.; Cortese, B.; Mowlem, M.C.; Morgan, H. Chemically resistant microfluidic valves from viton® membranes bonded to COC and PMMA. *Lab Chip* **2011**, *11*, 2455–2459. [CrossRef] [PubMed]

18. Goyal, S.; Desai, A.V.; Lewis, R.W.; Ranganathan, D.R.; Li, H.R.; Zeng, D.X.; Reichert, D.E.; Kenis, P.J.A. Thiolene and SIFEL-based microfluidic platforms for liquid-liquid extraction. *Sen. Actuators B Chem.* **2014**, *190*, 634–644. [CrossRef] [PubMed]

19. Renckens, T.J.A.; Janeliunas, D.; van Vliet, H.; van Esch, J.H.; Mul, G.; Kreutzer, M.T. Micromolding of solvent resistant microfluidic devices. *Lab Chip* **2011**, *11*, 2035–2038. [CrossRef] [PubMed]

20. Riche, C.T.; Zhang, C.C.; Gupta, M.; Malmstadt, N. Fluoropolymer surface coatings to control droplets in microfluidic devices. *Lab Chip* **2014**, *14*, 1834–1841. [CrossRef] [PubMed]

21. Rolland, J.P.; Van Dam, R.M.; Schorzman, D.A.; Quake, S.R.; DeSimone, J.M. Solvent-resistant photocurable "liquid teflon" for microfluidic device fabrication. *J. Am. Chem. Soc.* **2004**, *126*, 2322–2323. [CrossRef] [PubMed]

22. Vitale, A.; Quaglio, M.; Marasso, S.L.; Chiodoni, A.; Cocuzza, M.; Bongiovanni, R. Direct photolithography of perfluoropolyethers for solvent-resistant microfluidics. *Langmuir* **2013**, *29*, 15711–15718. [CrossRef] [PubMed]

23. De Marco, C.; Gaidukeviciute, A.; Kiyan, R.; Eaton, S.M.; Levi, M.; Osellame, R.; Chichkov, B.N.; Turri, S. A new perfluoropolyether-based hydrophobic and chemically resistant photoresist structured by two-photon polymerization. *Langmuir* **2013**, *29*, 426–431. [CrossRef] [PubMed]

24. Credi, C.; Levi, M.; Turri, S.; Simeone, G. Stereolithography of perfluoropolyethers for the microfabrication of robust omniphobic surfaces. *Appl. Surf. Sci.* **2017**, *404*, 268–275. [CrossRef]

25. Melchels, F.P.W.; Feijen, J.; Grijpma, D.W. A review on stereolithography and its applications in biomedical engineering. *Biomaterials* **2010**, *31*, 6121–6130. [CrossRef] [PubMed]

26. Urrios, A.; Parra-Cabrera, C.; Bhattacharjee, N.; Gonzalez-Suarez, A.M.; Rigat-Brugarolas, L.G.; Nallapatti, U.; Samitier, J.; DeForest, C.A.; Posas, F.; Garcia-Cordero, J.L. 3D-printing of transparent bio-microfluidic devices in peg-da. *Lab Chip* **2016**, *16*, 2287–2294. [CrossRef] [PubMed]

27. Ligon, S.C.; Liska, R.; Stampfl, J.; Gurr, M.; Mülhaupt, R. Polymers for 3D printing and customized additive manufacturing. *Chem. Rev.* **2017**, *117*, 10212–10290. [CrossRef] [PubMed]

28. Rapp, B.E. *Microfluidics: Modeling, Mechanics and Mathematics*; William Andrew: Norwich, NY, USA, 2016.

micromachines

MDPI

Technical Note

Characterization of 3D-Printed Moulds for Soft Lithography of Millifluidic Devices

Nurul Mohd Fuad [1], Megan Carve [1], Jan Kaslin [2] and Donald Wlodkowic [1,3,*]

[1] School of Science, RMIT University, Melbourne, VIC 3083, Australia; nfua256@aucklanduni.ac.nz (N.M.F.); s3437621@student.rmit.edu.au (M.C.)

[2] Australian Regenerative Medicine Institute, Monash University, Clayton, VIC 3800, Australia; jan.kaslin@monash.edu

[3] Centre for Additive Manufacturing, School of Engineering, RMIT University, Melbourne, VIC 3083, Australia

* Correspondence: donald.wlodkowic@rmit.edu.au; Tel.: +61-3-9925-7157

Received: 2 February 2018; Accepted: 7 March 2018; Published: 8 March 2018

Abstract: Increased demand for inexpensive and rapid prototyping methods for micro- and millifluidic lab-on-a-chip (LOC) devices has stimulated considerable interest in alternative cost-effective fabrication techniques. Additive manufacturing (AM)—also called three-dimensional (3D) printing—provides an attractive alternative to conventional fabrication techniques. AM has been used to produce LOC master moulds from which positive replicas are made using soft-lithography and a biocompatible elastomer, poly(dimethylsiloxane) (PDMS). Here we characterize moulds made using two AM methods—stereolithography (SLA) and material-jetting (MJ)—and the positive replicas produced by soft lithography and PDMS moulding. The results showed that SLA, more than MJ, produced finer part resolution and finer tuning of feature geometry. Furthermore, as assessed by zebrafish (*Danio rerio*) biotoxicity tests, there was no toxicity observed in SLA and MJ moulded PDMS replicas. We conclude that SLA, utilizing commercially available printers and resins, combined with PDMS soft-lithography, is a simple and easily accessible technique that lends its self particularly well to the fabrication of biocompatible millifluidic devices, highly suited to the in-situ analysis of small model organisms.

Keywords: stereolithography; material jetting; soft lithography; Lab-on-a-Chip; millifluidic; biodevices; biotests; polydimethylsiloxane

---

1. Introduction

Lab-on-a-chip (LOC) devices permit the manipulation of extremely small volumes of fluids and regents within networks of miniaturized channels, and are being readily adopted in the fields of biomedical and ecological testing [1–7]. Consequently, interest in the development of superior rapid and inexpensive prototyping and fabrication techniques has increased.

Traditionally, LOC devices have been prototyped and produced by methods such as laser cutting, micro milling, or soft lithography [5,8,9]. However, these techniques often require multiple stages of assembly, a high degree of technical expertise, and the use of clean room facilities. In the case of laser cutting, low resolution and heat-related substratum deformations may occur.

Additive manufacturing (AM) techniques such as stereolithography (SLA) and material jetting (MJ) provide an elegant and practical alternative LOC fabrication method [2–4,10]. By using computer-assisted design (CAD), SLA and MJ produce a solid monolithic three-dimensional (3D) object by the successive polymerization of a light-sensitive resin. The produced part requires minimal post-processing. The multiple stages of assembly and the high degree of technical proficiency required by conventional LOC manufacturing techniques is thus avoided. AM has been used to produce optically transparent LOC devices with integrated fluidic interconnects and functional elements such as pumps and valves [11–14].

The sub-millimeter resolution achieved by the market-dominant SLA and MJ machines makes them highly suited to application in millifluidic LOC device prototyping and production. SLA machines such as the FORM 2 and ProJet 7000 HD, are capable of printing layers as thin as ≥25 μm, and MJ machines such as the Objet J750 can print layers as thin as ≥14 μm. The minimum feature size achievable by SLA and MJ is dependent on accuracy, z-axis layer thickness, and laser-spot (SLA) or droplet (MJ) size. For SLA machines, features may be produced as small as ~70 μm, and for MJ, ~100 μm [3,14,15].

Millifluidic LOC devices typically require designs customized to the selected model metazoan organism which are typically ≤50 mm in size [10]. Consequently, LOC geometric features which are often required include channel cross-section dimensions ranging from 250–2000 μm and features with high aspect ratios (width-to-height). In general, millifluidic LOC devices are difficult to rapidly prototype, customize, and fabricate using traditional techniques and materials.

Biocompatibility of the LOC material is critical [10,16,17]. The material used in SLA and MJ is a liquid light-sensitive resin composed of a polymer–photoinitiator system, and auxiliary compounds for tuning material mechanical and aesthetic properties [18]. Companies recommend using their proprietary resins in their AM machines to optimize final print resolution and mechanical properties. However, studies have identified that quantities of toxic compounds may leach from proprietary AM materials, reviewed in [19]. For example, Zhu et al. [9] showed that leachate from several commonly available AM materials were toxic to a variety of model organisms, with exposure having both sublethal and lethal affects. Tsuda et al. [20] found that the leaching of unknown compounds from materials can unpredictably affect test organisms, and may compromise the outcomes of biological tests.

In some respects, the results found by Zhu et al. [9], and similar studies conducted by Oskui et al. [21], Alifui-Segbaya [22], and Macdonald et al. [17], were unexpected because some of the materials tested (e.g., VisiJet SL Clear) have been awarded a United States Pharmacopoeia (USP) Class VI certification [23]. However, it is important to note that this certification is highly dependent on strict adherence to post-processing cleaning procedures.

Nonetheless, these studies highlight a potential obstacle in the use of proprietary AM resins for fabricating LOC devices intended for biological applications. Since advances in AM LOC devices can enable a growing number of applications [6], ongoing research aims to develop the capability of AM systems capable of using biologically compatible substrata. Several proprietary resins are reported to be biocompatible, including the E-Shell series (Envision Tec. Inc., Dearborn, MI, USA). E-shell has been successfully applied in micro-needle fabrication with SLA-based two-photon polymerization systems [24,25].

Custom-designed open-source resins, primarily consisting of poly(ethylene glycol) diacrylate (PEG-DA), are also reported to be biocompatible [10,26–28]. Bhattacharjee et al. [10] used PEG-DA to fabricate biocompatible petri dishes. After printing, the petri dishes were rendered biocompatible by post-curing under ultraviolet light, followed by 24 h extraction in water to remove potential toxic compounds (i.e., uncreated monomers and photoinitiators). Rogers et al. [13] used a B9 Creator SLA printer and a PEG-DA custom resin to manufacture microfluidic valves and channels designed to be as narrow as 350 μm in diameter. Using the the AM technique of material extrusion, Tsuda et al. [20] created biocompatible LOC devices printed in thermoplastic materials, rendered biocompatible by post-treatment with oxygen plasma bonding of a silicone-based polymer.

Those involved in the design and application of AM for LOC in vitro bioassays should exercise caution in their approach, and aim to avoid the use of potentially toxic materials which may leach from AM-printed parts in aqueous media. Compared to the direct AM of LOC devices, transfer moulding has been explored as a more rapid and economic manufacturing method [29]. Transfer moulding involves the AM of moulds which are used to cast a relief in a material which has the desired properties, such as biocompatible poly(dimethylsiloxane) (PDMS). For example, Glick et al. [29] used transfer moulding to produce a biocompatible LOC with geometrically complex components and rounded channels as narrow as 100 μm. In this example, the ProJet 3000 SLA printer was used to produce the moulds, and PDMS was used for casting.

The main motivation of our work was to demonstrate an easily and widely accessible user-friendly and inexpensive fabrication method using "off-the-shelf" resins and commercially available AM printers to produce biocompatible millifluidic systems for application in bioengineering, biomedical and toxicological fields, and for use in laboratories that may or may not have direct access to AM facilities. This method utilizes standard technology and materials and requires minimum infrastructure investments, thus reducing the entry barrier to this technology when compared to conventional techniques.

In this work, we use two different AM methods—SLA (ProJet 7000 HD) and MJ (Objet350)—to produce moulds for casting PDMS, and compare these AM methods for application in PDMS transfer moulding for the production of biocompatible millifluidic LOC devices. We characterize both the moulds and PDMS replicas, and assess the toxicity of PDMS replicas by using zebrafish (*Danio rerio*) toxicity tests.

2. Materials and Methods

2.1. Metrology

Scanning electron micrographs were taken using a Quanta 200 SEM (FEI Inc., Hillsboro, OR, USA) operating at 20.0 kV and 5.0 Spot size in high vacuum. Samples were coated with Au/Pd using an in-house sputter coater operating at 15 mA for 2 min [9]. Optical profilometry was performed using a ContourGT-I 3D Optical Microscope (Bruker Daltonics Inc., Billerica, MA, USA) with stitching enabled, as described by Zhu et al. [9]. Images were analysed with Image J.

2.2. Additive Manufacturing

AM prototypes were designed using SolidWorks 2011 (Dassault Systems SolidWorks Corp., Waltham, MA, USA). The CAD was exported as a .stl file and processed by the AM machine software packages. SLA was performed with ProJet 7000 HD in VisiJet SL Clear resin (3D Systems Inc., Rock Hill, SC, USA), and MJ performed with Objet350 Connex, in VeroClear resin (Stratasys Inc., Eden Prairie, MN, USA) and as described earlier [9,30].

2.3. Soft-Lithography

Replica moulding was performed in PDMS (Sylgard 184; DowCorning Corp, Midland, MI, USA) according to a standard protocol as described earlier [31]. Briefly, the PDMS was mixed at a 10:1 weight-to-weight (w/w) ratio of elastomer base to curing agent, then degassed at 40 Torr to remove any residual air bubbles. PDMS was then poured on master moulds to achieve approximately 5 mm thickness and cured thermally at 75 °C for up to 1 h.

2.4. Non-Contact Laser Machining

Controls were fabricated using infra-red computer numerical control (CNC) laser machining of poly(methyl methacrylate) (PMMA) substrate by the methods described by Khoshmanesh et al. [32] and Akagi et al. [31]. The laser cutting system achieved x–y accuracy of up to 5 μm.

2.5. Toxicological Profiling

The OECD 236 Fish Embryo Acute Toxicity (FET) Assay [33] and fish behavioural toxicity test [9,30] were employed to assess toxicity as described previously [9,30]. Five days post-fertilisation (DPF) zebrafish (*Danio rerio*) embryos were used in all tests. Animals were treated according to Monash University Ethics Committee regulations and protocols.

3. Results and Discussion

A combination of AM negative relief moulds and soft lithography were used to produce a biocompatible millifluidic LOC. Preliminary validation experiments showed that this approach is highly suited to the design, prototyping, and production of LOC with cages designed to mirror the shape of the biological specimen.

The reproduction quality of geometric features of the printed moulds made by two different AM techniques was compared, and PDMS replicas were assessed using SEM and quantitative 3D topographic surface mapping using optical profilometry. The *x–y* and *z* resolution of the printed moulds were also evaluated and compared. Of the two techniques, SLA was found most suitable for this application.

The two AM machines used in this study were the Objet350 Connex (MJ) and ProJet 7000 HD (SLA), and both are claimed to have "ultra-high definition" capabilities. Positive relief test structures (moulds) were fabricated using the ProJet 7000 HD in VisiJet SL Clear resin, and the Objet350 Connex in VeroClear resin. Accuracy of the moulds was evaluated by comparing the fabricated dimensions to the dimensions in the CAD. There was variability between SLA and MJ in terms of accuracy (Figures 1 and 2).

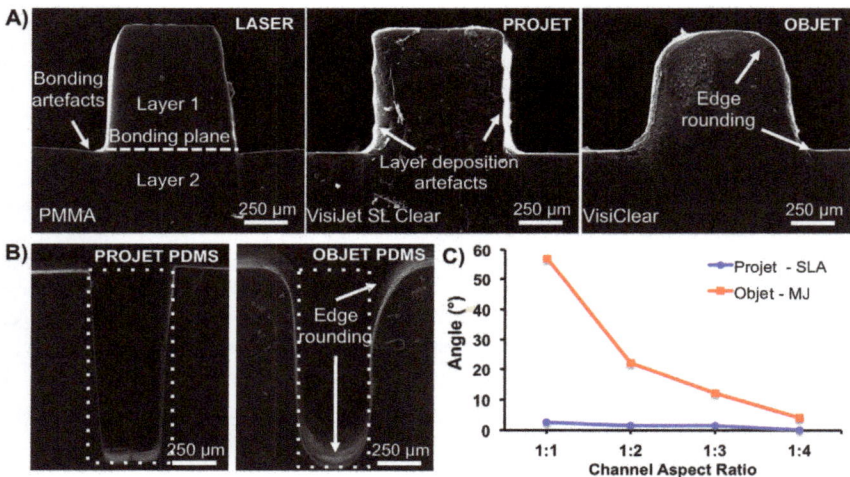

Figure 1. Physical characteristics of moulds fabricated using additive manufacturing processes: (**A**) SEM image of the cross-section of millifluidic positive relief pattern made with infrared laser cutting (and subsequent thermal bonding to obtain a complete mould), ProJet 7000 HD (in VisiJet SL Clear resin), and Objet350 Connex (in VeroClear resin). (**B**) Deviation from designed feature geometry of high aspect ratio poly(dimethylsiloxane) (PDMS) replicas. Dotted boxes represent the designed geometry superimposed on SEM images of PDMS channel cross-sections. (**C**) Quantitative analysis of AM moulds. Angle parameter denotes an average deviation from the designed geometry measured top plane edges of positive relief moulds. Measurements were performed on cross-sectional views obtained using SEM. Measurements were made with ImageJ software (*n* = 20).

In general, we found that the smallest feature reliably produced by both machines was ~300 μm in width and height. Features such as rectangular channel geometries with straight sidewalls fabricated by the ProJet 7000 HD more closely matched the theoretical CAD design than the features fabricated by the Objet350 Connex.

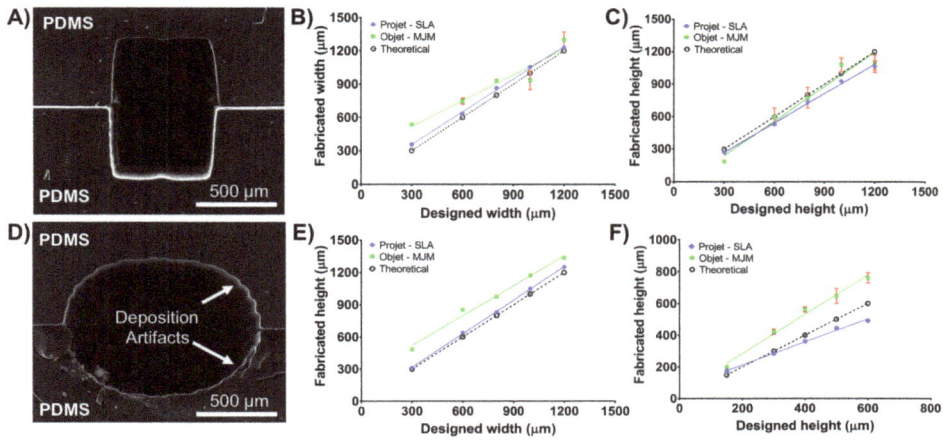

Figure 2. Physical characteristics of millifluidic channel replicas obtained from stereolithography (SLA; ProJet) or material jetting (MJ; Objet350 Connex) printed moulds: (**A**) SEM image of the channel with rectangular cross-section formed by moulding PDMS on a positive relief pattern made with ProJet 7000 HD (in VisiJet SL Clear resin). The channel was obtained from two identical and sandwiched PDMS replicas; (**B**) Relationship between the designed and actual width parameter of rectangular channels fabricated using SLA or MJ; (**C**) The relationship between the designed and actual height parameter of rectangular channels fabricated using SLA or MJ; (**D**) SEM micrographs of the channel with circular cross-section formed by moulding PDMS on a positive relief pattern made with ProJet 7000 HD in VisiJet SL Clear resin; (**E**) Relationship between the designed and actual feature width (defined here as a diameter in the centre plane) of circular channels fabricated using SLA or MJ; and (**F**) The relationship between the designed and actual height parameter (defined here as a diameter in the vertical plane) of circular channels fabricated using SLA or MJ. Measurements using ImageJ software were performed on cross-sectional views obtained using SEM (n = 20).

For positive relief test structures fabricated by the ProJet 7000 HD, deviation from the designed geometry was typically <10%, irrespective of feature dimension, and vertical walls exhibited an angular skew in the vertical plane of 6–9° compared to the design parameters (Figure 1). In contrast, those fabricated by the Objet350 Connex failed to reproduce straight edges and corners. Considerable edge rounding—up to 60°—occurred at the bottom and top planes of the channels. Rounding was more pronounced when fabricating features ≤600 μm in height, which resulted in deviations from the designed geometry >40%.

The Objet350 Connex tended to exceed the designed channel width; for example, when printing features designed to be smaller than 400 μm, the printed object feature was ∼650 μm (±25 μm) (Figure 2). Accuracy improved when fabricating channels designed to be larger than 900 μm in width. While the printer achieved excellent results in terms of height deposition, significant edge rounding artefacts prevented the printed channels having a rectangular shape as specified by design. Furthermore, the Objet350 Connex could not reproduce semicircular geometries reliably, and deviations from the designed geometry exceeded 40%.

We explored the fabrication of millifluidic channels with both rectangular and circular cross-sections (Figure 2, and Figure S1). When fabricating channels with both rectangular and circular cross-sections from two sandwiched PDMS replicas, the ProJet 7000 HD achieved the most consistency with theoretical design parameters. For instance, a channel with a designed geometry of 500 × 500 μm, width and height, was reproduced with dimensions of 520 ± 6 μm × 484 ± 5 μm (Figure 2). This corresponded to average deviations from a desired CAD geometry in x and y-axis of ∼5%. Overall, the ProJet 7000 HD tended to exceed channel width by 5–7%. The ProJet 7000 HD also failed to match the

CAD file with respect to height. The degree of inaccuracy varied depending on the designed height. Precision decreased for designed rectangular channel heights of 1000 μm and for circular cross-section heights of 400 μm (Figure 2).

The control method—infra-red laser cutting—was unable to fabricate an optically transparent positive relief pattern with a channel width <300 μm. Achievable laser x–y accuracy was up to 5 μm. The vertical walls exhibited an angular skew in the vertical plane of ∼8° compared to the specified design parameters (Figure 1). This is likely due to significant thermal deformation and melting damage of PMMA sheets (data not shown). These results show that these geometries are not easily fabricated by laser micromachining or photolithography methods.

A considerable advantage of using SLA to fabricate LOC moulds is that it permits fine control over complex topologies without any additional cost or complex protocols. Accordingly, we explored the capability of fabricating miniaturized 3D-cages that mirrored the shape of the selected biological specimen, larval zebrafish (*D. rerio*) (Figure 3). Such 3D geometries are typically difficult to achieve using conventional methods such as laser micromachining or photolithography. In addition, these methods require multiple stages of assembly, multiple masks, repeated coatings, photoresist re-flow, development, and alignment procedures.

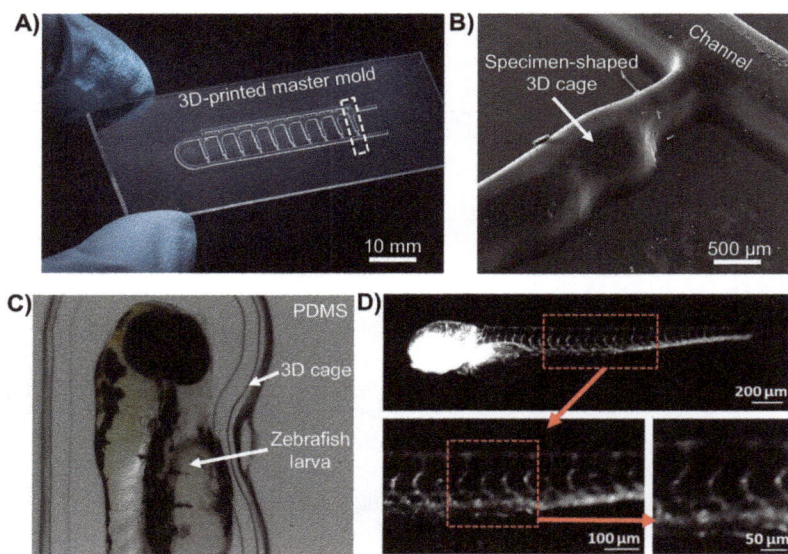

Figure 3. Fabrication of miniaturized 3D cages that mirror the shape of the biological specimens such as the larval stages of zebrafish (*Danio rerio*) using SLA moulds: (**A**) Microphotograph of a positive relief pattern fabricated using a ProJet 7000 HD in VisiJet SL Clear resin. The mould is the size of a standard microscope slide 25 mm × 75 mm. The dashed box denotes a single 3D cage; (**B**) Topographic surface analysis of a single 3D cage (marked as dashed box in (A) using scanning electron microscopy (SEM); (**C**) *D. rerio* larvae immobilized inside a PDMS cage, made from an SLA mould and that mirrors the organisms shape; and (**D**) High-resolution fluorescence imaging of a transgenic Fli1a:EGFP *D. rerio* larva showing green fluorescent protein (GFP)-expressing vasculature. Larger and smaller vessels are clearly distinguishable in the trunk region (boxed area, red outline).

Soft-lithography moulds the size of a standard microscope slide (25 mm × 75 mm; Figure 3) were fabricated using SLA process in under 40 min, and cost ∼US$2 per template. In comparison, laser machining soft-lithography moulds took ∼5 min, and cost ∼US$0.2 per mould. However, the control method was more labour intensive than SLA. It required the cutting of independent PMMA layers, subsequent alignment for assembly in 3D structures, and then thermal bonding for >60 min. Whereas AM required transferring the CAD to the AM software.

PDMS cured well on the VisiJet SL Clear moulds and VisiJet SL Clear moulds were geometrically stable during the PDMS curing process at 70 °C for 1 h. In addition, the releasing of PDMS replicas did not require any application of releasing agents. However, the curing of PDMS on VeroClear resin experienced several polymerization issues, and were more difficult to release.

Importantly, the SLA mould could be used multiple times without any degradation or deformation of pattern geometries observed during the repeated PDMS curing cycles.

Quantitative 3D topographic surface mapping using optical profilometry revealed that PDMS replicas had low surface roughness (Figure 4). Median surface roughness of PDMS replicas made from VisiJet SL Clear resin moulds was higher than the control PMMA moulds (Ra = 850 nm vs. 200 nm, respectively). The SLA moulds had had an average surface roughness of 1.7 ± 0.4 μm (taking into account both the tops of channels and bases of troughs). Consequently, following plasma treatment, PDMS replicas achieved proper hydraulic sealing to both glass and PDMS surfaces.

The PDMS millifluidic LOC produced from the SLA mould was optically transparent, thus enabling high-resolution fluorescence imaging (Figure 3).

Several studies have shown that photoinitiator and short-chain polymer residues that may leach from SLA devices exhibit a level of toxicity to test organisms [17,21,22,30,34]. This indicates that compounds remaining in the SLA-printed parts are not irreversibly trapped or adequately removed in standard manufacturer recommended post-processing protocols (e.g., solvent extraction and post-curing) [23]. This has cast doubt on the appropriateness of using SLA to produce biomicrofluidic LOC devices intended for in vitro bioassays [9,16,17,35]. However, using AM to create moulds, and the final LOC being produced by soft-lithography in PDMS side-steps these concerns.

Biocompatibility of PDMS replicas fabricated from SLA moulds was evaluated by using the well-established and very sensitive standard OECD FET test on developing *D. rerio* embryos [9,17]. The embryonic developmental stage is considered to be the most sensitive to environmental perturbations, and can be readily applied to assess impacts of any potential toxic effects of chemicals and solid phases [9,33].

The FET bioassay showed no observable toxicity of PDMS replicas fabricated using SLA moulds (Figure 4). At 24 h, all *D. rerio* embryos exposed to PDMS replicas survived without any developmental abnormalities. In contrast, at 24 h, the survival rate of embryos exposed to test parts fabricated by SLA and post-processed using standard cleaning procedures was ∼0% (data not shown). Behavioural toxicity bioassay revealed a lack of any behavioural abnormalities in zebrafish larvae exposed to PDMS replicas. In contrast, *D. rerio* larvae exposed to parts made with SLA exhibited sudden and complete paralysis. This suggests that toxic compounds did not transfer from SLA moulds to PDMS during the casting and curing process.

One of the main limitations of conventional soft lithography for millifluidic devices is the low availability of inexpensive master-mould fabrication techniques, and AM presents an attractive solution to overcoming this obstacle. Current-generation SLA technologies are generally able to print with resolutions >100 μm in the x–y dimension and >25 μm in the z dimension. Consequently, this technology is particularly suitable for fabricating millifluidic devices such as those designed with channel minimum cross-section dimensions ranging from 500 to 1500 μm for use with metazoan model organisms. While AM cannot match the performance and resolution of photolithography, it does permit the design and manufacture of complex geometries. SLA, rather than MJ, was shown to be a better choice for the fabrication of the millifluidic devices used in our study.

Despite some practical limitations, the significance and convenience of PDMS replica moulding and PDMS-on-glass-LOC is still high. Due to high metabolic rate, gas diffusion through PDMS layer is particularly advantageous for devices aimed at bioassays on model metazoan organisms. Moreover, the elastic properties of PDMS can alleviate mechanical damage during the loading and immobilization process of fragile embryonic stages. The replicas achieved a suitable level of pattern fidelity, with only a relatively insignificant deviation from the desired geometry of the fluidic channels. PDMS replicas also permitted bright field and fluorescent imaging at relatively high resolution. In addition, we showed that there was no transfer of toxic compounds from AM moulds to PDMS replicas.

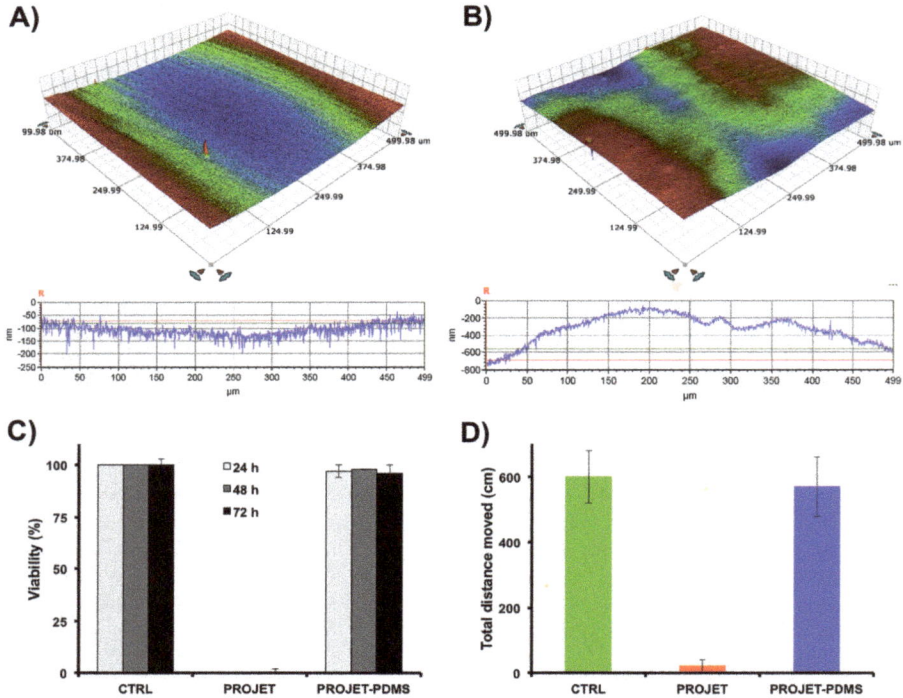

Figure 4. Characterization of PDMS replicas obtained from both control (CTRL) laser-cutting methods and SLA-AM (ProJet 7000 HD) positive relief moulds: (**A**) Quantitative topographic surface analysis of a representative 500 μm × 500 μm section of a millifluidic channel fabricated in PDMS from moulds made with laser cut and thermally bonded PMMA moulds; PMMA surface topography such as valleys (peak-to-peak value of 200 ± 20 nm) are thermal deformations formed during laser cutting and subsequent oven bonding. (**B**) Surface roughness renders of PDMS replicas obtained from ProJet 7000 HD printed moulds. Surface topography characteristic of SLA was peak-to-peak = 850 ± 70 nm. Measurements consisting of at least three independent samples were taken using ContourGT-I 3D optical profilometery. (**C,D**) Toxicity profiling of PDMS test wells casted on a mould fabricated using ProJet 7000 HD in VisiJet SL Clear resin. PDMS replicas were tested using OECD 236 Fish Embryo Acute Toxicity (FET) Assay with zebrafish (*Danio rerio*). Comparative analysis between negative control polystyrene wells (CTRL), SLA moulds printed in VisiJet Clear SL (PROJET), and wells made from moulding PDMS with Visijet Clear SL moulds (PROJET-PDMS): (C) survival of *D. rerio* larvae, and (D) behavioural responses of *D. rerio* larvae, measured as the change in total distance travelled (±S.E) after 5 min in wells. Results represent data from at least three independent experiments performed in triplicate.

Here we present a combinatorial fabrication approach utilizing SLA and soft lithography in PDMS as a viable method for the inexpensive production of high-definition biocompatible millifluidic LOC devices. These devices are easily adaptable for use in a range of studies using small to large metazoan model organisms. The presented technique offers the possibility of prototyping millifluidic devices and iterating multiple designs per day to support fast optimization cycles and achieve significant advancement in the definition of manufactured structures. Due to the growing availability of AM facilities, which enables the outsourced fabrication of small batches of prototypes at a very low cost, our described manufacturing method is readily accessible to any biological laboratory. In addition, issues previously observed with polymer toxicity were avoided.

Supplementary Materials: The following are available online at www.mdpi.com/2072-666X/9/3/116/s1, Figure S1: Physical characteristics of semi-circular positive relief patterns fabricated using additive manufacturing processes. (A) Pattern fabricated in VisJet Clear material using the ProJet 7000 HD system; (B) PDMS replica obtained from the master depicted in (A); (C) Pattern fabricated in Vero Clear using the Objet350 Connex system; (D) PDMS replica obtained from the master depicted in (C). Dotted lines represent the designed CAD geometry superimposed on SEM images of representative cross-sections.

Acknowledgments: This work was supported by: the Australian Research Council DECRA [DE130101046] (D.W.); Vice-Chancellor's Senior Research Fellowship, RMIT University, Australia (D.W.), National Health and Medical Research Council [GNT1068411] (J.K.), Eva and Les Erdi foundation (J.K.). We acknowledge the assistance of: RMIT Micro Nano Research Facility (MNRF) during optical profilometry; RMIT Microscopy and Microanalysis Facility (RMMF) during SEM imaging experiments; Monash University FishCore in zebrafish breeding and embryo culture. Zebrafish husbandry and biotests were performed according to the regulations of Monash University Animal Ethics Committee.

Author Contributions: D.W. and N.M.F. conceived and designed the experiments; N.M.F. performed the experiments; N.M.F. analyzed the data; J.K. contributed materials; D.W. and M.C. wrote the paper.

Conflicts of Interest: The authors declare no conflict of interest. The funding sponsors had no role in the design of the study; in the collection, analyses, or interpretation of data; in the writing of the manuscript, and in the decision to publish the results.

Abbreviations

The following abbreviations are used in this manuscript:

LOC	Lab-on-a-Chip
AM	A dditive manufacturing
MJ	Material jetting
3D	Three-dimensional
PDMS	Poly(dimethylsiloxane)
SLA	Stereolithography
PEG-DA	poly(ethylene glycol) diacrylate
USP	United States Pharmacopeia
PMMA	Poly(methyl methacrylate)
CNC	Computer numerical control
OECDv	Organisation for Economic Co-operation and Development
CAD	Computer-assisted design
DPF	Days post-fertilisation
SEM	Scanning electron microscope
GFP	Green fluorescent protein

References

1. Waheed, S.; Cabot, J.M.; Macdonald, N.P.; Lewis, T.; Guijt, R.M.; Paull, B.; Breadmore, M.C. 3D printed microfluidic devices: Enablers and barriers. *Lab Chip* **2016**, *16*, 1993–2013.

2. Au, A.K.; Huynh, W.; Horowitz, L.F.; Folch, A. 3D-Printed Microfluidics. *Angew. Chem. Int. Ed.* **2016**, *55*, 3862–3881.

3. Ho, C.M.B.; Ng, S.H.; Li, K.H.H.; Yoon, Y.J. 3D printed microfluidics for biological applications. *Lab Chip* **2015**, *15*, 3627–3637.

4. Chen, C.; Mehl, B.T.; Munshi, A.S.; Townsend, A.D.; Spence, D.M.; Martin, R.S. 3D-printed microfluidic devices: Fabrication, advantages and limitations—A mini review. *Anal. Methods* **2016**, *8*, 6005–6012.

5. Cartlidge, R.; Nugegoda, D.; Wlodkowic, D. Millifluidic Lab-on-a-Chip technology for automated toxicity tests using the marine amphipod Allorchestes compressa. *Sens. Actuators B Chem.* **2017**, *239*, 660–670.

6. Campana, O.; Wlodkowic, D. The undiscovered country: Ecotoxicology meets microfluidics. *Sens. Actuators B Chem.* **2018**, *257*, 692–704.

7. Wlodkowic, D.; Khoshmanesh, K.; Akagi, J.; Williams, D.E.; Cooper, J.M. Wormometry-on-a-chip: Innovative technologies for in situ analysis of small multicellular organisms. *Cytom. Part A* **2011**, *79*, 799–813.

8. Zhu, F.; Wigh, A.; Friedrich, T.; Devaux, A.; Bony, S.; Nugegoda, D.; Kaslin, J.; Wlodkowic, D. Automated Lab-on-a-Chip Technology for Fish Embryo Toxicity Tests Performed under Continuous Microperfusion (μFET). *Environ. Sci. Technol.* **2015**, *49*, 14570–14578.

9. Zhu, F.; Skommer, J.; MacDonald, N.P.; Friedrich, T.; Kaslin, J.; Wlodkowic, D. Three-dimensional printed millifluidic devices for zebrafish embryo tests. *Biomicrofluidics* **2015**, *9*, doi:10.1063/1.4927379.

10. Bhattacharjee, N.; Urrios, A.; Kang, S.; Folch, A. The upcoming 3D-printing revolution in microfluidics. *Lab Chip* **2016**, *16*, 1720–1742.

11. Gong, H.; Woolley, A.T.; Nordin, G.P. High density 3D printed microfluidic valves, pumps, and multiplexers. *Lab Chip* **2016**, *16*, 2450–2458.

12. Au, A.K.; Bhattacharjee, N.; Horowitz, L.F.; Chang, T.C.; Folch, A. 3D-printed microfluidic automation. *Lab Chip* **2015**, *15*, 1934–1941.

13. Rogers, C.I.; Qaderi, K.; Woolley, A.T.; Nordin, G.P. 3D printed microfluidic devices with integrated valves. *Biomicrofluidics* **2015**, *9*, doi:10.1063/1.4905840.

14. Macdonald, N.P.; Cabot, J.M.; Smejkal, P.; Guijt, R.M.; Paull, B.; Breadmore, M.C. Comparing Microfluidic Performance of Three-Dimensional (3D) Printing Platforms. *Anal. Chem.* **2017**, *89*, 3858–3866.

15. Bhushan, B.; Caspers, M. An overview of additive manufacturing (3D printing) for microfabrication. *Microsyst. Technol.* **2017**, *23*, 1117–1124.

16. Zhu, F.; Skommer, J.; Friedrich, T.; Kaslin, J.; Wlodkowic, D. 3D printed polymers toxicity profiling: A caution for biodevice applications. *Proc. SPIE* **2015**, *9668*, doi:10.1117/12.2202392.

17. Macdonald, N.P.; Zhu, F.; Hall, C.J.; Reboud, J.; Crosier, P.S.; Patton, E.E.; Wlodkowic, D.; Cooper, J.M. Assessment of biocompatibility of 3D printed photopolymers using zebrafish embryo toxicity assays. *Lab Chip* **2016**, *16*, 291–297.

18. Ligon, S.C.; Liska, R.; Stampfl, J.; Gurr, M.; Mulhaupt, R. Polymers for 3D Printing and Customized Additive Manufacturing. *Chem. Rev.* **2017**, *117*, 10212–10290.

19. Carve, M.; Wlodkowic, D. 3D-Printed Chips: Compatibility of Additive Manufacturing Photopolymeric Substrata with Biological Applications. *Micromachines* **2018**, *9*, 91.

20. Tsuda, S.; Jaffery, H.; Doran, D.; Hezwani, M.; Robbins, P.J.; Yoshida, M.; Cronin, L. Customizable 3D printed 'Plug and Play' millifluidic devices for programmable fluidics. *PLoS ONE* **2015**, *10*, e0141640.

21. Oskui, S.M.; Diamante, G.; Liao, C.; Shi, W.; Gan, J.; Schlenk, D.; Grover, W.H. Assessing and Reducing the Toxicity of 3D-Printed Parts. *Environ. Sci. Technol. Lett.* **2016**, *3*, 1–6.

22. Alifui-Segbaya, F.; Varma, S.; Lieschke, G.J.; George, R. Biocompatibility of Photopolymers in 3D Printing. *3D Print. Addit. Manuf.* **2017**, *4*, 185–191.

23. 3D Systems Inc. *VisiJet SL Clear Cleaning Procedure for USP Class VI*; Library Technology Reports; 3D Systems Inc.: Valencia, CA, USA, 2013; pp. 1–2.

24. Miller, P.R.; Skoog, S.A.; Edwards, T.L.; Lopez, D.M.; Wheeler, D.R.; Arango, D.C.; Xiao, X.; Brozik, S.M.; Wang, J.; Polsky, R.; et al. Multiplexed microneedle-based biosensor array for characterization of metabolic acidosis. *Talanta* **2012**, *88*, 739–742,

25. Gittard, S.D.; Ovsianikov, A.; Monteiro-Riviere, N.A.; Lusk, J.; Morel, P.; Minghetti, P.; Lenardi, C.; Chichkov, B.N.; Narayan, R.J. Fabrication of polymer microneedles using a two-photon polymerization and micromolding process. *J. Diabetes Sci. Technol.* **2009**, *3*, 304–311.

26. Rogers, C.I.; Pagaduan, J.V.; Nordin, G.P.; Woolley, A.T. Single-monomer formulation of polymerized polyethylene glycol diacrylate as a nonadsorptive material for microfluidics. *Anal. Chem.* **2011**, *83*, 6418–6425,

27. Urrios, A.; Parra-Cabrera, C.; Bhattacharjee, N.; Gonzalez-Suarez, A.M.; Rigat-Brugarolas, L.G.; Nallapatti, U.; Samitier, J.; DeForest, C.A.; Posas, F.; Garcia-Cordero, J.L.; et al. 3D-printing of transparent bio-microfluidic devices in PEG-DA. *Lab Chip* **2016**, *16*, 2287–2294.

28. Hinton, T.J.; Hudson, A.; Pusch, K.; Lee, A.; Feinberg, A.W. 3D Printing PDMS Elastomer in a Hydrophilic Support Bath via Freeform Reversible Embedding. *ACS Biomater. Sci. Eng.* **2016**, *2*, 1781–1786.

29. Glick, C.C.; Srimongkol, M.T.; Schwartz, A.J.; Zhuang, W.S.; Lin, J.C.; Warren, R.H.; Tekell, D.R.; Satamalee, P.A.; Lin, L. Rapid assembly of multilayer micro fl uidic structures via 3D-printed transfer molding and bonding. *Microsyst. Nanoeng.* **2016**, *2*, 16063.

30. Zhu, F.; Friedrich, T.; Nugegoda, D.; Kaslin, J.; Wlodkowic, D. Assessment of the biocompatibility of three-dimensional-printed polymers using multispecies toxicity tests. *Biomicrofluidics* **2015**, *9*, doi:10.1063/1.4939031.

31. Akagi, J.; Khoshmanesh, K.; Evans, B.; Hall, C.J.; Crosier, K.E.; Cooper, J.M.; Crosier, P.S.; Wlodkowic, D. Miniaturized embryo array for automated trapping, immobilization and microperfusion of zebrafish embryos. *PLoS ONE* **2012**, *7*, e36630.

32. Khoshmanesh, K.; Akagi, J.; Hall, C.J.; Crosier, K.E.; Crosier, P.S.; Cooper, J.M.; Wlodkowic, D. New rationale for large metazoan embryo manipulations on chip-based devices. *Biomicrofluidics* **2012**, *6*, doi:10.1063/1.3699971.

33. OECD. Test No. 236: Fish Embryo Acute Toxicity (FET) Test Section 2: Effects on Biotic Systems. In *OECD Guidelines for the Testing of Chemicals*; OECD: Paris, France, 2013; Chapter 2.

34. Cebe, M.A.; Cebe, F.; Cengiz, M.F.; Cetin, A.R.; Arpag, O.F.; Ozturk, B. Elution of monomer from different bulk fill dental composite resins. *Dent. Mater.* **2015**, *31*, e141–e149.

35. Zhu, F.; Macdonald, N.; Skommer, J.; Wlodkowic, D. Biological implications of lab-on-a-chip devices fabricated using multi-jet modelling and stereolithography processes. *SPIE Microtechnol.* **2015**, *9518*, doi:10.1117/12.2180743.

micromachines

MDPI

Communication

Digital Manufacturing of Selective Porous Barriers in Microchannels Using Multi-Material Stereolithography

Yong Tae Kim *,†, Kurt Castro †, Nirveek Bhattacharjee and Albert Folch

Department of Bioengineering, University of Washington, Seattle, WA 98195, USA; castrokb@uw.edu (K.C.); nirveek@uw.edu (N.B.); afolch@uw.edu (A.F.)

* Correspondence: ytkim82@uw.edu; Tel.: +1-206-685-2257

† These authors contributed equally to this work.

Received: 1 February 2018; Accepted: 13 March 2018; Published: 14 March 2018

Abstract: We have developed a sequential stereolithographic co-printing process using two different resins for fabricating porous barriers in microfluidic devices. We 3D-printed microfluidic channels with a resin made of poly(ethylene glycol) diacrylate (MW = 258) (PEG-DA-258), a UV photoinitiator, and a UV sensitizer. The porous barriers were created within the microchannels in a different resin made of either PEG-DA (MW = 575) (PEG-DA-575) or 40% (*w/w* in water) PEG-DA (MW = 700) (40% PEG-DA-700). We showed selective hydrogen ion diffusion across a 3D-printed PEG-DA-575 porous barrier in a cross-channel diffusion chip by observing color changes in phenol red, a pH indicator. We also demonstrated the diffusion of fluorescein across a 3D-printed 40% PEG-DA-700 porous barrier in a symmetric-channel diffusion chip by measuring fluorescence intensity changes across the porous barrier. Creating microfluidic chips with integrated porous barriers using a semi-automated 3D printing process shortens the design and processing time, avoids assembly and bonding complications, and reduces manufacturing costs compared to micromolding processes. We believe that our digital manufacturing method for fabricating selective porous barriers provides an inexpensive, simple, convenient and reproducible route to molecule delivery in the fields of molecular filtration and cell-based microdevices.

Keywords: multi-material stereolithography; porous barrier; diffusion; microfluidics

---

1. Introduction

Precise control of small molecule diffusion through porous materials and its application to analyze cell behavior is an important technology in fundamental biology, biomedicine, and pharmaceutics [1–5]. Lab-on-a-chip research has progressed over the last decades by adapting microfabrication technology from semiconductor fabrication processes that bring advantages such as miniaturization, uniformity, accuracy, reproducibility, and fluid/cell/tissue manipulations [6–12]. Hydrogels, which are three-dimensional (3D) polymer networks with high porosity, have been employed in the field of tissue engineering, drug delivery systems, biomaterials, and size-selective separation, among others [13–18]. The pore size of hydrogels can be easily tuned by manipulating their crosslinking density.

Researchers have used microfluidic architectures in combination with hydrogel materials for delivery of chemicals in cell and tissue engineering applications [19–22]. For example, Lee et al. developed a macroporous polyacrylamide and poly(ethylene glycol) diacrylate (PEG-DA) hydrogel array within a microchannel to monitor solute transport into the hydrogel; they fabricated macroporous hydrogels composed of polyacrylamide and analyzed the diffusion of both glutathione *S*-transferase-green fluorescent protein and dextran-fluorescein isothiocyanate into the hydrogel; they also produced PEG-DA (MW = 700) (PEG-DA-700) hydrogels and tested the penetration of

250 kDa dextran into the hydrogel [23]. Huang et al. reported a 3D microfluidic platform, which was fabricated using typical photolithography methods, to observe cell to extracellular matrix and cell to cell interactions; they filled multiple adjacent channels with type-I collagen and Matrigel by balancing the capillary forces and surface tension; the different patterned gels in the microchannels enabled real-time observation of intercellular interactions in co-cultures using MDA-MB-231 cell and RAW cells [24]. Sung et al. announced a microfluidic device for testing drug toxicity in a pharmacokinetic-based manner that included multiple hydrogel-based cell culture chambers connecting to microfluidic channels to mimic multi-organ interactions [25]. However, fabrication of porous microstructures in a microfluidic chip usually requires the assembly and bonding of an additional membrane or hydrogel structure, a process that is complex, costly, and time-consuming, and often requires the intervention of a trained specialist.

3D printing has emerged as an attractive alternative technique for manufacturing microfluidic devices because of its attributes, i.e., 3D design, semi-automated and assembly-free 3D fabrication, rapidly decreasing costs, and faster production time [26,27]. 3D printing has also been extensively used to build hydrogel scaffolds [28,29]. For example, Mohanty et al. fabricated scalable scaffolds using a combined 3D-fused deposition modeling printing and salt leaching process for precise control of the geometry and dimensions of the scaffolds [30,31]. He et al. investigated the printability of a sodium alginate and gelatin mixture hydrogel using an extrusion-based 3D bioprinting system and demonstrated the viability of L929 mouse fibroblasts on the 3D-printed hydrogel scaffold [32]. Hinton et al. recently reported a complex 3D biological structure using a thermoreversible gelatin support bath that enabled freeform reversible embedding of hydrogel to suspend alginate, fibrin, collagen type I, and Matrigel [33]. However, multi-material printing has not been used yet for the production of porous barriers in microfluidics.

Integrating the fabrication of a porous microstructure within a microfluidic device using an automated process such as 3D printing allows for materials and performance modeling prior to fabrication, shortens the duration of design iterations, and reduces overall manufacturing costs. In this work, we developed a stereolithographic co-printing process to enable the inexpensive microfabrication of selective porous barriers in order to control the diffusion of small ions or molecules within a microchannel. We explored two different hydrogels—PEG-DA (MW = 575) (PEG-DA-575) and 40% (*w/w* in water) PEG-DA-700 (40% PEG-DA-700)—for printing the porous barriers separating two adjacent microchannels. The microchannels were fabricated with PEG-DA (MW = 258) (PEG-DA-258) [34]. Diffusion of hydrogen ions and fluorescein was demonstrated with a 3D-printed cross-channel diffusion chip (PEG-DA-575 barrier) and a 3D-printed symmetric-channel diffusion chip (40% PEG-DA-700 barrier), respectively.

2. Materials and Methods

2.1. Resin Composition for Multi-Material Stereolithography Printing

Photocurable resins for stereolithography printing were developed for the microfluidic chips that have selective porous barriers within their microchannels. The resin for creating the microchannel floor, walls, and roof is based on PEG-DA-258 (Sigma Aldrich, St. Louis, MO, USA) mixed with 0.6% (*w/w*) Irgacure 819 (IRG, BASF Corporation, Florham Park, NJ, USA) as a photoinitiator and 0.6% (*w/w*) 2-isopropylthioxanthone (ITX) as a photosensitizer. The resin for the porous barrier of the 3D-printed cross-channel diffusion chip contains PEG-DA-575 (Sigma Aldrich), 0.6% IRG, and 0.6% ITX. 40% PEG-DA-700 (Sigma Aldrich) in distilled water conjugated with 0.6% IRG is used for printing the porous barrier of the 3D-printed symmetric-channel diffusion chip.

2.2. Stereolithography—Setup and Printing

Both the cross-channel and symmetric-channel diffusion chip were designed with Autodesk Inventor® and saved in STL formats. In order to slice the designed object, we used Autodesk

Print Studio (Autodesk, Inc., San Rafael, CA, USA), which has a function for automatic mesh analysis and healing. After slicing the designed object into 25 μm-thick layers, the sliced images and the UV exposure settings for each layer outlined in a CSV file named layersettings.csv were compressed into a zipped folder. The folder was then sent to the 3D printer to fabricate the object. The fabrication of the multi-material diffusion microfluidic chip was carried out in bat configuration with an Ember DLP® 3D Printer (Autodesk, Inc.), which has a 405 nm wavelength LED, 50 μm in-plane resolution, and 10~100 μm out-of-plane resolution controlled by a digital light projector (DLP). To make transparent prints, we used a glass slide (75 mm (L) × 50 mm (W) × 1.0 mm (T)) as the substrate on which the designed object was built [34]. Prior to using the glass slide, the glass surface was thoroughly cleaned with acetone, isopropyl alcohol, and deionized (D.I.) water and dried at 70 °C overnight. The surface of the glass slide was plasma-activated with oxygen plasma for 180 s and treated with 3-(trimethoxysilyl) propyl methacrylate (TMSPMA) (Sigma-Aldrich) in an 85 °C chamber for 8 h. The silanized glass slide was attached to the aluminum build plate by coating one side of it with uncured PEG-DA-258 resin and exposing it to UV light using a broadband UV lamp (B-100 A, UVP). For the microchannel fabrication, the PEG-DA-258 resin was poured into a resin tray and then the build plate was lowered until it reached the resin tray for calibration. After calibration, the object was printed with repeated cycles of UV curing, resin tray separation from the current print position, Z-Axis overlift, and resin tray approach to the new print position. This process was repeated until the whole object was fabricated. For multi-material printing, the printer was paused at the desired layer and the PEG-DA-258 resin tray was changed to the other material (PEG-DA-575 or 40% PEG-DA-700), or vice versa. The glass slide with the printed object was removed from the build plate and cleaned using D.I. water and pressurized air.

2.3. Molecule Diffusion Analysis

To test the hydrogen ion diffusion through the porous barrier printed using PEG-DA-575 in the 3D-printed cross-channel diffusion chip, phenol red (Sigma) was employed to observe color changes corresponding to changes in acidity. Then 0.1 M HCl was injected into the bottom channel of the cross-channel diffusion chip and phenol red was loaded into the top channel of the diffusion chip. The color change of the phenol red was recorded using a Nikon SMZ1500 stereomicroscope (Nikon, Tokyo, Japan) fitted with a Canon Rebel EOS 5D Mark II DSLR camera (Canon, Tokyo, Japan).

The 3D-printed symmetric-channel diffusion microchip was utilized to transport fluorescein molecules through its porous barrier made of 40% PEG-DA-700 resin. The 0.1 M fluorescein was inserted into Channel 1 of the symmetric-channel diffusion chip while PBS buffer was loaded into Channel 2. The diffusion of fluorescein across the porous barrier from one channel to the other was measured for 1 hour using an inverted epifluorescence microscope (Nikon Eclipse Ti, Plan Apo λ 2× objective).

3. Results and Discussion

3.1. Selection of Resin Components

We have fabricated selective porous barriers in microchannels using multi-material stereolithography. We chose PEG-DA-258 to print microfluidic channels due to its characteristics such as photo-curability, water impermeability, transparency, and low cost [34]. We added 0.6% IRG to the PEG-DA-258 resin because of its ability to absorb energy and initiate free radical polymerization at longer UV wavelengths (405 nm), corresponding to that of the printer DLP projector. We also added ITX at 0.6% to increase the Z-resolution of 3D-printed microchannels. ITX, which is a sulfur-type photosensitizer, is widely used in UV curing to promote the polymerization of photopolymers because of its UV absorption characteristics. PEG-DA with higher molecular weights were chosen as porous barrier materials since it had been shown before that PEG-DA hydrogel networks (MW = 575 to 20,000) have a mesh size of 0.1 to 10 nm [35–37]. A porous barrier in the cross-channel diffusion chip was printed using PEG-DA-575 mixed with 0.6% IRG and 0.6% ITX for small ion diffusion such as hydrogen ions. We also used 40% PEG-DA-700 in D.I. water containing 0.6% IRG to make a porous barrier for the transport of larger molecules since adding 60% water to PEG-DA-700 resin increases the pore size in a hydrogel polymer matrix [16].

3.2. Fabrication of a 3D-Printed Cross-Channel Diffusion Chip

Figure 1A shows the fabrication process of a cross-channel diffusion chip. We developed two different resins to print the 3D-printed cross-channel diffusion chip: a resin to build the porous barrier (PEG-DA-575) and a microfluidic channel resin (PEG-DA-258). Prior to the printing of the 3D-printed cross-channel diffusion chip, the designed object was sliced into 25 µm-thick layers. A PEG-DA-258 resin tray and a PEG-DA-575 resin tray were prepared to build the microchannel part and the porous barrier part, respectively. First, the PEG-DA-258 resin tray was inserted into the Ember DLP® 3D Printer and the first layer was created with 6 s UV exposure to aid the adhesion of the device to the glass slide. After printing a 100 µm-thick bottom layer, a 1 mm high bottom channel was fabricated using the PEG-DA-258 resin with a 0.3 s exposure setting for each layer. Prior to build the porous barrier, PEG-DA-258 residue was washed using D.I. water. The PEG-DA-575 resin tray was inserted into the printer for fabricating the porous barrier. The 100 µm-thick porous barrier was printed by irradiating each layer for 0.8 s. Compared to PEG-DA-258, PEG-DA-575 needs longer exposure times to fabricate polymer structures because, for the same polymer mass, PEG-DA-575 has less active sites (diacrylate groups) for photopolymerization than PEG-DA-258; the smaller number of active sites results in a slower reaction speed in PEG-DA-575. The lower molecular weight monomers in PEG-DA-258 have a higher mobility in the resin, which increases the probability of radical initiation on the carbon-carbon double bond (this radical initiation is the first step of the polymerization process). Moreover, due to its higher mobility compared to PEG-DA-575, the PEG-DA-258 molecule has a greater chance of encountering other molecules, which also results in a faster reaction speed. To fabricate a 1 mm-high top channel, the resin tray was switched back to the PEG-DA-258 resin tray after rinsing the PEG-DA-575 residue and the top channel was fabricated with the same conditions as the bottom channel. In order to prevent the microchannel from clogging with PEG-DA-258 residue due to overexposure, the resin in the microchannels was removed after printing a 100 µm-thick top channel roof and the fabrication continued until the diffusion chip was completed. At the end of the print process, the microchannels were flushed in a dark room with distilled water to remove any PEG-DA resin from the channel.

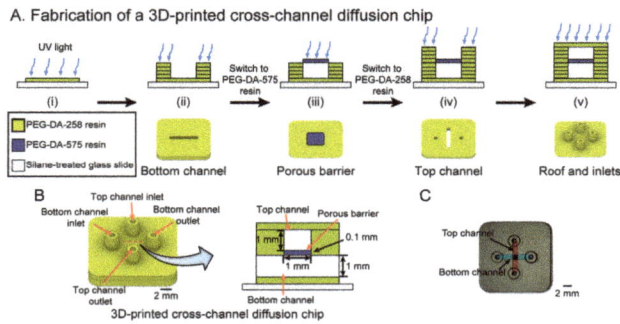

Figure 1. (**A**) Fabrication process in cross-section schematics (top row) and 3D view (bottom row) of a 3D-printed cross-channel diffusion microchip containing a PEG-DA-575 porous barrier. The process depicts the fabrication of (i) the bottom layer; (ii) the bottom channel; (iii) the porous barrier; (iv) the top channel; and (v) the roof and inlets; (**B**) CAD representation of the finished 3D-printed cross-channel diffusion chip (left) and its cross-sectional schematic (right); (**C**) photograph (top view) of the 3D-printed cross-channel diffusion chip; the top channel is filled with red dye and the bottom channel is filled with blue dye for visualization purposes.

Figure 1B illustrates a CAD design of the 3D-printed cross-channel diffusion chip and its cross-sectional view. A photo image of the 3D-printed cross-channel diffusion chip is displayed in Figure 1C. The bottom channel is filled with blue dye while the top channel is filled with red dye.

3.3. Hydrogen Ion Diffusion Test

We confirmed the selective diffusion of hydrogen ions through a PEG-DA-575 barrier by observing pH shifts as color changes with the pH-sensitive phenol red dye. The phenol red color was shown to gradually shift from yellow to red over the pH range 6.8 to 8.2 [38]. 0.1 M HCl solution was injected into the bottom channel of the cross-channel diffusion chip and then 0.5% phenol red in PBS was loaded into the top channel. Hydrogen ion diffusion was recognized by observing the color change of phenol red as shown in Figure 2. The phenol red solution at 0 s gradually took on a yellow color at 10 s at the porous barrier area, corresponding to the overlap between the two microchannels. The phenol red turned completely yellow after 70 s. The color change happened only in the top channel, which means hydrogen ions can pass through the porous barrier from the bottom channel to the top channel, while phenol red remains in the top channel, as illustrated in the cross-section view of Figure 2.

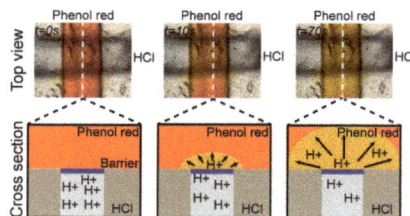

Figure 2. Hydrogen ion diffusion experiment (**top row**) and cross-section schematic (**bottom row**) using the 3D-printed cross-channel diffusion chip. The phenol red in the channel ($t = 0$ s) gradually changed to yellow at the channel intersection ($t = 10$ s) until it turned completely yellow ($t = 70$ s), while the bottom channel did not change color.

3.4. Fabrication of a 3D-Printed Symmetric-Channel Diffusion Chip

Small-ion diffusion was successfully carried out with the 3D-printed cross-channel diffusion chip (see Section 3.3). However, some biomedical applications may require the delivery of larger molecules such as dyes and drugs. To make a hydrogel barrier with larger pores, we used 40% PEG-DA-700 mixed with 60% water. The higher molecular weight of PEG-DA can produce larger pores in the hydrogel because of its longer chain length, which increases the mesh size of the polymer network. To polymerize a dilute PEG-DA-700 resin, a longer exposure time is required, which increases the risk of clogging channels under the porous barrier. We changed the diffusion chip design to a symmetric-channel chip having a porous barrier in between two channels, which ensured that the longer exposure time required to polymerize 40% PEG-DA-700 did not occlude the channels.

The fabrication process of a 3D-printed symmetric-channel diffusion chip is described in Figure 3A. PEG-DA-258 resin and 40% PEG-DA-700 resin in D.I. water were used to print the symmetric-channel and porous barrier, respectively. Prior to printing, the 3D designed symmetric-channel diffusion chip was sliced into 25 μm thick layers similar to the cross-channel chip. As explained before, a 100 μm-thick bottom layer was printed with 6 s exposure and then eight layers of the channel (200 μm) were fabricated with 0.3 s UV irradiation for each layer. Next, the PEG-DA-258 residue was washed using D.I. water and the 40% PEG-DA-700 resin tray was inserted. The porous barrier was created by a single 6.5 s exposure of the resin, which allowed the PEG-DA-700 hydrogel structure to grow down and attach to the glass slide. The porous barrier is designed as the shape of an extruded rectangle between two microchannels and is, by design, 200 μm tall, 300 μm wide and 2 mm long. The PEG-DA-258 resin tray was re-installed in the 3D printer to build the last of the channel layers after cleaning the PEG-DA-700 residue. To avoid clogging of the microchannels, they were cleaned after printing a 100 μm-thick roof layer and then the diffusion chip fabrication was completed. The microchannels were washed using distilled water to remove uncured PEG-DA resin. Figure 3B,C show a schematic cross section and a photograph, respectively, of the device.

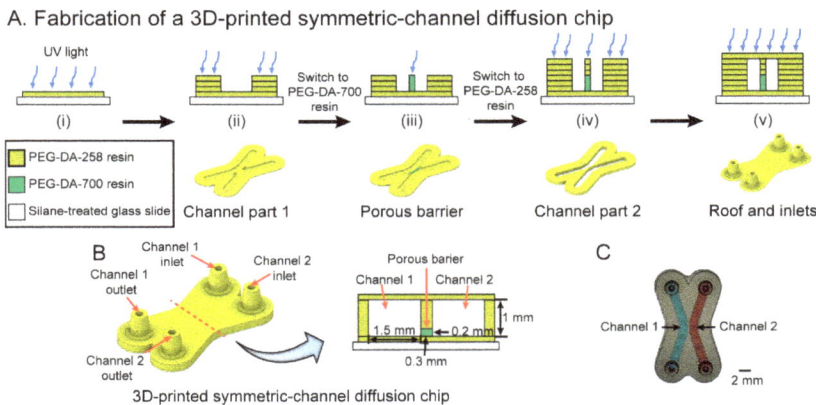

Figure 3. (A) Fabrication process in cross-section schematics (**top row**) and 3D view (**bottom row**) of a 3D-printed symmetric-channel diffusion microchip with a 40% *w/w* PEG-DA-700 porous barrier. The process depicts the fabrication of (i) the bottom layer; (ii) the channel part 1; (iii) the porous barrier; (iv) the channel part 2; and (v) the roof and inlets; (**B**) CAD representation of the 3D-printed symmetric-channel diffusion chip (**left**) and its cross-sectional schematic (**right**); (**C**) photograph of the 3D-printed symmetric-channel diffusion chip; Channel 1 is filled with blue dye and Channel 2 is filled with red dye for visualization purposes.

3.5. Fluorescein Diffusion Test

Molecular diffusion across the 40% PEG-DA-700 barrier in the 3D-printed symmetric-channel chip was tested by observing the transport of fluorescein (as a model of a small drug) using an inverted epifluorescence microscope. 0.1 mM fluorescein was loaded into Channel 1 while PBS buffer was loaded into Channel 2. The change in fluorescence intensity as fluorescein diffused from Channel 1 to Channel 2 was measured for 1 h. The results of the fluorescein diffusion are shown in Figure 4. The green signal was observed only in Channel 1 when fluorescein was loaded into it (Figure 4A). Then, fluorescein molecules transferred to Channel 2 through the 40% PEG-DA-700 hydrogel barrier (Figure 4B). After 60 min, the green color was clearly observed in Channel 2 as seen with the fluorescence image in Figure 4C. The fluorescence intensity was quantitatively analyzed and plotted in Figure 4D. The fluorescence intensity gradually increased with time from 0 to 60 min in both the 40% PEG-DA-700 hydrogel barrier and Channel 2.

Fick's second law of diffusion was employed to calculate the diffusivity of fluorescein molecules into the 40% PEG-DA-700 barrier [39]:

$$\frac{\partial C}{\partial t} = D\frac{\partial^2 C}{\partial x^2}, \tag{1}$$

where D represents the diffusivity of fluorescein in 40% PEG-DA-700 and C indicates the concentration of fluorescein.

The solution to Equation (1) can be expressed as

$$C(x,t) = \frac{C_x - C_0}{C_s - C_0} = 1 - erf\left(\frac{x}{2\sqrt{Dt}}\right) = erfc\left(\frac{x}{2\sqrt{Dt}}\right), \tag{2}$$

where $C(x=0) = C_s$ is the concentration of fluorescein at the junction of the porous barrier and the channel containing fluorescein ("Channel 1"), and $C(x=\infty) = C_0 = 0$ corresponds to the initial concentration of fluorescein in PEG-DA [39]. We assume that C_s and C_0 remain constant over time.

The characteristic diffusion length (L) at a given time (t) is defined as the distance at which the concentration of the diffusing species reaches 50% of the source concentration (C_s), and can be approximated as [39]:

$$L \approx (Dt)^{1/2}. \tag{3}$$

To determine the diffusivity of fluorescein in the 40% PEG-DA-700 barrier, we acquired time-lapse images of fluorescein diffusion through the hydrogel from 0 min to 10 min using an inverted epifluorescence microscope. The fluorescent time-lapse images were analyzed using ImageJ to obtain the diffusion profiles, which show that the fluorescein molecules diffuse into the porous barrier (Figure 4E). For the duration of the experiment ($t = 10$ min), since $C(x = 200 \ \mu m) = 0$, we can assume that there is negligible mass loss from the porous barrier into Channel 2. We plotted the experimentally observed characteristic diffusion length (L) as a function of the square root of time ($t^{1/2}$) and performed a linear regression analysis to determine the diffusivity (D). The diffusivity of fluorescein in 40% PEG-DA-700 was estimated to be ~4.6 × 10^{-7} cm^2/s from the slope ($D^{1/2}$) of the fitted curve ($L = 6.77951t^{1/2}$, adjusted $R^2 = 0.99343$) (Figure 4E inset).

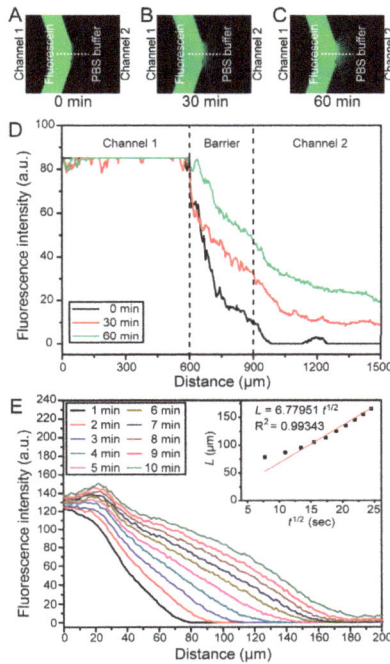

Figure 4. Fluorescein diffusion test using the 3D-printed symmetric-channel diffusion microchip. (**A–C**) Fluorescein diffused through the 40% PEG-DA-700 hydrogel barrier from 0 min to 60 min; (**D**) graph of the fluorescein intensity profiles obtained across the microchannel in the three images above, showing that the fluorescein intensity gradually increased in Channel 2; (**E**) concentration profile of fluorescein in the 40% PEG-DA-700 hydrogel barrier from 1 min to 10 min. (Inset) Plot of the characteristic diffusion length as a function of the square root of time (dots); the red line is a linear fit of the dots, $L = 6.77951t^{1/2}$ (adjusted $R^2 = 0.99343$).

4. Conclusions

Our simple 3D-printed microfluidic channel with an integrated porous barrier is the first step toward 3D-printed devices capable of more advanced functionalities. We developed a stereolithographic co-printing process to demonstrate the selective diffusion of small ions or fluorescein molecules through a 3D-printed porous barrier. The 3D-printed cross-channel chip and the 3D-printed symmetric-channel chip were successfully printed using PEG-DA-575 and 40% PEG-DA-700, respectively, for the porous barrier and PEG-DA-258 for the microchannels. We were able to confirm the selective hydrogen ion diffusion through the PEG-DA-575 hydrogel barrier by observing the color change of phenol red. Fluorescein diffusion was demonstrated using the 3D-printed symmetric-channel chip, which has a 40% PEG-DA-700 hydrogel barrier in the microchannel. The proposed stereolithographic co-printing process for fabricating selective porous barriers will provide an inexpensive, simple, convenient, and reproducible molecule delivery platform that can be applied in the fields of tissue engineering, drug delivery systems, and biomaterials. 3D-printed hydrogels with larger porosities could potentially be used for filtering particles from bodily fluids such as circulating tumor DNA, exosomes, bacteria, and circulating tumor cells, and in general 3D-printed hydrogels could be used for immobilizing biomolecules within microfluidic devices for biosensing applications. Moreover, the biocompatibility of 3D-printed PEG-DA-258 microdevices as well as

PEG-DA-700 structures makes our strategy of printing porous barriers inside a microchannel amenable to the development of cell-based assays and organ-on-chip platforms [40,41].

Acknowledgments: This research was partially supported by the National Cancer Institute (Grant No. 5R01CA181445) and the BioNano Health-Guard Research Center funded by the Ministry of Science, ICT (MSIT) of Korea as Global Frontier Project (Grant No. H-GUARD_2014M3A6B2060302).

Author Contributions: Yong Tae Kim fabricated the diffusion chips, tested the ion and molecule diffusion using the diffusion chips, and wrote the paper. Kurt Castro developed the stereolithographic co-printing methods and PEG-DA-575 hydrogel printing conditions. Nirveek Bhattacharjee discussed the details of experiments and advanced 3D printing using PEG-DA-258 resin. Albert Folch advised on and enabled funding for the project.

Conflicts of Interest: The authors declare no conflict of interest.

References

1. Novikov, D.S.; Fieremans, E.; Jensen, J.H.; Helpern, J.A. Random walks with barriers. *Nat. Phys.* **2011**, *7*, 508–514. [CrossRef] [PubMed]
2. Nance, E.A.; Woodworth, G.F.; Sailor, K.A.; Shih, T.-Y.; Xu, Q.; Swaminathan, G.; Xiang, D.; Eberhart, C.; Hanes, J. A dense poly(ethylene glycol) coating improves penetration of large polymeric nanoparticles eithin brain tissue. *Sci. Transl. Med.* **2012**, *4*, 149ra119. [CrossRef] [PubMed]
3. Verhulsel, M.; Vignes, M.; Descroix, S.; Malaquin, L.; Vignjevic, D.M.; Viovy, J.L. A review of microfabrication and hydrogel engineering for micro-organs on chips. *Biomaterials* **2014**, *35*, 1816–1832. [CrossRef] [PubMed]
4. Gizzatov, A.; Key, J.; Aryal, S.; Ananta, J.; Cervadoro, A.; Palange, A.L.; Fasano, M.; Stigliano, C.; Zhong, M.; Di Mascolo, D.; et al. Hierarchically structured magnetic nanoconstructs with enhanced relaxivity and cooperative tumor accumulation. *Adv. Funct. Mater.* **2014**, *24*, 4584–4594. [CrossRef] [PubMed]
5. Park, H.B.; Jung, C.H.; Lee, Y.M.; Hill, A.J.; Pas, S.J.; Mudie, S.T.; Wagner, E.V.; Freeman, B.D.; Cookson, D.J. Polymers with cavities tuned for fast selective transport of small molecules and ions. *Science* **2007**, *318*, 254–258. [CrossRef] [PubMed]
6. El-Ali, J.; Sorger, P.K.; Jensen, K.F. Cells on chips. *Nature* **2006**, *442*, 403–411. [CrossRef] [PubMed]
7. Unger, M.A. Monolithic microfabricated valves and pumps by multilayer soft lithography. *Science* **2000**, *288*, 113–116. [CrossRef] [PubMed]
8. Sackmann, E.K.; Fulton, A.L.; Beebe, D.J. The present and future role of microfluidics in biomedical research. *Nature* **2014**, *507*, 181–189. [CrossRef] [PubMed]
9. Ren, K.; Zhou, J.; Wu, H. Materials for microfluidic chip fabrication. *Acc. Chem. Res.* **2013**, *46*, 2396–2406. [CrossRef] [PubMed]
10. Hsu, C.H.; Folch, A. Spatio-temporally-complex concentration profiles using a tunable chaotic micromixer. *Appl. Phys. Lett.* **2006**, *89*, 144102. [CrossRef]
11. Scott, A.; Weir, K.; Easton, C.; Huynh, W.; Moody, W.J.; Folch, A. A microfluidic microelectrode array for simultaneous electrophysiology, chemical stimulation, and imaging of brain slices. *Lab Chip* **2013**, *13*, 527–535. [CrossRef] [PubMed]
12. Lai, H.; Folch, A. Design and dynamic characterization of "single-stroke" peristaltic PDMS micropumps. *Lab Chip* **2011**, *11*, 336–342. [CrossRef] [PubMed]
13. Seliktar, D. Designing cell-compatible hydrogels. *Science* **2012**, *336*, 1124–1128. [CrossRef] [PubMed]
14. Vagias, A.; Košovan, P.; Koynov, K.; Holm, C.; Butt, H.J.; Fytas, G. Dynamics in stimuli-responsive poly(*N*-isopropylacrylamide) hydrogel layers as revealed by fluorescence correlation spectroscopy. *Macromolecules* **2014**, *47*, 5303–5312. [CrossRef]
15. Nguyen, K.T.; West, J.L. Photopolymerizable hydrogels for tissue engineering applications. *Biomaterials* **2002**, *23*, 4307–4314. [CrossRef]
16. Zustiak, S.P.; Boukari, H.; Leach, J.B. Solute diffusion and interactions in cross-linked poly(ethylene glycol) hydrogels studied by fluorescence correlation spectroscopy. *Soft Matter* **2010**, *6*, 3609–3618. [CrossRef] [PubMed]
17. Hubbell, J.A. Hydrogel systems for barriers and local drug delivery in the control of wound healing. *J. Control. Release* **1996**, *39*, 305–313. [CrossRef]
18. Tourovskaia, A.; Figueroa-Masot, X.; Folch, A. Long-term microfluidic cultures of myotube microarrays for high-throughput focal stimulation. *Nat. Protoc.* **2006**, *1*, 1092–1104. [CrossRef] [PubMed]

19. Hou, X.; Zhang, Y.S.; De Santiago, G.T.; Alvarez, M.M.; Ribas, J.; Jonas, S.J.; Weiss, P.S.; Andrews, A.M.; Aizenberg, J.; Khademhosseini, A. Interplay between materials and microfluidics. *Nat. Rev. Mater.* **2017**, *2*, 17016. [CrossRef]

20. Khademhosseini, A.; Langer, R. Microengineered hydrogels for tissue engineering. *Biomaterials* **2007**, *28*, 5087–5092. [CrossRef] [PubMed]

21. Peppas, N.A.; Hilt, J.Z.; Khademhosseini, A.; Langer, R. Hydrogels in biology and medicine: From molecular principles to bionanotechnology. *Adv. Mater.* **2006**, *18*, 1345–1360. [CrossRef]

22. Cosson, S.; Lutolf, M.P. Hydrogel microfluidics for the patterning of pluripotent stem cells. *Sci. Rep.* **2014**, *4*, 4462. [CrossRef] [PubMed]

23. Lee, A.G.; Arena, C.P.; Beebe, D.J.; Palecek, S.P. Development of macroporous poly(ethylene glycol) hydrogel arrays within microfluidic channels. *Biomacromolecules* **2010**, *11*, 3316–3324. [CrossRef] [PubMed]

24. Huang, C.P.; Lu, J.; Seon, H.; Lee, A.P.; Flanagan, L.A.; Kim, H.-Y.; Putnam, A.J.; Jeon, N.L. Engineering microscale cellular niches for three-dimensional multicellular co-cultures. *Lab Chip* **2009**, *9*, 1740–1748. [CrossRef] [PubMed]

25. Sung, J.H.; Kam, C.; Shuler, M.L. A microfluidic device for a pharmacokinetic–pharmacodynamic (PK–PD) model on a chip. *Lab Chip* **2010**, *10*, 446–455. [CrossRef] [PubMed]

26. Au, A.K.; Huynh, W.; Horowitz, L.F.; Folch, A. 3D-Printed Microfluidics. *Angew. Chem. Int. Ed.* **2016**, *55*, 3862–3881. [CrossRef] [PubMed]

27. Bhattacharjee, N.; Urrios, A.; Kang, S.; Folch, A. The upcoming 3D-printing revolution in microfluidics. *Lab Chip* **2016**, *16*, 1720–1742. [CrossRef] [PubMed]

28. Kirchmajer, D.M.; Gorkin, R., III; in het Panhuis, M. An overview of the suitability of hydrogel-forming polymers for extrusion-based 3D-printing. *J. Mater. Chem. B* **2015**, *3*, 4105–4117. [CrossRef]

29. Pantani, R.; Turng, L.S. Manufacturing of advanced biodegradable polymeric components. *J. Appl. Polym. Sci.* **2015**, *132*. [CrossRef]

30. Mohanty, S.; Larsen, L.B.; Trifol, J.; Szabo, P.; Burri, H.V.R.; Canali, C.; Dufva, M.; Emnéus, J.; Wolff, A. Fabrication of scalable and structured tissue engineering scaffolds using water dissolvable sacrificial 3D printed moulds. *Mater. Sci. Eng. C* **2015**, *55*, 569–578. [CrossRef] [PubMed]

31. Mohanty, S.; Sanger, K.; Heiskanen, A.; Trifol, J.; Szabo, P.; Dufva, M.; Emnéus, J.; Wolff, A. Fabrication of scalable tissue engineering scaffolds with dual-pore microarchitecture by combining 3D printing and particle leaching. *Mater. Sci. Eng. C* **2016**, *61*, 180–189. [CrossRef] [PubMed]

32. He, Y.; Yang, F.; Zhao, H.; Gao, Q.; Xia, B.; Fu, J. Research on the printability of hydrogels in 3D bioprinting. *Sci. Rep.* **2016**, *6*, 29977. [CrossRef] [PubMed]

33. Hinton, T.J.; Jallerat, Q.; Palchesko, R.N.; Park, J.H.; Grodzicki, M.S.; Shue, H.-J.; Ramadan, M.H.; Hudson, A.R.; Feinberg, A.W. Three-dimensional printing of complex biological structures by freeform reversible embedding of suspended hydrogels. *Sci. Adv.* **2015**, *1*, e1500758. [CrossRef] [PubMed]

34. Urrios, A.; Parra-Cabrera, C.; Bhattacharjee, N.; Gonzalez-Suarez, A.M.; Rigat-Brugarolas, L.G.; Nallapatti, U.; Samitier, J.; DeForest, C.A.; Posas, F.; Garcia-Cordero, J.L.; et al. 3D-printing of transparent bio-microfluidic devices in PEG-DA. *Lab Chip* **2016**, *16*, 2287–2294. [CrossRef] [PubMed]

35. Cruise, G.M.; Scharp, D.S.; Hubbell, J.A. Characterization of permeability and network structure of interfacially photopolymerized poly(ethylene glycol) diacrylate hydrogels. *Biomaterials* **1998**, *19*, 1287–1294. [CrossRef]

36. Mellott, M.B.; Searcy, K.; Pishko, M.V. Release of protein from highly cross-linked hydrogels of poly(ethylene glycol) diacrylate fabricated by UV polymerization. *Biomaterials* **2001**, *22*, 929–941. [CrossRef]

37. Russell, R.J.; Axel, A.C.; Shields, K.L.; Pishko, M.V. Mass transfer in rapidly photopolymerized poly(ethylene glycol) hydrogels used for chemical sensing. *Macromolecules* **2001**, *42*, 4893–4901. [CrossRef]

38. Pick, E.; Keisari, Y. A simple colorimetric method for the measurement of hydrogen peroxide produced by cells in culture. *J. Immunol. Methods* **1980**, *38*, 161–170. [CrossRef]

39. Kirby, B.J. *Micro- and Nanoscale Fluid Mechanics: Transport in Microfluidic Devices*; Cambridge University Press: Cambridge, UK, 2010; ISBN 978-0-521-11903-0.

40. Cvetkonic, C.; Raman, R.; Chan, V.; Williams, B.J.; Tolish, M.; Bajaj, P.; Sakar, M.S.; Asada, H.H.; Saif, M.T.; Bashir, R. Three-dimensionally printed biological machines powered by skeletal muscle. *Proc. Natl. Acad. Sci. USA* **2014**, *111*, 10123–10130.

41. Khademhosseini, A.; Yeh, J.; Jon, S.; Eng, G.; Suh, K.Y.; Burdick, J.A.; Langer, R. Molded polyethylene glycol microstructures for capturing cells within microfluidic channels. *Lab Chip* **2014**, *4*, 425–430. [CrossRef] [PubMed]

micromachines

MDPI

Article

Open Design 3D-Printable Adjustable Micropipette that Meets the ISO Standard for Accuracy

Martin D. Brennan, Fahad F. Bokhari and David T. Eddington *

Department of Bioengineering, University of Illinois at Chicago, Chicago, IL 60607, USA;
mbrenn3@uic.edu (M.D.B.); fbokha2@uic.edu (F.F.B.)
* Correspondence: dte@uic.edu; Tel.: +1-312-355-3278

Received: 30 March 2018; Accepted: 17 April 2018; Published: 18 April 2018

Abstract: Scientific communities are drawn to the open source model as an increasingly utilitarian method to produce and share work. Initially used as a means to develop freely-available software, open source projects have been applied to hardware including scientific tools. Increasing convenience of 3D printing has fueled the proliferation of open labware projects aiming to develop and share designs for scientific tools that can be produced in-house as inexpensive alternatives to commercial products. We present our design of a micropipette that is assembled from 3D-printable parts and some hardware that works by actuating a disposable syringe to a user-adjustable limit. Graduations on the syringe are used to accurately adjust the set point to the desired volume. Our open design printed micropipette is assessed in comparison with a commercial pipette and meets the ISO 8655 standards.

Keywords: open source labware; 3D printing; functional prototyping

1. Introduction

The open source development model, initially applied to software, is thriving in the development of open source scientific equipment due in part to increasing access of 3D printing [1,2]. Additive manufacturing methods have existed for decades although the recent availability of inexpensive desktop printers [3,4] has made it feasible for consumers to design and print prototypes and even functional parts, as well as consumer goods [5,6]. The proliferation of free CAD software [7–10] and design sharing sites [11–14] has also supported the growth and popularity of open designed parts and projects. Open design 3D-printable lab equipment is an attractive idea because, like open source software, it allows free access to technology that is otherwise inaccessible due to proprietary and/or financial barriers. Open design tools create the opportunity for scientists and educational programs in remote or resource-limited areas to participate with inexpensive and easy to make tools [15–20]. Open source development also enables the development of custom solutions to meet unique applications not met by commercial products that are shared freely and are user modifiable [5,21–26]. Some advanced, noteworthy, open source scientific equipment include a PCR device [27], a tissue scaffold printer [28] and a two-photon microscope [29], although simple tools have the potential to be impactful as they can serve a wider community.

Some simple and clever printable parts that have emerged are ones that give a new function to a ubiquitous existing device, such as a drill bit attachment designed to hold centrifuge tubes, allowing a dremel to be used as a centrifuge [30]. Although this may make a rather crude centrifuge, it may be an adequate solution for a fraction of the cost of a commercial centrifuge. Other examples of open source research tools that utilize 3D-printed parts include optics equipment [31], microscopes [32,33], syringe pumps [34,35], reactionware [36–39] and microfluidics [40,41], and the list continues to grow [42].

One example of an everyday scientific tool that provides the opportunity for an open design solution is the micropipette. An open design micropipette that can be made inexpensively affords more options for labs and educational settings. Micropipettes are an indispensable tool used routinely in lab tasks and can easily cost $1000 USD for a set. Often, a lab will require several sets each for a dedicated task. Some pipettes may even be re-calibrated for use with liquids of different properties.

Air displacement pipettes use a piston operating principle to draw liquid into the pipette [43]. In a typical commercial pipette. the piston is made to be gas-tight with a gasketed plunger inside a smooth barrel. Consumer-grade fused deposition modeling (FDM) printers are unable to build a smooth surface due to the formation of ridges that occur as each layer is deposited [44]. The ridges formed by FDM make it impractical to form a gas-tight seal between moving parts, even with a gasket. Existing printable open-design micropipettes get around this limitation by stretching a membrane over one end of a printed tube, which when pressed causes the displacement. The displacement membrane can be made from any elastic material such as a latex glove. A few open design micropipettes exist including a popular one, which in addition to the printed parts, uses parts scavenged from a retractable pen [45]. A major limitation of this design is that there is no built-in feature such as a readout for the user to set the displacement to a desired volume. This requires the user to validate the volumes dispensed with a high precision scale. Without verification with a scale, the volumes dispensed can only be estimated based on the calculations of the deflection of the membrane, which is not a practical protocol.

We submit a new design whose major strength is the ability to adjust to any volume aided by the built-in scale. Our open design 3D-printable micropipette works by actuating a disposable syringe to a user-adjustable set point. This allows the user to set the pipette to a volume by reading the graduations on the syringe barrel. We have also designed an adjusted graduation scale, to be used in place of the printed on scale, that corrects for the compressibility of air, which allows the pipette to be set accurately. Additionally, our pipette offers a simplified assembly requiring no glue, tape or permanent connections.

2. Materials and Methods

Our printed pipette is designed to actuate a 1-mL or 3-mL syringe to a user-set displacement. The core of the design is two printed parts, the body and the plunger, which are able to be printed on a consumer-grade FDM printer (Figure 1), in this case a Makerbot Replicator.

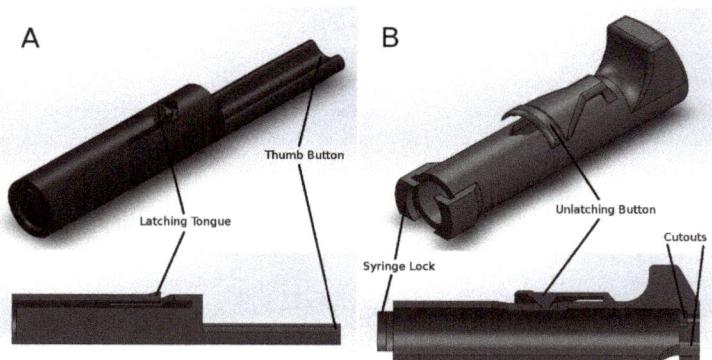

Figure 1. CAD renderings of printable parts and cross-sections. (**A**) The printed plunger part is a shaft that pushes on the syringe plunger and slides in the printed body part. The printed plunger has a latching tongue and button, which interfaces with the printed body part. (**B**) The printed body part holds the syringe and interfaces with the printed plunger part. The body part features the unlatching button and slots to hold the syringe in place by its flanges. The body part also has two cutouts in the top for the plunger button and for the hex nut and bolt. STL files of the printable parts are available in the Supplementary Materials.

A 1-mL or 3-mL syringe twists to lock in the body part and is held into place by the syringe flanges. The 30–300-μL configuration uses a 1-mL pipette, and the 100–1000-μL configuration uses a 3-mL pipette. The plunger part slides freely in the body part and actuates the syringe by pushing the thumb button (Figure 2). The pipette is spring loaded towards a set point, which is adjustable by a set-screw. When the thumb button is pressed, the system locks when it reaches the latched position, where it is ready to draw in fluid. The plunger is held in place with a latching mechanism, which when released, draws in fluid. The displacement is equal to the distance between the set position and the latched position. The pipette can also be pressed past the latched position to 'blow-out' the transferred fluid completely from the pipette tip. For a demonstration of the assembly and operation, see the videos in the Supplementary Materials. Additional materials required for assembly include two springs, a nut and a bolt (Tables A1 and A2). Attempts to make a printable luer lock adapter for tips was abandoned as the surface of printed parts is too rough to make an air-tight seal with the luer or pipette tip. Instead, a combination of a barbed luer adapter and elastic tubing is used to attach the pipette tips. Our printed pipette mimics commercial pipettes in design, function and user operation, making it intuitive to use.

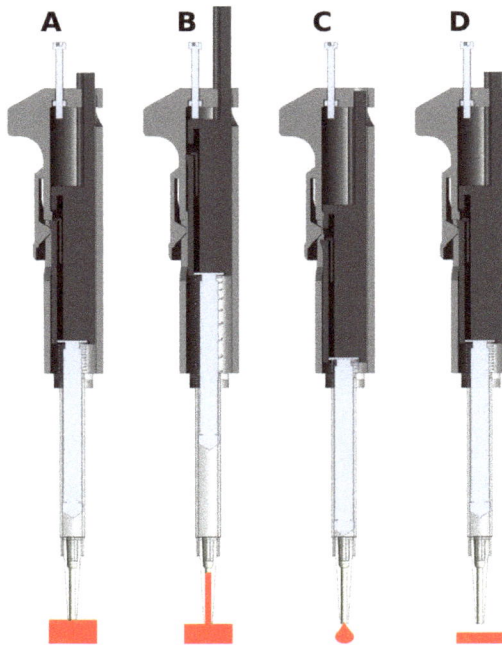

Figure 2. CAD renderings of assembled pipette and function. The pipette actuates the syringe to three positions. (**A**) The latched position. When the plunger is pressed, the pipette locks at this position. The tip is then placed in a liquid, and the unlatching button is pressed to release the pipette back to the set position (**B**), drawing in liquid. (**B**) The set position. The position of the screw determines the total displacement that the plunger moves. The pipette is spring loaded to return to this position. (**C**) The blow-out position. The fluid is transferred by pressing the plunger past the latched position to blow-out all the liquid. (**D**) Return to the latched position. The pipette returns to the latched position ready to perform another transfer.

Our pipette uses the air-displacement method, where a vacuum is applied to a pocket of air to draw liquid into the pipette. As air is a compressible fluid, this pocket of air expands due to the weight of the liquid pulling on it. Due to this effect, the graduations on the syringe are not accurate, as they are designed for measuring liquid within the syringe. At larger volumes, this effect is more pronounced, resulting in the volume measured being greater than the amount of liquid pulled into the syringe. We remedied this by creating a new scale to account for the expansion. The scale is printed on a transparency sheet and is taped onto the syringe for accurate measurements (the scales appear in Appendix A: Figures A1 and A2).

2.1. Fabrication and Assembly

Two parts are printed at normal resolution with rafts on a Makerbot Replicator (STL files are available in the Supplementary Materials). A disposable syringe and a few extra parts are used to assemble the pipette (Figure 3). A small amount of paraffin wax is applied to the screw to prevent slop from causing the set point to drift after each actuation. The nut is sunk into the hex inset in the printed body part. The bolt is threaded in from the top of the body into the nut. Two springs are threaded onto the plunger of a 1-mL syringe for the 30–300-µL configuration. Springs are placed inside a 3-mL syringe for the 100–1000-µL configuration. The plunger part is inserted in the body, and the syringe assembly is pushed in and locked from the syringe flanges to the body part to complete assembly (Figure 4). Parts and cost are listed in Tables A1 and A2 in Appendix A.

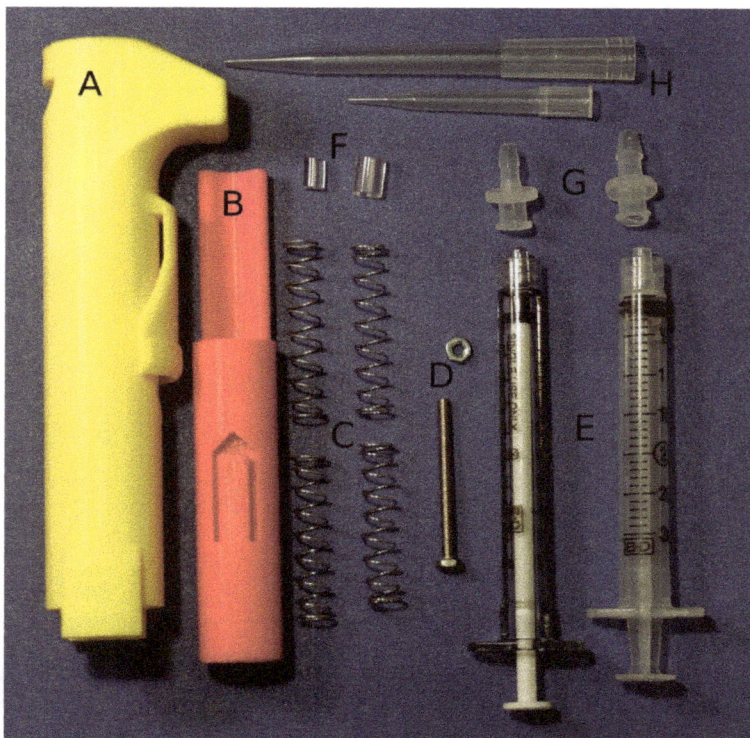

Figure 3. Photo of the parts to make the pipette. (**A**) Printed body part. (**B**) Printed plunger part. (**C**) Springs. (**D**) M3 hex nut and M3 bolt. (**E**) Disposable syringes. (**F**) Two sizes of tubing to adapt pipette tips to luer barbs. (**G**) Luer lock syringe-to-barb adapters. (**H**) Pipette tips. A list of parts is in Appendix A: Tables A1 and A2.

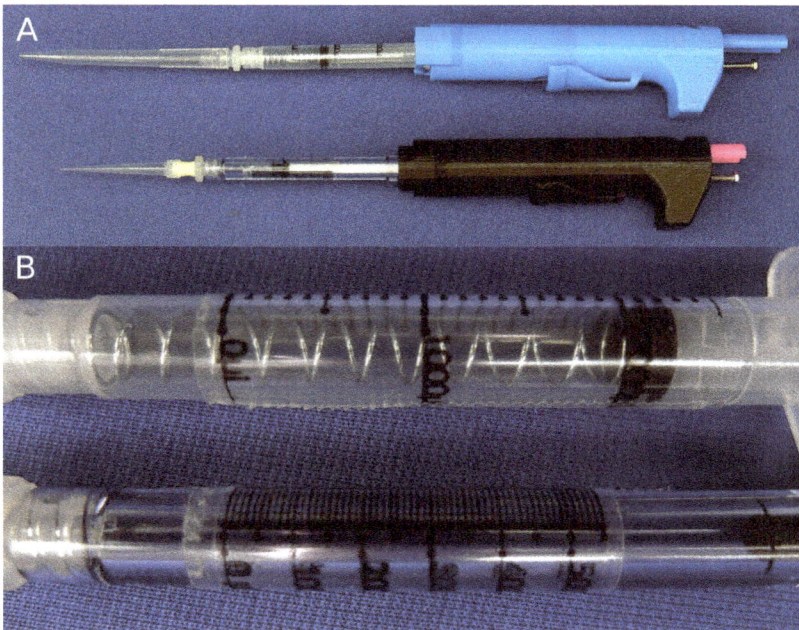

Figure 4. Photos of assembled pipettes. (**A**) Two assemblies of the pipette: the 100–1000-µL configuration (top) and the 30–300-µL configuration (bottom). (**B**) Close-up photo of the taped on scale for each of the syringes. Scales are available in the Appendix A: Figures A1 and A2.

2.2. Validation

The printed pipette's accuracy and precision were characterized and compared to a commercial pipette, as well as ISO 8655. The printed pipette was adjusted to the target volume by eye from the syringe graduations. Deionized water was transferred and measured with a high precision scale. Five transfers were recorded and averaged to account for random variability. Data were taken for printed pipettes with existing syringe graduations, as well as with our adjusted scale. Data were taken with commercial pipettes of 30–300 µL and 100–1000 µL to compare to the printed pipette. Accuracy and precision are expressed as systematic error and random error, respectively, and are calculated according to ISO 8655 [43] (Tables 1 and 2). The systematic error, or accuracy, is calculated according to the following equations where accuracy (A) is the difference of the mean volume ($\bar{V}$) and the nominal volume (V_0):

$$A = \bar{V} - V_0, A\% = 100\% \times A/V_0 \tag{1}$$

The precision or random error is the standard deviation (s) of the measurements, and (cv) is the coefficient of variation:

$$s = \sqrt{\frac{\sum_{i=1}^{n}(V_i - \bar{V})^2}{n-1}}, cv = 100\% \times s/\bar{V} \tag{2}$$

3. Results

Using the adjusted syringe graduation scale, our printed pipette meets ISO 8655 standards, but using the existing syringe graduation scale does not meet the ISO standard for accuracy. According to ISO 8655 for a pipette with the maximum nominal volume of 1000 µL, the systematic error cannot exceed 8 µL, and the random error cannot exceed 3 µL [43]. For the maximum nominal volume of 300 µL, the systemic error cannot exceed 4 µL, and the random error cannot exceed 1.5 µL. The commercial pipettes met these standards handily, but in our initial tests with our printed pipette, we noticed large negative systematic error, suggesting that we were missing a biasing factor. This error was exaggerated while transferring larger volumes (Figures 5 and 6). After further investigation, we realized that the water in the tip pulls on the air due to gravity, which causes the air to expand, making measuring with the graduations inaccurate. The graduations are intended to measure incompressible fluids within the syringe, while we were using them to measure air while under vacuum. The syringe graduations could not be expected to accurately indicate the volume of water pulled into the tip, because the air had expanded under vacuum. This explains the negative systematic error we were experiencing.

Our solution was to replace the built-in scale with a scale that accounted for the expansion. We compared expected volumes with actual measurements and noticed a linear relationship. We scaled up the graduations by a factor of 1:1.027 for both the 100–1000-µL, and 30–300-µL configurations. These adjusted scales are printed on paper or a transparency sheet and taped onto the syringe with the zero at the plunger when the pipette is in the latched position (scales appear in Appendix A: Figures A1 and A2). With the new scale taped on the syringe, our validation testing met the ISO 8655 standard. Replacing the built-in scale with our scale corrected this effect. Our printed pipette with the adjusted scale meets ISO standards for accuracy and precision (Tables 1 and 2).

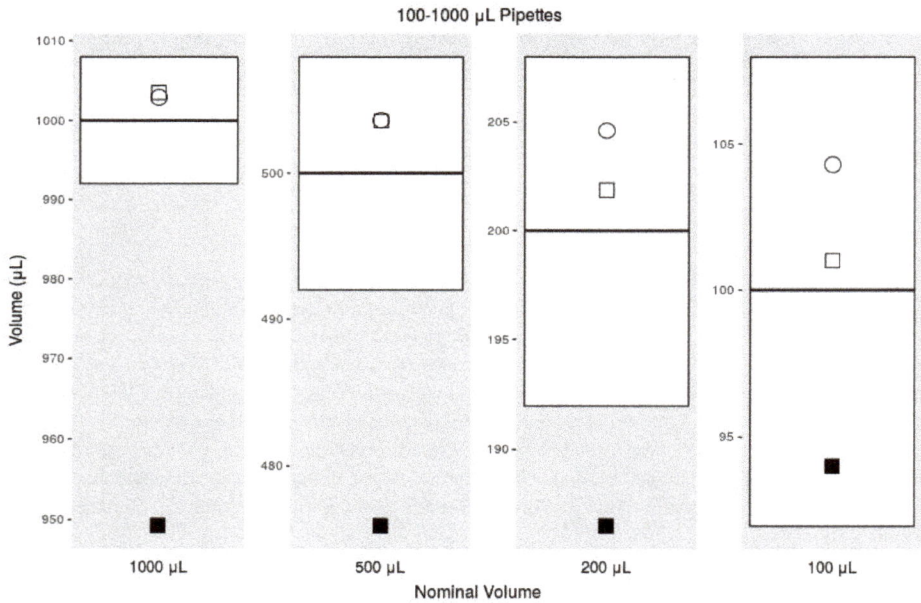

Figure 5. Comparison of the accuracy, or systematic error, among 100–1000-µL pipettes. The box plot indicates the range of the ISO standard; the plotted circle indicates the commercial pipette; and the plotted squares are the printed pipette. Empty squares are with the printed pipette with our adjusted graduation scale, and the filled squares are measurements taken with the syringe's existing graduations (data from Table 1.)

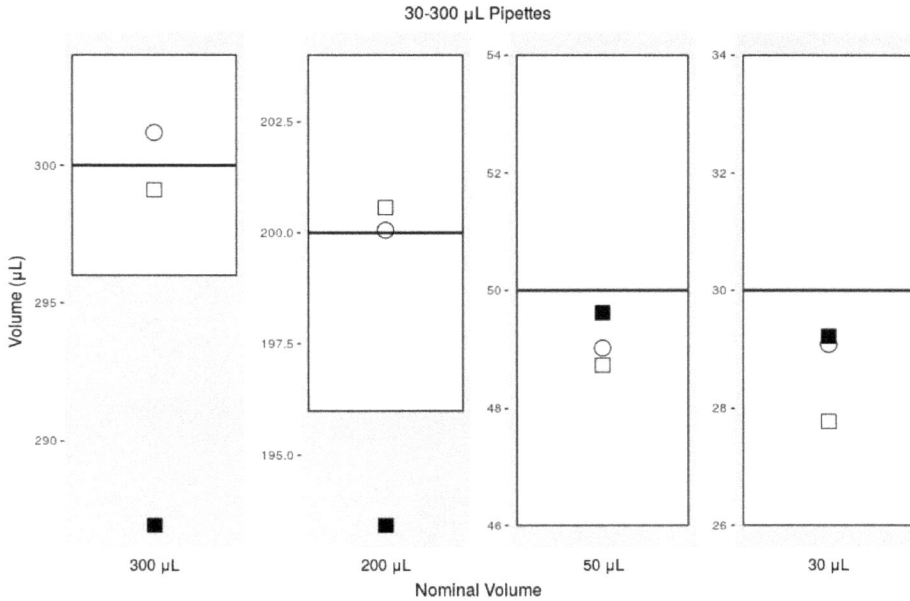

Figure 6. Comparison of the accuracy, or systematic error, among 30–300-μL pipettes. The box plot indicates the range of the ISO standard; the plotted circle indicates the commercial pipette; and the plotted squares are the printed pipette. Empty squares are with the printed pipette with our adjusted graduation scale, and the filled squares are measurements taken with the syringe's existing graduations (data from Table 2.)

Table 1. ISO 8655 for 100–1000-μL comparing a commercial pipette with our printed pipette used with the existing 3-mL syringe scale and an adjusted scale.

		Mean	Systematic Error	% Sys. Err.	Random Error	% Rand. Err.
1000 μL	ISO 8655, 100–1000 μL	1000	8.00	0.80	3.00	0.30
	Commercial Pipette	1002.98	2.98	0.30	1.72	0.17
	Printed Pipette	949.29	−50.71	−5.07	0.60	0.06
	Printed Pipette Scale	1003.57	3.57	0.36	0.89	0.09
500 μL	ISO 8655, 100–1000 μL	500	8.00	1.60	3.00	0.60
	Commercial Pipette	503.67	3.67	0.73	0.49	0.10
	Printed Pipette	475.99	−24.01	−4.80	4.75	1.00
	Printed Pipette Scale	503.62	3.62	0.72	1.64	0.33
200 μL	ISO 8655, 100–1000 μL	200	8.00	4.00	3.00	1.50
	Commercial Pipette	204.61	4.61	2.30	0.15	0.07
	Printed Pipette	186.55	−13.45	−6.72	1.31	0.70
	Printed Pipette Scale	201.87	1.87	0.94	1.47	0.73
100 μL	ISO 8655, 100–1000 μL	100	8.00	8.00	3.00	3.00
	Commercial Pipette	104.29	4.29	4.29	1.65	1.58
	Printed Pipette	94.02	−5.98	−5.98	4.81	5.12
	Printed Pipette Scale	101.00	1.00	1.00	1.05	1.04

Table 2. ISO 8655 for 30–300-µL comparing a commercial pipette with our printed pipette used with the existing 3-mL syringe scale and an adjusted scale.

		Mean	Systematic Error	% Sys. Err.	Random Error	% Rand. Err.
300 µL	ISO 8655, 30–300 µL	300	4.00	1.33	1.50	0.50
	Commercial Pipette	301.19	1.19	0.40	0.53	0.18
	Printed Pipette	286.91	−13.09	−4.36	0.42	0.15
	Printed Pipette Scale	299.11	−0.89	−0.30	0.48	0.16
200 µL	ISO 8655, 30–300 µL	200	4	2	1.5	0.75
	Commercial Pipette	200.06	0.06	0.03	0.46	0.23
	Printed Pipette	193.40	−6.60	−3.30	2.86	1.48
	Printed Pipette Scale	200.57	0.57	0.28	0.86	0.43
50 µL	ISO 8655, 30–300 µL	50	4	8	1.5	3
	Commercial Pipette	49.02	−0.98	−1.96	0.10	0.20
	Printed Pipette	49.62	−0.38	−0.76	1.26	2.53
	Printed Pipette Scale	48.73	−1.27	−2.54	1.11	2.27
30 µL	ISO 8655, 30–300 µL	30	4	13.3	1.5	5
	Commercial Pipette	29.08	−0.92	−3.06	0.09	0.31
	Printed Pipette	29.22	−0.78	−2.59	0.31	1.07
	Printed Pipette Scale	27.78	−2.22	−7.41	1.37	4.93

4. Discussion and Conclusions

Our printed pipette improves on existing open design pipettes in several ways. Most significantly, the user is able to adjust the syringe accurately without verifying the volume with a scale. The pipette can be adjusted to any discrete volume within range. In addition, the assembly requires no permanent connections using tape or glue, which allows for re-configuration and easy replacement of parts. Conveniently, our design can also reach to the bottom of a 15-mL conical tube allowing tasks such as aspirating supernatant fluid from a cell pellet. Unlike existing designs, our design does not use a membrane that may wear out and require replacement. The only major limitation of this design compared to the biropipette [45] is the option for a pipette tip ejector. The biropipette also uses arguably easier to source parts (a pen vs. a syringe) relative to what is required for our design. The biropipette also was developed in OpenSCAD, which is free and open source CAD software, whereas our design was created in Solidworks, a proprietary software. This creates a limitation for users who may not have access to the proprietary software used to make the design. Fortunately, the produced STL files are universal and can be imported and modified by any CAD software.

The luer to pipette tip connection is a good candidate for a 3D-printed solution. Unfortunately, the current state of FDM cannot make this adapter with high enough resolution or, critically, surface smoothness [44]. Other printing techniques, such as stereolithography (SLA) and multijet printing (MJP), provide higher resolution and smoother finishing, yet FDM is the the most widely available and affordable of the printing technologies. SLA printers are coming down in price, but are still significantly more expensive, especially considering the higher cost of the consumable resin vs. FDM filament. Ultimately, the goal of this project is to make this pipette inexpensive and easy to make.

Our design relies heavily on the syringe, a specialized part, for the core function, as well as accuracy. Fortunately, as disposable syringes are a regulated medical device, they must meet high standards, making them a reliable part for this design. As good as the performance of our printed pipette is, a commercial pipette should be expected to be more reliable and preferred for critical procedures. Adjusting our printed pipette to the desired volume requires more time and attention, which would be tedious for procedures requiring many adjustments. On the other hand, due to the $6 cost to make it, many printed pipettes could be assembled and pre-set, each for a planned transfer, rather then re-adjusting one commercial pipette for each step. The printable pipette can also be used as a disposable, sparing the commercial micropipette when damage from volatile reagents is a concern or in situations were cross-contamination and resterilization is required. This pipette would also be ideal for high school and teaching labs or any resource-limited program.

The open design nature of this project encourages anyone to use and submit changes to improve and add features to this design. All of the working files and documentation are kept in a public repository [46]. Future directions for this project include developing a tip ejection system and additional configurations for transferring volumes below 30 µL, as well as increasing overall user friendliness, such as making the adjusting bolt easier to grip and improving ergonomics. It is our hope that this project will be under continuous development where improvements and compatibility with different syringes and printers can be crowd-sourced from community contributors.

Supplementary Materials: The following are available online at https://zenodo.org/record/1220094# .WtcQ2FKSCJl: File S1: body-clip.STL, File S2: plunger.STL, Video S1: Assembly of the pipette, Video S2: Operation of the pipette.

Acknowledgments: We would like to acknowledge the UIC Bioengineering Senior Design team of 2014 for initial work on this project: Oluwafemi Aboloye, Pedro Hurtado, Rafael Romero and Vivian Sandoval.

Author Contributions: M.D.B. wrote the manuscript. M.D.B. and F.F.B. designed the device, collected and analyzed data and made the figures. D.T.E. advised and reviewed the manuscript.

Conflicts of Interest: Martin Brennan and David Eddington are founders of Go Go Micro LLC and have equity interest in the company.

Appendix A. Additional Materials

Table A1. Parts and cost for the 30–300-µL pipette.

30–300 mL Configuration:	Source	Unit Price	Part Number
Filament	Makerbot	$1.63	NA
1-mL Syringe	BD Biosciences	$0.15	309628
M3 Bolt, 35 mm	McMaster-Carr	$0.12	91287A026
M3 Nut	McMaster-Carr	$0.01	90591A121
Music Wire Compression Springs * (2)	Jones Spring Co	$1.23	C10-022-048
Female Luer to 1/8" Hose Barb Adapter	Cole-Parmer	$0.40	EW-30800-08
Tygon Tubing, 1/16" ID × 3/16" OD	Cole-Parmer	$0.03	EW-06407-72
300-µL Tips (10)	Fisher Scientific	$0.34	02-707-447
	Total	$3.91	

* Spring specifications: Overall length 38 mm, OD 7.62 mm, wire diameter 0.56 mm, Max load 0.86 kg, with closed and flat ends. Also available from McMaster-Carr P/N: 9657K347.

Table A2. Parts and cost for the 100–1000-µL pipette.

100–1000 mL Configuration:	Source	Unit Price	Part Number
Filament	Makerbot	$1.63	NA
3-mL Syringe	BD Biosciences	$0.73	309657
M3 Bolt, 35 mm	McMaster-Carr	$0.12	91287A026
M3 Nut	McMaster-Carr	$0.01	90591A121
Music Wire Compression Springs * (2)	Jones Spring Co	$1.23	C10-022-048
Female luer to 5/32" Hose Barb Adapter	Cole-Parmer	$0.66	EW-45508-06
Tygon Tubing, 1/8" ID × 1/4" OD	Cole-Parmer	$0.05	EW-06407-76
1000-µL Tips (10)	Fisher Scientific	$0.41	02-707-400
	Total	$4.84	

* Spring specifications: Overall length 38 mm, OD 7.62 mm, wire diameter 0.56 mm, Max load 0.86 kg, with closed and flat ends. Also available from McMaster-Carr P/N: 9657K347.

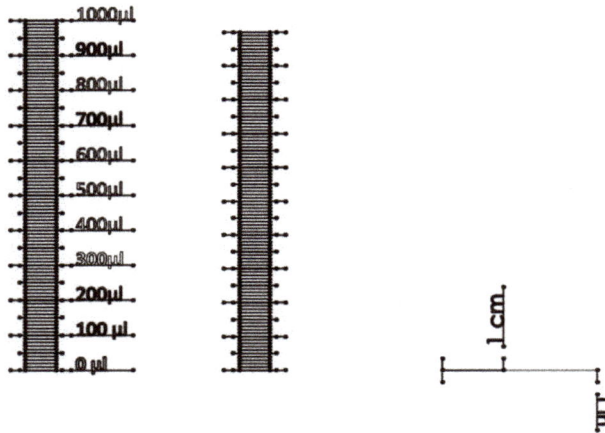

Figure A1. Adjusted scale graduations for the 30–300-µL pipette. Adjusted scale graduations that can be printed on paper or a transparency sheet and taped onto a 1-mL syringe. Using these graduations will correct for expansion of air and allow ISO standards to be met. One-inch and 1-cm marks to check the scale are included, as well as a reproduction of the existing 1-mL syringe graduation marks.

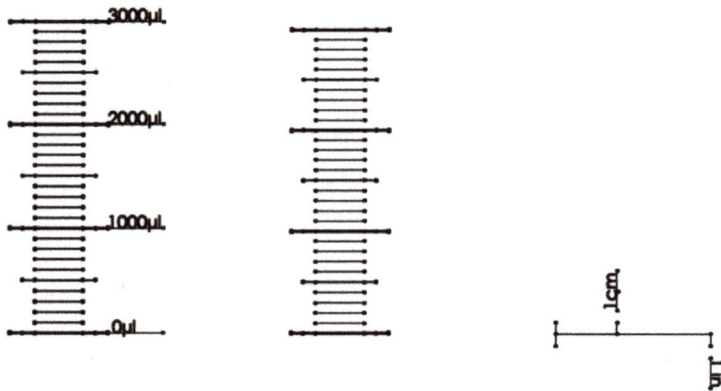

Figure A2. Adjusted scale graduations for the 100–1000-µL pipette. Adjusted scale graduations that can be printed on paper or a transparency sheet and taped onto a 3-mL syringe. Using these graduations will correct for expansion of air and allow ISO standards to be met. One-inch and 1-cm marks to check the scale are included, as well as a reproduction of the existing 3-mL syringe graduation marks. Print this image to 100% scale.

References

1. Baden, T.; Chagas, A.M.; Gage, G.; Marzullo, T.; Prieto-Godino, L.L.; Euler, T. Open Labware: 3-D Printing Your Own Lab Equipment. *PLoS Biol.* **2015**, *13*, e1002086. [CrossRef]
2. Pearce, J. M. *Open-Source Lab: How to Build Your Own Hardware and Reduce Research Costs*; Newnes: New South Wales, Australia, 2013.
3. Makerbot Industries. Makerbot. 2018. Available online: https://www.makerbot.com/ (accessed on 18 April 2018).
4. RepRap Project. RepRap. 2018. Available online: reprap.org (accessed on 18 April 2018).

5.   Fullerton, J.N.; Frodsham, G.C.M.; Day, R.M. 3D printing for the many, not the few. *Nat. Biotechnol.* **2014**, *32*, 1086–1087. [CrossRef]

6.   Wittbrodt, B.; Glover, A.G.; Laureto, J.; Anzalone, G.C.; Oppliger, D.; Irwin, J.L.; Pearce, J.M. Life-cycle economic analysis of distributed manufacturing with open-source 3-D printers. *Mechatronics* **2013**, *23*, 713–726. [CrossRef]

7.   Kintel, M. OpenSCAD. 2018. Available online: http://www.openscad.org/ (accessed on 18 April 2018).

8.   Blender Foudation. Blender. 2018. Available online: www.blender.org (accessed on 18 April 2018).

9.   Trimble Inc. SketchUp. 2016. Available online: www.sketchup.com (accessed on 18 April 2018).

10.   Autodesk Inc. Autodesk 123D. 2018. Available online: www.123dapp.com (accessed on 18 April 2018).

11.   Makerbot Industries. Thingiverse. 2018. Available online: https://www.thingiverse.com/ (accessed on 18 April 2018).

12.   National Institutes of Health. NIH 3D Print Exchange. 2018. Available online: http://3dprint.nih.gov/ (accessed on 18 April 2018).

13.   Stratasys. GrabCAD. Available online: https://grabcad.com/ (accessed on 18 April 2018).

14.   GitHub Inc. GitHub. 2018. Available online: https://github.com/ (accessed on 18 April 2018).

15.   Marzullo, T.C.; Gage, G.J. The SpikerBox: A Low Cost, Open-Source BioAmplifier for Increasing Public Participation In Neuroscience Inquiry. *PLoS ONE* **2012**, *7*, e30837. [CrossRef]

16.   Lang, T. Advancing global health research through digital technology and sharing data. *Science* **2011**, *331*, 714–717. [CrossRef]

17.   Fobel, R.; Fobel, C.; Wheeler, A.R. DropBot: An open-source digital microfluidic control system with precise control of electrostatic driving force and instantaneous drop velocity measurement. *Appl. Phys. Lett.* **2013**, *102*, 1–5. [CrossRef]

18.   Baker, E. Open source data logger for low-cost environmental monitoring. *Biodivers. Data J.* **2014**, e1059. [CrossRef]

19.   Cybulski, J.S.; Clements, J.; Prakash, M. Foldscope: origami-based paper microscope. *PLoS ONE* **2014**, *9*, e98781. [CrossRef]

20.   Bhamla, M.S.; Benson, B.; Chai, C.; Katsikis, G.; Johri, A.; Prakash, M. Hand-powered ultralow-cost paper centrifuge. *Nat. Biomed. Eng.* **2017**, *1*, 0009. [CrossRef]

21.   Pearce, J.M. Building Research Equipment with Free, Open-Source Hardware. *Science* **2012**, *337*, 1303–1304. [CrossRef]

22.   Rankin, T.M.; Giovinco, N.A.; Cucher, D.J.; Watts, G.; Hurwitz, B.; Armstrong, D.G. Three-dimensional printing surgical instruments: are we there yet? *J. Surg. Res.* **2014**, *189*, 193–197. [CrossRef]

23.   Sulkin, M.S.; Widder, E.; Shao, C.; Holzem, K.M.; Gloschat, C.; Gutbrod, S.R.; Efimov, I.R. Three-dimensional printing physiology laboratory technology. *Am. J. Physiol.-Heart Circ. Physiol.* **2013**, *305*, H1569–H1573. [CrossRef]

24.   Dryden, M.D.M.; Wheeler, A.R. DStat: A versatile, open-source potentiostat for electroanalysis and integration. *PLoS ONE* **2015**, *10*, 1–17. [CrossRef]

25.   Costa, E.T.; Mora, M.F.; Willis, P.A.; Lago, C.L.; Jiao, H.; Garcia, C.D. Getting started with open-hardware: Development and control of microfluidic devices. *Electrophoresis* **2014**, *35*, 2370–2377. [CrossRef]

26.   Tek, P.; Chiganos, T.C.; Mohammed, J.S.; Eddington, D.T.; Fall, C.P.; Ifft, P.; Rousche, P.J. Rapid prototyping for neuroscience and neural engineering. *J. Neurosci. Methods* **2008**, *172*, 263–269. [CrossRef]

27.   Chai Biotechnologies Inc. Open PCR. 2017. Available online: http://openpcr.org/ (accessed on 18 April 2018).

28.   Trachtenberg, J.E.; Mountziaris, P.M.; Miller, J.S.; Wettergreen, M.; Kasper, F.K.; Mikos, A.G. Open-source three-dimensional printing of biodegradable polymer scaffolds for tissue engineering. *J. Biomed. Mater. Res. Part A* **2014**, *102*, 4326–4335. [CrossRef]

29.   Rosenegger, D.G.; Tran, C.H.T.; LeDue, J.; Zhou, N.; Gordon, G.R. A High Performance, Cost-Effective, Open-Source Microscope for Scanning Two-Photon Microscopy that Is Modular and Readily Adaptable. *PLoS ONE* **2014**, *9*, e110475. [CrossRef]

30.   Garvey, C. DremelFuge. 2009. Available online: https://www.thingiverse.com/thing:1483 (accessed on 18 April 2018).

31.   Zhang, C.; Anzalone, N.C.; Faria, R.P.; Pearce, J.M. Open-Source 3D-Printable Optics Equipment. *PLoS ONE* **2013**, *8*, e59840. [CrossRef]

32. Baden, T. Raspberry Pi Scope. 2014. Available online: http://3dprint.nih.gov/discover/3dpx-000609 (accessed on 18 April 2018).

33. Walus, K. A Fully Printable Microscope. 2014. Available online: http://3dprint.nih.gov/discover/3dpx-000304 (accessed on 18 April 2018).

34. Wijnen, B.; Hunt, E.J.; Anzalone, G.C.; Pearce, J.M. Open-Source Syringe Pump Library. *PLoS ONE* **2014**, *9*, e107216. [CrossRef]

35. Patrick, W.G.; Nielsen, A.A.; Keating, S.J.; Levy, T.J.; Wang, C.W.; Rivera, J.J.; Mondragón-Palomino, O.; Carr, P.A.; Voigt, C.A.; Oxman, N.; et al. DNA assembly In 3D printed fluidics. *PLoS ONE* **2015**, *10*, 1–18. [CrossRef]

36. Symes, M.D.; Kitson, P.J.; Yan, J.; Richmond, C.J.; Cooper, G.J.; Bowman, R.W.; Vilbrandt, T.; Cronin, L. Integrated 3D-printed reactionware for chemical synthesis and analysis. *Nat. Chem.* **2012**, *4*, 349–354. [CrossRef]

37. Kitson, P.J.; Rosnes, M.H.; Sans, V.; Dragone, V.; Cronin, L. Configurable 3D-Printed millifluidic and microfluidic 'lab on a chip' reactionware devices. *Lab Chip* **2012**, *12*, 3267–3271. [CrossRef]

38. Mathieson, J.S.; Rosnes, M.H.; Sans, V.; Kitson, P.J.; Cronin, L. Continuous parallel ESI-MS analysis of reactions carried out In a bespoke 3D printed device. *Beilstein J. Nanotechnol.* **2013**, *4*, 285–291. [CrossRef]

39. Kitson, P.J.; Marshall, R.J.; Long, D.; Forgan, R.S.; Cronin, L. 3D Printed high-throughput hydrothermal reactionware for discovery, optimization, and scale-up. *Angew. Chem. Int. Ed.* **2014**, *53*, 12723–12728. [CrossRef]

40. Brennan, M.D.; Rexius-Hall, M.L.; Eddington, D.T. A 3D-Printed Oxygen Control Insert for a 24-Well Plate. *PLos ONE* **2015**, *10*, e0137631. [CrossRef]

41. Dragone, V.; Sans, V.; Rosnes, M.H.; Kitson, P.J.; Cronin, L. 3D-printed devices for continuous-flow organic chemistry. *BeilsteIn J. Org. Chem.* **2013**, *9*, 951–959. [CrossRef]

42. Plos Collections. Open Source Toolkit Hardware. 2017. Available online: http://collections.plos.org/open-source-toolkit-hardware (accessed on 18 April 2018).

43. International Organization for Standardization. *Piston-Operated Volumetric Apparatus—Part 1: Terminology, General Requirements and User Recommendations;* ISO 8655-1:2002(en); ISO: Geneva, Switzerland, 2002.

44. Takagishi, K.; Umezu, S. Development of the Improving Process for the 3D Printed Structure. *Sci. Rep.* **2017**, *7*, 39852. [CrossRef]

45. Baden, T. Biropette: Customisable, High Precision Pipette. 2014. Available online: http://www.thingiverse.com/thing:255519 (accessed on 18 April 2018).

46. Eddington Lab. 3D Printable Micropipette. 2017. Available online: https://github.com/Biological-Microsystems-Laboratory/micropipette (accessed on 18 April 2018).

micromachines

MDPI

Perspective

Emerging Anti-Fouling Methods: Towards Reusability of 3D-Printed Devices for Biomedical Applications

Eric Lepowsky [1] and Savas Tasoglu [1,2,3,4,5,*]

1   Department of Mechanical Engineering, University of Connecticut, Storrs, CT 06269, USA;
    eric.lepowsky@uconn.edu
2   Department of Biomedical Engineering, University of Connecticut, Storrs, CT 06269, USA
3   Institute of Materials Science, University of Connecticut, Storrs, CT 06269, USA
4   Institute for Collaboration on Health, Intervention, and Policy, University of Connecticut,
    Storrs, CT 06269, USA
5   The Connecticut Institute for the Brain and Cognitive Sciences, University of Connecticut,
    Storrs, CT 06269, USA
*   Correspondence: savas.tasoglu@uconn.edu; Tel.: +1-860-486-5919

Received: 21 March 2018; Accepted: 19 April 2018; Published: 20 April 2018

Abstract: Microfluidic devices are used in a myriad of biomedical applications such as cancer screening, drug testing, and point-of-care diagnostics. Three-dimensional (3D) printing offers a low-cost, rapid prototyping, efficient fabrication method, as compared to the costly—in terms of time, labor, and resources—traditional fabrication method of soft lithography of poly(dimethylsiloxane) (PDMS). Various 3D printing methods are applicable, including fused deposition modeling, stereolithography, and photopolymer inkjet printing. Additionally, several materials are available that have low-viscosity in their raw form and, after printing and curing, exhibit high material strength, optical transparency, and biocompatibility. These features make 3D-printed microfluidic chips ideal for biomedical applications. However, for developing devices capable of long-term use, fouling—by nonspecific protein absorption and bacterial adhesion due to the intrinsic hydrophobicity of most 3D-printed materials—presents a barrier to reusability. For this reason, there is a growing interest in anti-fouling methods and materials. Traditional and emerging approaches to anti-fouling are presented in regard to their applicability to microfluidic chips, with a particular interest in approaches compatible with 3D-printed chips.

Keywords: 3D printing; microfluidic chips; anti-fouling; surface coatings

---

1. Introduction

Microfluidic devices are widely used in numerous fields. Pertaining specifically to biotechnology, microfluidic devices have been used for cancer screening [1–3], microphysiological system engineering [4,5], high-throughput drug testing [6,7], and point-of-care diagnostics [8–10]. Within the broad field of microfluidics, novel technologies have been reported, such as paper-based microfluidics and three-dimensional (3D) printed devices. Each of these alternative methods addresses different drawbacks of the conventional soft lithography fabrication of poly(dimethylsiloxane) (PDMS) microfluidic devices. Using paper as a substrate alleviates the need for external pumps, in addition to the added benefits of low-cost fabrication and disposability [11–16]. Three-dimensional (3D) printing is a relatively new technique in microfluidics that offers rapid prototyping of a variety of materials, enabling fabrication of high resolution molds, as well as direct fabrication of 3D-printed chips [17–24]. The latter is of particular interest, as microfluidic chips can be designed using computer aided design (CAD) software and printed with high resolution. Compared to the laborious fabrication involved with PDMS chips, 3D-printed chips are low-cost, efficient, and highly customizable. Microfluidic devices, of all forms and methods, are indispensable tools in biomedical engineering. Accordingly, there is a need for reusable, long-term use of microfluidic chips for applications such as long-term health monitoring. Efforts towards designing long-term use microfluidic chips can be aided by the low-cost, high-throughput capabilities offered by 3D-printing.

Fostered by biomedical research, a transformation of the healthcare industry is imminent. There is an unmet need for transformative technologies that are essential for enabling the shift from hospital-centered, reactive care to proactive, person-centered care which focuses on individuals' well-being [25]. Next-generation technologies are the vital factor in developing affordable and accessible care, while also lowering the costs of healthcare. A promising solution to this challenge is low-cost continuous health monitoring; this approach allows for effective screening, analysis, and diagnosis and facilitates proactive medical intervention. In a study reported by Health Affairs, it was shown that a 90% increase in specific preventative screenings back in 2006 would have saved more than 2 million lives without a significant increase in healthcare costs—in fact, a 0.2% decrease in costs was estimated [26]. Not only does public health benefit, but savings in healthcare costs could be drastically increased by further research focused on the development and implementation of preventative care methods and tools. To support this much-needed approach, there is an impending demand for low-cost, compact, and innovational technologies to perform routine health measurements across the population. Microfluidic devices have a proven record for being effective analytical devices, capable of controlling the flow of fluid samples, containing reaction and detection zones, and displaying results, all within a compact footprint [27–30]. The next crucial step within microfluidic technologies is to address the need for microfluidic chips capable of long-term use, with a particular interest in designing microfluidic chips that can be embedded within sophisticated medical devices without the need for replacement by the user [31,32].

A primary challenge in producing microfluidic devices for long-term use is the biofouling that often occurs on the surface of integrated channels and features [33,34]. This phenomenon occurs due to surface interactions between the walls of the channels and the biological sample flowing through the channels. As a result, there is an accumulation of unwanted substances on the surface of the microfluidic channels, thereby adversely affecting the performance of the device. Specifically, protein fouling, or nonspecific protein absorption, is the cause of failure for many microfluidic devices [31]. Furthermore, once a protein layer is formed, bacterial attachment and growth of biofilms is facilitated. To address the long-term use of microfluidic devices, there has been a growing research interest in developing new anti-fouling methods and materials [31,32,35–39]. If fouling can be reduced to miniscule levels, a device may be considered reusable for the purpose of point-of-care, long-term health monitoring performed *ex vivo*. Possible solutions for reducing or preventing biofouling have been explored, including dynamic coating and chemical modification. Additionally, with the advent of 3D-printed microfluidic devices, there is a newfound interest in studying the protein absorption of 3D-printed materials [40–42]. Directly associated with protein absorption are possible anti-fouling techniques which may be implemented to improve the surface properties of 3D-printed devices. With continued research, these anti-fouling methods applied to 3D-printed microfluidic devices will work towards the future of reusable, long-term use, low-cost, point-of-care biomedical devices. Herein, we assess the reusability of 3D-printed chips for biomedical applications through a review of 3D-printed microfluidic fabrication techniques, 3D-printable materials, and traditional and emerging anti-fouling methods and materials.

2. 3D-Printed Microfluidics

2.1. Fabrication Methods for 3D-Printed Microfluidics

There are several 3D printing techniques, including the following: fused deposition modeling (FDM), stereolithography (SLA), photopolymer inkjet printing, selection laser sintering (SLS), binder deposition, laminated object manufacturing [19], two-photon polymerization (2PP), and micromirror-controlled projection printing [20]. In FDM, depicted in Figure 1a, a thermoplastic filament is fed through a nozzle [17,19,20,43]. Within the nozzle, rollers force the solid filament through heaters which melt the filament prior to extrusion. The melted filament is then deposited onto the print-bed where it solidifies post-extrusion. SLA 3D printers use light to cure layers of photosensitive resin, as seen in Figure 1b [17,19,20,43]. The resin is cured by photopolymerization caused by a laser or digital projector that is directed towards the surface of the resin by motorized mirrors. SLA offers high resolution, as defined by the size of the laser spot and the type of resin. Photopolymer inkjet printing, modeled in Figure 1c, also prints with a photosensitive polymer, but functions on the concept behind inkjet printing: the liquid photopolymer is deposited onto the print-bed by an inkjet printing head [17,19,20,43,44]. By printing layer-by-layer, each layer is deposited and subsequently cured by UV light. Two powder-based 3D printing methods are SLS and binder jetting. In SLS, a laser is used to thermally bind precursor polymer powders in a layer-by-layer fashion [19,45]. In binder jetting, a liquid adhesive, or binder, is deposited by a print head onto a bed of powder [19,43]. For both SLS and binder deposition, a roller is used to form layers of powder. Another 3D printing method used for microfabrication is 2PP, which uses a near-IR light with high peak power within a UV-photopolymerizable resin. The photo-initiator within the resin absorbs one UV photon and two near-IR photons, causing it to radicalize, thereby breaking its current monomer bonds and forming polymer bonds [19,46,47]. In terms of application in microfluidic device fabrication, FDM, SLA, and photopolymer inkjet printing are the most commonly implemented [18].

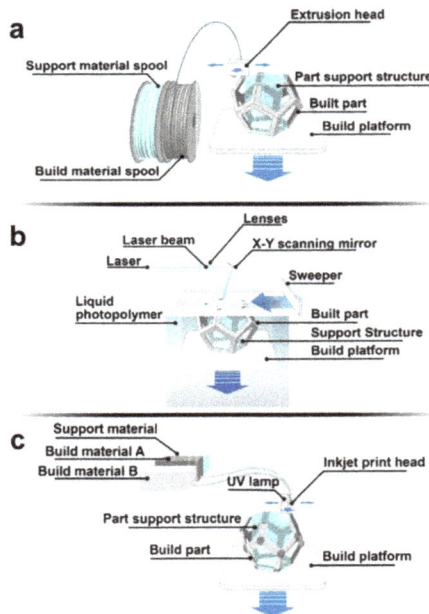

Figure 1. Schematic representation of 3D printing methods commonly applied in the fabrication of microfluidic devices. (**a**) Fused deposition modeling (FDM) [48]: a solid filament is fed from an external spool through the extrusion head, in which the filament is heated and extruded; (**b**) Stereolithography (SLA) [49]: a laser is directed at a scanning mirror which focuses the laser on a pool of photo-sensitive resin; (**c**) Photopolymer inkjet printing [50]: photopolymer material and support material are fed into an inkjet printing head which deposits the material in layers while an attached UV lamp cures the printed material. Illustrations courtesy of [48–50].

2.2. Materials for 3D-Printed Microfluidics

Beyond the fabrication methods, another important aspect of 3D-printed microfluidic devices is the material itself. Certain mechanical properties are very relevant to 3D printability, including the elastic modulus and tensile strength of the post-printed material, and the viscosity of the pre-printed material (Table 1) [17]. The elastic modulus is representative of the mechanical relationship between stress and strain. The 3D-printed materials must withstand high internal pressures from the fluid flow; due to the high fluid-induced stress, the resulting degree of strain is an important factor in choosing a material [51]. Likewise, tensile strength also helps to characterize the mechanical properties of the material in terms of the forces it can withstand. Additionally, the viscosity of the material before printing (for resins) is an important factor in determining the printability and resolution of the print. The material must be thin enough (low viscosity) to be precisely deposited in minute quantities but must also be sufficiently thick (high viscosity) such that it does not spread, migrate, or deform during the printing process. Viscosity also plays a role in the ease of cleaning the microfluidic channels post-printing—a lower viscosity is preferred for easier and quicker emptying and rinsing of channels and semi-enclosed geometries.

Additional material properties that are necessary for microfluidics include high optical transparency, hydrophobicity or hydrophilicity, and biocompatibility (Table 1). Optical transparency is extremely useful for microfluidic applications since visual observation and imaging are often necessary. In particular, a transparent microfluidic chip allows for qualitative observations—a visual check on proper fluid flow and reagent interactions—as well as quantitative, through-chip imaging using either visible transmitted light or fluorescence detection. PDMS is widely used for microfluidic device fabrication and has excellent transparency [52]. Several 3D-printable resins have comparable transparency. However, these resins may suffer from fogginess or discoloration over time due to excessive UV-light exposure after fabrication. In some cases, such as for Formlabs clear resin, a thin layer of isopropyl alcohol, oil, or PDMS can be applied to the exterior surfaces of the printed chip to facilitate transparency. Additionally, some commonly used materials in 3D printing exhibit natural fluorescence due to the photo-initiators used for curing, contributing to fluorescent background noise, which could be problematic in fluorescent imaging systems—a challenge which can be mitigated by proper calibration of the imaging process [43,53,54]. Another important property of materials in microfluidics is their intrinsic hydrophobicity or hydrophilicity. Hydrophobic surfaces are more prone to suffering from fouling due to surface interactions, whereas hydrophilic surfaces can facilitate a protective hydration layer [31,55]. Since practically all 3D-printable resins are hydrophobic, either copolymerization, material grafting, or a surface treatment is necessary for anti-fouling purposes. While it should not absorb or react with biological agents in the fluid sample, such as proteins or bacteria, the material should be biocompatible. In biomedical applications, it is likely that living cells may come into contact with the 3D-printed material. Fortunately, the majority of 3D-printable resins are reportedly biocompatible or can be UV-treated to reduce cytotoxicity [20,40,56].

Table 1. Materials for 3D printing.

| Material/Example | 3D Printing Properties | | | Important Properties for Microfluidics | | |
	Elastic Modulus (GPa)	Tensile Strength (MPa)	Viscosity (cps)	Optical Transparency	Hydro-Phobicity/Philicity	Bio-Compatibility
Poly(dimethylsiloxane) (PDMS) (for comparison) [52,57]	0.00132–0.00297	3.51–7.65	N/A	High transparency (standard)	Hydrophobic	Biocompatible
MakerBot polylactic acid (PLA) [40,58,59]	3.368	56.6	N/A	Semi-transparent	Hydrophobic, easily modified	Biocompatible, biodegradable
MakerBot acrylonitrile butadiene styrene (ABS) [40,58]	1.807	28.5	N/A	Opaque	Hydrophobic	Biocompatible
Formlabs proprietary methacrylate [40,43,60–62]	2.7	61.5	850–900	Transparent, discolors	Hydrophobic	N/A
Asiga PlasCLEAR polypropylene/ABS [43,63,64]	N/A	52.6	342	Semi-transparent	N/A	N/A

Table 1. *Cont.*

Material/Example	3D Printing Properties			Important Properties for Microfluidics		
	Elastic Modulus (GPa)	Tensile Strength (MPa)	Viscosity (cps)	Optical Transparency	Hydro-Phobicity/Philicity	Bio-Compatibility
Stratasys Object acrylates and acrylics [19,43,65]	2–3	50–65	N/R	Transparent, discolors	Hydrophobic	Biocompatible
3DSystems VisiJet Clear Class [40,66–69]	0.866–2.168	20.5–49	150–260	Semi-transparent	N/A	Biocompatible
Somos WaterShed XC [19,40,43,70–73]	2.77	50.4	260	Transparent, discolors	Hydrophobic	Biocompatible
MiiCraft acrylates (BV-007 Clear Resin) [43,71]	N/A	N/A	N/A	Semi-transparent	N/A	Biocompatible available
DWS Lab Vitra 429 & DS3000 [74]	1.38	32–35	600–850	Transparent	N/A	Short-term biocompatible

2.3. Benefits and Applications of 3D-Printed Microfluidics

Given the variety of fabrication methods and materials available for 3D printing, it is also important to consider the benefits and applications of 3D-printed microfluidics. The conventional method for fabricating microfluidic devices is soft lithography, which involves casting PDMS based on a semi-3D mold [20]. A channel pattern, designed using CAD software, is used to fabricate a master negative-mold. The mold is fabricated, most often by a photoresist process, consisting of spin coating SU-8 on a wafer and exposing it to an ultraviolet source filtered by a photomask. The master mold is then filled with PDMS, which is then allowed to cure. After curing, the PDMS is peeled away from the master, cut into the shape of the desired device, and inlet ports are punched. The cast PDMS is then oxygen plasma treated to improve adhesion to a glass substrate, resulting in the final microfluidic device, in which one of the four channel walls is provided by the glass. This conventional fabrication process is extremely time consuming, involves extensive manual operation, and requires new molds for different designs [75–77]. Additionally, the channel height of traditional PDMS microfluidic chips cast using soft lithography, determined by the material properties of the photoresist, is extremely limited. Although photoresist-based molds can have a very high resolution, fabricating complex designs is a tedious and exacting process, requiring photomasks to be perfectly aligned, to expose sequential layers of photoresist [78]. Even with multiple layers of photoresist, three-dimensional PDMS chips are not feasible, as 3D PDMS designs are produced by layering and thus cannot be arbitrarily complex. Three-dimensional (3D) printing offers a solution to these challenges, yet it comes with its own unique set of limitations [19]. The resolution achievable by 3D printing cannot match that of soft lithography. Similarly, the smooth surface of PDMS cannot be recreated by 3D-printed materials. 3D printing can be implemented in two ways: (1) to fabricate a mold, and (2) to directly print the device.

While conventional photoresist-based molds yield smoother surface topology and smaller feature sizes, this method is often economically unfeasible, inaccessible, and involves a lengthy prototyping process [79]. On the other hand, molds can be fabricated by 3D printing, which enables low-cost production of more complex designs. Fabrication of PDMS microfluidic chips using 3D-printed master molds provides many of the advantages of 3D printing fabrication while maintaining the desirable material properties of PDMS, such as biocompatibility and gas permeability [17]. The mold benefits from rapid prototyping techniques, reducing the overall cost and time required for fabrication and enabling a simple method for printing new mold designs. Additionally, 3D-printed molds enable considerably more complex channel geometries: circular cross-section channels [80], non-planar channels such as helices and pillars [81], and even complex intermingling internal patterns supported by sacrificial scaffolds which can be dissolved or melted away post-production [82–84].

Three-dimensional (3D) printing can also be implemented to directly fabricate microfluidic devices. Rather than designing a negative-mold for the chip, the chip is printed using the 3D-printed material. This method reduces fabrication costs, increases the speed of production, and is ideal for rapid prototyping and low-cost microfluidic devices. Since the whole device is directly printed, 3D printing eliminates the assembly steps required by conventional molding processes and allows for quick changes to the design, simply requiring a few hours for the new design to print. Another advantage of progressing beyond mold-based fabrication is the ability to produce arbitrarily defined structures in a fully 3D space, with no significant increase in fabrication complexity and time [78,85]. Furthermore, a wide variety of microfluidic components can be fabricated by 3D printing, including passive, which rely on external actuation or capillary forces, and active, which apply energy to manipulate fluid flow components. Examples of passive 3D-printed components include a micromixer, gradient generator, droplet generator, reaction zones, and check valves [17,86–89]. Active 3D-printed components which have been demonstrated include active valves, flow switches, and pumps [17,54,70].

Fully-functional 3D-printed microfluidic chips have also been designed for biomedical applications (Figure 2), such as a chip for inertial focusing and separation of bacteria [90], high-throughput drug transport and cell viability testing [91], electrochemical detection [92], pathogen detection of bacteria and viruses [93,94], biological assays and cell observations [95,96], and cancer assays and studies on cell migration [97]. Figure 2a demonstrates a mixing scheme which has broad applicability to high-throughput combinatorial testing applications, such as drug screening, biochemical assays, lab-on-chip devices, and biosensors [98]. Figure 2b shows an bacteria detection device which separates captured bacteria using inertial focusing [90]. The fabrication of this device's complex helical geometry and trapezoidal cross-section channels was only made possible by the capabilities of 3D printing. Figure 2c depicts a single-valve 3D-printed microfluidic device which allows for the controlled opening and closing of channels [70]. While soft-lithographic automation generally involves PDMS layering, alignment, and bonding, the pictured device offers a rapid, prototyped alternative, which is completely fabricated by stereolithography. Additionally, two-valve and four-valve switches have also been demonstrated. Such actuated switches can be pivotal in microfluidic sample handling for various assays. Figure 2d shows a three-dimensional microfluidic device built from identical, individual elements [99]. As another example of enhanced control of fluid flow, these modular 3D-printed elements allow for a wide variety of fluidic circuits to be constructed. Figure 2e offers an example of a 3D-printed mini-bioreactor (MBRA) for cultivating fecal microbial communities to study epidemic *Clostridium difficile* strains [100]. The MBRAs used in this study benefitted from the simple, high-throughput fabrication process offered by 3D printing. Figure 2f demonstrates another 3D-printed microfluidic mixing device; however, instead of solely relying on two-dimensional serpentines and splitters, the pictured device utilizes three-dimensional channel patterns to improve the gradient generation [86]. Finally, Figure 2g,h detail a multi-material 3D-printed device for the microphysiological study of cardiac tissues [101]. The functionality of the device relies on contractions of cardiac tissue to deflect a cantilever substrate, thereby stretching an embedded strain gauge which generates a resistance chance proportional to the stress of the tissue (Figure 2g). The device itself was printed with multiple materials to form the cantilever base, strain gauge wire, wire cover, tissue-guiding microfilaments, electrical leads, and surrounding wells. The inset images of Figure 2h show (1) a confocal microscopy image of the immunostained cardiac tissues tested using this device, (2) an image of an active cantilevered device, and (3) an example of the resistance signal from the cantilever deflection. These examples demonstrate the wide applicability of 3D-printed microfluidics to biomedical applications.

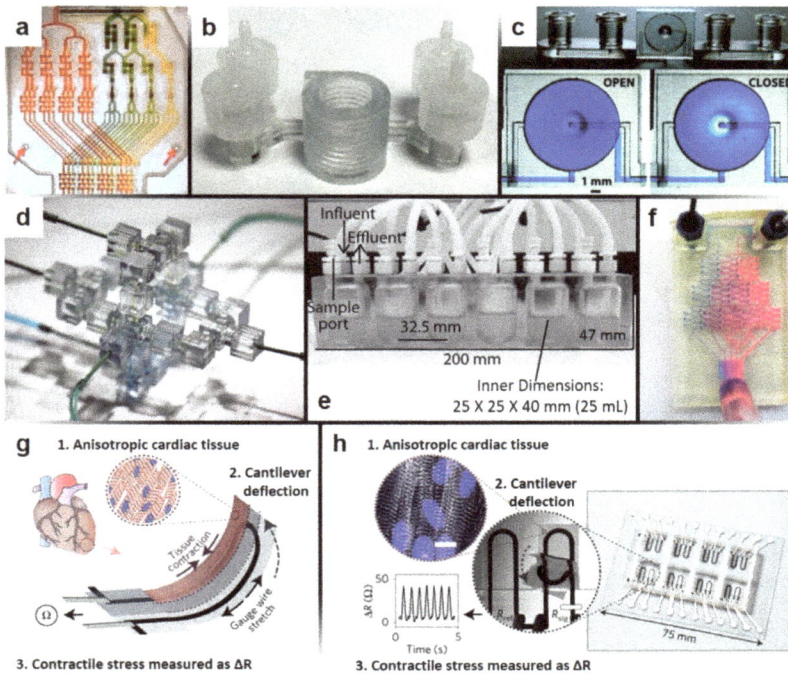

Figure 2. Biomedical applications of 3D-printed microfluidic devices; (**a**) A combinatorial mixer which generates four titrations of two dye solutions and produces combinatorial mixes of the dye titrations to deliver sixteen mixture combinations into separate outlet microchannels (reproduced, with permission, from [98]); (**b**) a helical-shaped 3D microfluidic device with trapezoidal-shaped channels for the detection of pathogenic bacteria by inertial focusing (reproduced, with permission, from [90]); (**c**) automated 3D-printed microfluidic single-valve device. Below are micrographs of the valve unit in its open and closed states (reproduced, with permission, from [70]); (**d**) single-outlet sub-circuit elements are connected to form a four-outlet mixer. Each sub-circuit element is identical, constituted by a single inlet splitter (reproduced, with permission, from [99]); (**e**) example of a simple, high-throughput mini-bioreactor array (MBRA) used for the cultivation of microbial communities (reproduced, with permission, from [19,100]); (**f**) Three-dimensional gradient generator for the mixing of two dyes, consisting of three levels of combining, mixing, and splitting (reprinted, with permission, from [86], Copyright 2014 American Chemical Society); (**g**,**h**) instrumented cardiac microphysiological device fabricated by multi-material 3D printing; (**g**) illustration of the working principles of the microphysiological device; (**h**) images of the fully-printed device (reproduced, with permission, from [101]).

3. Anti-Fouling Methods and Materials

3.1. Traditional Anti-Fouling Methods and Materials

Attributed to the hydrophobicity of PDMS, microfluidic devices fabricated using this traditional material exhibit low wettability and are prone to biofouling from nonspecific protein and analyte absorption and cell and bacterial adhesion [102]. To counteract the fouling, coatings aim to minimize intermolecular forces and interactions between the surface and samples. Pertaining to protein fouling, absorption into the substrate can be due to hydrogen bonding, electrostatic forces, charge-transfers, or hydrophobic interactions [55].

Considerable work has been done to create a hydrophilic PDMS surface with anti-fouling properties [31,36]. Physical methods to accomplish this adaptation rely on changing the state of the PDMS surface via physical processes. For instance, a coating material may be absorbed by the PDMS by hydrophobic or electrostatic interaction. Alternatively, the surface can be treatment-activated by plasma, ozone, or ultraviolet exposure. There are also chemical modification methods that have been studied, such as covalent bonding between a coating material and the PDMS. Both physical and chemical methods have their respective drawbacks—physical modification results in only temporary anti-fouling, with the effectiveness decreasing over time, while chemical modification requires complex processes with several reagents and procedural steps. Consequently, these methods may not be ideal for commercialization, particularly for fabricating long-term use or low-cost devices, since they are often ephemeral and laborious. Polymer chain length is an important factor when considering anti-fouling mechanisms [31]. Protein resistance can be achieved by a hydration layer of short chain length polymers, which forms due to the hydrophilicity of the short chains. In addition to forming a hydrophilic hydration layer, long chain length materials have the added benefit of steric repulsion from the flexible polymer chains. When tuning the protein resistance of a material, longer chain lengths are therefore preferred since they exhibit both a hydration layer and steric repulsion. Beyond hydrophilicity, anti-fouling mechanisms should be electrically neutral and should only have hydrogen bond acceptors—no hydrogen bond donors [103].

A widely used anti-fouling surface treatment is a poly(ethylene oxide) (PEO) or poly(ethylene glycol) (PEG) coating. PEO and PEG are hydrophilic and nontoxic [104,105]. They can be added as a coating on PDMS by physical or chemical absorption, direct covalent attachment, or graft copolymerization. The addition of PEG chains to the surface of the microfluidic channels offers anti-fouling capabilities due to a hydration layer and steric repulsion, the strength of which is dependent on factors such as surface density and the length of the PEG chains [106–110]. Another class of materials that has been demonstrated for anti-fouling purpose is zwitterions. Polyzwitterion-based coatings have an equal amount of positively and negatively charged groups and thus exhibit electrical neutrality, which is a desirable characteristic [31]. Zwitterionic materials are able to form thick hydration layers by bonding with water molecules, forming a shield across the surface to prevent protein absorption [111–116]. Other types of anti-fouling coatings to prevent the absorption of nonspecific proteins include saccharide-based coatings [117–120], polyhydroxy-polymer-based coatings [121,122], amide-containing-hydrophilic-polymer-based coatings [36,123,124], and fluorinated-polymer-based coatings [125–128].

3.2. Emerging Anti-Fouling Methods and Materials

The numerous methods proposed for anti-fouling of PDMS suggest that reusable microfluidic devices are plausible. However, other materials and methods are needed for anti-fouling coatings of 3D-printed microfluidic devices. Novel advancements from the Wyss Institute at Harvard University have yielded SLIPS (Slippery Liquid-Infused Porous Surfaces). SLIPS is an ultra-repellent, self-healing, transparent surface coating for industrial and medical materials [129]. While most repellent surface materials can repel aqueous substances, they often fail to repel other liquids, fail or degrade under physical stress, cannot self-heal after repeated use, and may be very costly to implement [129–131]. The SLIPS technology relies on nano- and micro-structured porous material infused with lubricating fluid [129]. The commonality between the traditional methods and materials described above and SLIPS is the implementation of a lubrication or hydration layer. High optical transparency and compatibility with a myriad of substrates are what differentiate SLIPS from other coatings. These features also make SLIPS an ideal anti-fouling material for 3D-printed microfluidic chips for biomedical applications, in addition to a variety of previously reported applications [130–133].

As a transparent, anti-fouling material, SLIPS was demonstrated for implementation on an endoscope lens to prevent vision loss after repeated submersions in blood and mucus [37]. The coating exhibited conformability, mechanical adhesion, high transparency, biocompatibility, and high resistance to fogging and fouling by bodily fluids (Figure 3a,b) [134,135]. These material characteristics were attributed to the liquid-infused structure of the coating, which consisted of a porous silica particle network infiltrated with various oils, such as silicone oil which is already used in a variety of medical applications due to its biocompatibility and proven clinical performance [37,130,136–138]. By applying the coating to the lens, cleaning procedures in between uses of the endoscope were unnecessary or, in the worst cases, reduced to ten-to-fifteen times shorter than for an untreated lens. Additionally, the surface coating maintained a clear view through the lens, which is also extremely important for 3D-printed microfluidic chips for diagnostic applications, as diagnostics often rely on through-chip imaging.

As opposed to the stationary lens of an endoscope, in an environment with interfaces under flow, such as the channels of a microfluidic chip, it is also important to consider the stability and longevity of immobilized hydration layers as surface coatings. The surface hydration layer provides a low-adhesion interface that deters the attachment of fouling materials. However, as fouling material-laden fluid flows over the hydration layer, there is an increase in the instability of the interface [139]. Experimental results and mathematical modeling by Howell et al. provided several important findings: (1) initial conditioning to flush out excess lubricant from the surface reduces further losses; (2) surface structure and lubricant viscosity do not cause significant differences in performance; and (3) the formation of an air–water interface within the channel causes disruptions in the lubricant layer.

SLIPS technology has also been applied as a coating for medical materials that can effectively suppress thrombosis, even under high-pressure and high-shear flow [35]. An anti-thrombogenic surface coating was formed by covalently binding a flexible tethered perfluorocarbon (TP) layer on the material surface, which was then coated by a layer of a liquid perfluorocarbon (LP). The resulting bilayer is referred to as a tethered-liquid perfluorocarbon (TLP) surface. The TLP is capable of repelling blood and preventing the adhesion of blood components and bacteria, thereby reducing thrombosis without the need for anti-coagulants (Figure 3c). Even when subjected to a flowing fluid, the TP maintains the LP hydration layer. The TLP is also simple to fabricate—virtually any substrate can be modified by low-plasma surface modification, a procedure which is commonly used for the commercial modification of materials and is independent of surface geometry and properties [140]. The TLP coating is ideal for treating the surfaces of 3D-printed microfluidic channels as it is compatible with a large variety of materials and exhibits anti-thrombogenic, anti-bacterial, and anti-fouling properties.

Another lubricant-infused coating that has been applied to polymeric biomedical devices is a new, advanced type of the traditional fluorinated-polymer-based coatings. Badv et al. recently demonstrated an omniphobic, lubricant-infused coating which was applied to catheters to prevent thrombosis [141]. The proposed omniphobic coating functioned on a similar liquid lubricant concept as SLIPS—a biocompatible liquid lubricant is held to the surface of the catheter to form an anti-fouling coating. However, instead of forming a hydrophilic hydration layer as SLIPS aims to accomplish, the coating developed by Badv et al. was based on a self-assembled monolayer (SAM) of hydrophobic organosilanes which attenuates clotting. Recently, omniphobic, lubricant-infused coatings have been developed as a new class of coatings created by tethering biocompatible, perfluorocarbon lubricants onto SAMs of hydrophobic organosilanes [35,130]. In greater detail, fluorous molecules can be physically adsorbed onto fluorous-containing surfaces, facilitated by the strong intermolecular interaction between the fluorinated lubricant and the fluorosilane layer. Such omniphobic coatings are excellent for resisting blood clot formation and are more effective than traditional anti-fouling methods, such as PEG grafting, for blocking non-specific adhesion of cells and bacteria [142,143]. In addition to the great effectiveness of omniphobic coatings, they are stable and durable when exposed to shear stress [35,139].

A feature that sets the work of Badv et al. apart from other traditional omniphobic, fluorinated -polymer-based coatings is the method of application. The most common technique for applying SAMs of fluorine-based silanes is liquid phase deposition (LPD). LPD has a few notable drawbacks and limitations. LPD is not commercially feasible due to the high volume of solvent waste that it produces which is harmful to the environment [144]. Another limitation is that the self-polymerization of silanes in the liquid may prevent homogenous silane layer formation [145]. Additionally, LPD-treated surfaces are likely to be exposed to impurities and reaction side products which may have detrimental effects on the surface itself [144]. Conversely, the proposed surface coating was applied via chemical vapor deposition (CVD), which has the following benefits: a simplified application process, a lesser effect on the topography of the receiving substrate, and greater effectiveness. CVD was performed by oxygen plasma treatment of the catheters, followed by silanization, and completed by the addition of a biocompatible liquid lubricant.

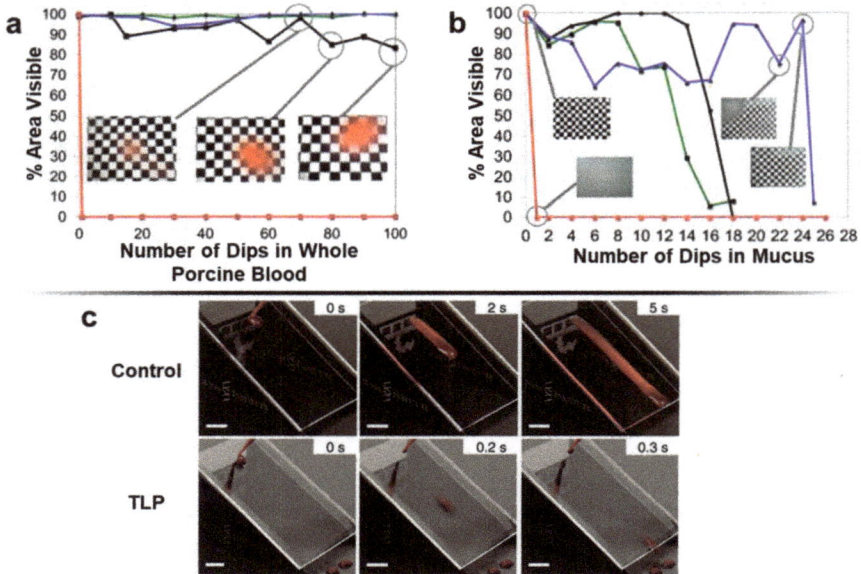

Figure 3. Emerging anti-fouling methods and materials: SLIPS (Slippery Liquid-Infused Porous Surfaces). (**a,b**) Evolution of the reduction in the visible area of an endoscope surface coated by SLIPS as a function of the number of dips. Silicone oil of low viscosity (10 cSt) was used as the lubricating liquid. Red corresponds to an uncoated endoscope, which fails immediately after a single dip. Green, blue, and black correspond to three replicates of SLIPS coated endoscopes; (**a**) endoscope dipped in whole porcine blood. Inset images show the visibility of the field of view at 70, 08, and 100 dips for the poorest performing sample; (**b**) endoscope dipped in mucus. Insets show the visibility of a coated endoscope after repeated dips compared to an untreated endoscope (reproduced, with permission, from [37]). (**c**) Comparison of the repellency of a tethered-liquid perfluorocarbon (TLP) treated surface to an untreated surface (reproduced, with permission, from [35]).

3.3. Anti-Fouling Methods Applied to 3D-Printed Materials

Biofouling from nonspecific protein and analyte absorption and cell and bacterial adhesion is not a detriment of only PDMS; biofouling can occur on the surfaces of all microfluidic chips, of all materials. While extensive work has been conducted to address the anti-fouling of PDMS, disproportionately less work has been reported for 3D printing. Nonetheless, recent studies and reviews have considered the biofouling of 3D-printed materials, demonstrating its significance and importance in the field. The objective of a study done by Siddiqui et al. was to propose a strategy for developing, characterizing, and testing 3D-printed feed spacers, both numerically by membrane fouling simulator (MFS) studies and experimentally [42]. Although the biofouling of the 3D-printed device was not directly assessed, the results showed that a 3D-printed feed spacer was similar. Feed spacers are commonly used devices in spiral-wound reverse osmosis and nanofiltration membrane systems to enhance water mixing and to reduce biofouling. As reported by Baker et al., the surface of feed spacers is prone to the initial deposition of fouling which can accumulate and spread [146]. Van Paassen et al. then showed that biofouling accumulated on feed spacers causes an exponential increase in the pressure drop across the membrane module containing the feed spacer [147]. To remedy the biofouling of feed spacers, various solutions have been proposed [148,149]. More recently, researchers have focused on the surface chemistry of the feed spacers. Multiple studies have considered anti-fouling coatings; however, these are more likely to wear out over time [150–154]. The solution proposed by Siddiqui et al. considered the native anti-fouling properties of 3D-printed materials. Their findings demonstrated the benefits of 3D-printing—feed spacers can be designed and printed with complex geometries to reduce the effects of biofouling. This studied demonstrates an example of the importance and application of studying the fouling of 3D-printed materials.

In reporting on the 3D printing of biopolymers for tissue engineering applications, Li et al. discussed the biocompatibility of common 3D-printed materials, such as polycaprolactone (PCL), poly-lactic acid (PLA), and acrylonitrile butadiene styrene (ABS) [41]. While PCL and PLA are natively biocompatible, ABS is not as ideally suited for biomedical applications. For this reason, there is considerable interest in possible methods for modifying the surface of printed ABS to render it hydrophilic and biocompatible. A traditional anti-fouling method which is compatible with a wide variety of materials, including ABS and other 3D-printed materials, is surface modification by the grafting of PEG [41,110]. As an alternative, McCullough and Yadavlli utilized an acetone solution to transform a porous ABS device fabricated by fused deposition modeling into a water-tight microfluidic device [155]. The acetone-treated microfluidic device retained the structural fidelity of microstructures as small as 250 μm, while benefiting from improved water impermeability, hydrophilicity, and biocompatibility. In addition to the acetone soaking, a traditional process of photo-induced PEG grafting was performed to further facilitate the formation of a stable, biocompatible surface by reducing biofouling behavior [155].

Anti-fouling methods have also been applied to 3D-printed molds for microfluidic device fabrication. Villegas et al. utilized omniphobic, lubricant-infused coatings, similar to the previously coating implemented by Badv et al., to fabricate 3D-printed molds for fabricating smooth PDMS microfluidic channels [79,141]. While the use of 3D-printed molds for casting PDMS microfluidic devices has been previously reported and described above, methods to improve the surface topology of the mold have not been previously studied [81,82]. The surface topology of a 3D-printed mold is likely to exhibit high variability and roughness, thereby increasing the difficulty of creating smooth, defined microfluidic channels from the mold [80]. In the recent study by Villegas et al., they addressed this challenge by investigating the use of lubricant-infused surfaces as a method for decreasing surface roughness—in addition to the more common use of lubricant-infusion for creating omniphobic slippery surfaces, which can be used in a multitude of applications including anti-biofouling. In this particular study, the mold was 3D-printed using a multi-jet modeling printer equipped with high-resolution nozzles [156]. The mold was then sonicated in ethanol and oxygen plasma treated before a self-assembled monolayer of fluorosilane was applied by chemical vapor deposition. Finally, a fluorocarbon lubricant was applied to form the smooth, lubricant-infused interface. This study, therefore, demonstrated the compatibility of 3D-printed material with omniphobic lubricant-infused coatings applied via chemical vapor deposition.

4. Conclusions

Microfluidic devices are invaluable tools in biomedical engineering. Microfluidics have a proven record of being effective analytical and diagnostic devices. Presently, there is an ongoing evolution in the field of microfluidics as researchers explore 3D printing as a viable fabrication method. 3D printing is revolutionizing the fabrication workflow of microfluidics. As opposed to the time-

and resource-intensive process of soft lithography for PDMS chip fabrication, 3D printing offers a low-cost, customizable platform for the direct fabrication of microfluidic chips. There are several 3D printing methods, of which fused deposition modeling, stereolithography, and photopolymer inkjet printing are applicable to 3D-printed microfluidic chips. Additionally, many materials are offered that may be compatible, boasting features such as high mechanical strength post-printing, low viscosity pre-printing, optical transparency, and biocompatibility.

The diagnostic capabilities of microfluidic chips and the benefits of 3D printing can be leveraged to address the unmet need for the long-term use of microfluidic chips for biomedical applications, such as for long-term health monitoring. Long-term, routine health monitoring has the capacity to improve public health while also reducing healthcare costs. A noteworthy obstacle in developing long-term diagnostic microfluidic chips is the fouling that is common on the surfaces of integrated microchannels and features. Similar to traditional PDMS, 3D-printed resins are most often hydrophobic, which means they are prone to surface interactions between the walls of the channels and the biological sample flowing through the channels. The resulting accumulation of unwanted substances on the surface of the microfluidic channels adversely affects the performance of the device. To combat fouling, there has been a growing research interest in developing new anti-fouling methods and materials. Traditional anti-fouling relies on chemical and/or physical surface modification. As an emerging solution, SLIPS (Slippery Liquid-Infused Porous Surfaces) offers a transparent, durable and robust surface treatment that can be applied to practically any surface, including the channels of a 3D-printed microfluidic chip. Continued research and characterization will result in reusable 3D-printed microfluidic chips that are low-cost, robust, transparent, and biocompatible.

Acknowledgments: S.T. acknowledges the American Heart Association Scientist Development Grant (15SDG25080056), Connecticut Innovations Biopipeline Award, and the University of Connecticut Research Excellence Program award for financial support of this research.

Conflicts of Interest: S.T. is founder of, and have an equity interest in mBiotics, LLC, a company that is developing microfluidic technologies for point-of-care diagnostic solutions. S.T.'s interests were viewed and managed in accordance with the conflict of interest policies of the University of Connecticut. The authors have no other relevant affiliations or financial involvement with any organization or entity with a financial interest in or financial conflict with the subject matter or materials discussed in the manuscript apart from those disclosed.

References

1. Wankhede, S.P.; Du, Z.; Berg, J.M.; Vaughn, M.W.; Dallas, T.; Cheng, K.H.; Gollahon, L. Cell detachment model for an antibody-based microfluidic cancer screening system. *Biotechnol. Prog.* **2006**, *22*, 1426–1433. [CrossRef] [PubMed]
2. Leoncini, E.; Ricciardi, W.; Cadoni, G.; Arzani, D.; Petrelli, L.; Paludetti, G.; Brennan, P.; Luce, D.; Stucker, I.; Matsuo, K.; et al. Lab-on-a-chip for oral cancer screening and diagnosis. *Head Neck* **2014**, *36*, 1391. [CrossRef]
3. Kim, L. *Overview of the Microfluidic Diagnostics Commercial Landscape*; Humana Press: Totowa, NJ, USA, 2013; Volume 949, ISBN 978-1-62703-133-2.
4. Srinivasan, V.; Pamula, V.K.; Fair, R.B. An integrated digital microfluidic lab-on-a-chip for clinical diagnostics on human physiological fluids. *Lab Chip* **2004**, *4*, 310–315. [CrossRef] [PubMed]
5. Hsu, Y.-H.; Moya, M.L.; Hughes, C.C.W.; George, S.C.; Lee, A.P. A microfluidic platform for generating large-scale nearly identical human microphysiological vascularized tissue arrays. *Lab Chip* **2013**, *13*, 2990. [CrossRef] [PubMed]
6. Toh, Y.-C.; Lim, T.C.; Tai, D.; Xiao, G.; van Noort, D.; Yu, H. A microfluidic 3D hepatocyte chip for drug toxicity testing. *Lab Chip* **2009**, *9*, 2026. [CrossRef] [PubMed]
7. Yu, L.; Chen, M.C.W.; Cheung, K.C. Droplet-based microfluidic system for multicellular tumor spheroid formation and anticancer drug testing. *Lab Chip* **2010**, *10*, 2424. [CrossRef] [PubMed]
8. Chin, C.D.; Linder, V.; Sia, S.K. Commercialization of microfluidic point-of-care diagnostic devices. *Lab Chip* **2012**, *12*, 2118. [CrossRef] [PubMed]
9. Sia, S.K.; Kricka, L.J. Microfluidics and point-of-care testing. *Lab Chip* **2008**, *8*, 1982. [CrossRef] [PubMed]

10. Foudeh, A.M.; Fatanat Didar, T.; Veres, T.; Tabrizian, M. Microfluidic designs and techniques using lab-on-a-chip devices for pathogen detection for point-of-care diagnostics. *Lab Chip* **2012**, *12*, 3249. [CrossRef] [PubMed]

11. Amin, R.; Ghaderinezhad, F.; Li, L.; Lepowsky, E.; Yenilmez, B.; Knowlton, S.; Tasoglu, S. Continuous-Ink, Multiplexed Pen-Plotter Approach for Low-Cost, High-Throughput Fabrication of Paper-Based Microfluidics. *Anal. Chem.* **2017**, *89*. [CrossRef] [PubMed]

12. Cate, D.M.; Adkins, J.A.; Mettakoonpitak, J.; Henry, C.S. Recent developments in paper-based microfluidic devices. *Anal. Chem.* **2015**, *87*, 19–41. [CrossRef] [PubMed]

13. Martinez, A.W.; Phillips, S.T.; Whitesides, G.M.; Carrilho, E. Diagnostics for the developing world: Microfluidic paper-based analytical devices. *Anal. Chem.* **2010**, *82*, 3–10. [CrossRef] [PubMed]

14. Yetisen, A.K.; Akram, M.S.; Lowe, C.R. Paper-based microfluidic point-of-care diagnostic devices. *Lab Chip* **2013**, *13*, 2210. [CrossRef] [PubMed]

15. Lepowsky, E.; Ghaderinezhad, F.; Knowlton, S.; Tasoglu, S. Paper-based assays for urine analysis. *Biomicrofluidics* **2017**, *11*. [CrossRef] [PubMed]

16. Ghaderinezhad, F.; Amin, R.; Temirel, M.; Yenilmez, B.; Wentworth, A.; Tasoglu, S. High-throughput rapid-prototyping of low-cost paper-based microfluidics. *Sci. Rep.* **2017**, *7*, 1–9. [CrossRef] [PubMed]

17. Amin, R.; Knowlton, S.; Hart, A.; Yenilmez, B.; Ghaderinezhad, F.; Katebifar, S.; Messina, M.; Khademhosseini, A.; Tasoglu, S. 3D-printed microfluidic devices. *Biofabrication* **2016**, *8*, 022001. [CrossRef] [PubMed]

18. Amin, R.; Joshi, A.; Tasoglu, S. Commercialization of 3D-printed microfluidic devices. *J. 3D Print. Med.* **2017**, *1*, 85–89. [CrossRef]

19. Au, A.K.; Huynh, W.; Horowitz, L.F.; Folch, A. 3D-Printed Microfluidics. *Angew. Chemie Int. Ed.* **2016**, *55*, 3862–3881. [CrossRef] [PubMed]

20. Ho, C.M.B.; Ng, S.H.; Li, K.H.H.; Yoon, Y.-J. 3D printed microfluidics for biological applications. *Lab Chip* **2015**, *15*, 3627–3637. [CrossRef] [PubMed]

21. Sochol, R.D.; Sweet, E.; Glick, C.C.; Wu, S.; Yang, C.; Restaino, M.; Lin, L. Microelectronic engineering 3D printed microfluidics and microelectronics. *Microelectron. Eng.* **2018**, *189*, 52–68. [CrossRef]

22. Knowlton, S.; Yu, C.H.; Ersoy, F.; Emadi, S.; Khademhosseini, A.; Tasoglu, S. 3D-printed microfluidic chips with patterned, cell-laden hydrogel constructs. *Biofabrication* **2016**, *8*, 25019. [CrossRef] [PubMed]

23. Knowlton, S.; Yenilmez, B.; Tasoglu, S. Towards Single-Step Biofabrication of Organs on a Chip via 3D Printing. *Trends Biotechnol.* **2016**, *34*, 685–688. [CrossRef] [PubMed]

24. Knowlton, S.; Joshi, A.; Syrrist, P.; Coskun, A.F.; Tasoglu, S. 3D-Printed Smartphone-Based Point of Care Tool for Fluorescence- and Magnetophoresis-Based Cytometry. *Lab Chip* **2017**, *17*, 2839–2851. [CrossRef] [PubMed]

25. National Science Foundation. Smart and Connected Health (SCH). Available online: https://www.nsf.gov/pubs/2016/nsf16601/nsf16601.htm (accessed on 27 November 2017).

26. Ward, B.W.; Clarke, T.C.; Nugent, C.N.; Schiller, J.S. Early Release of Selected Estimates Based on Data From the 2015 National Health Interview Survey (05/2016). National Health Interview Survey Early Release Program. Available online: https://www.cdc.gov/nchs/data/nhis/earlyrelease/earlyrelease201605.pdf (accessed on 27 November 2017).

27. Ashraf, M.W.; Tayyaba, S.; Afzulpurkar, N. Micro Electromechanical Systems (MEMS) based microfluidic devices for biomedical applications. *Int. J. Mol. Sci.* **2011**, *12*, 3648–3704. [CrossRef] [PubMed]

28. Jivani, R.R.; Lakhtaria, G.J.; Patadiya, D.D.; Patel, L.D.; Jivani, N.P.; Jhala, B.P. Biomedical microelectromechanical systems (BioMEMS): Revolution in drug delivery and analytical techniques. *Saudi Pharm. J.* **2016**, *24*, 1–20. [CrossRef] [PubMed]

29. Fujii, T. PDMS-based microfluidic devices for biomedical applications. *Microelectron. Eng.* **2002**, *61–62*, 907–914. [CrossRef]

30. Dario, P.; Carrozza, M.C.; Benvenuto, A.; Menciassi, A. Micro-systems in biomedical applications. *J. Micromech. Microeng.* **2000**, *10*, 235–244. [CrossRef]

31. Zhang, H.; Chiao, M. Anti-fouling coatings of poly(dimethylsiloxane) devices for biological and biomedical applications. *J. Med. Biol. Eng.* **2015**, *35*, 143–155. [CrossRef] [PubMed]

32. Picher, M.M.; Küpcü, S.; Huang, C.-J.; Dostalek, J.; Pum, D.; Sleytr, U.B.; Ertl, P. Nanobiotechnology advanced antifouling surfaces for the continuous electrochemical monitoring of glucose in whole blood using a lab-on-a-chip. *Lab Chip* **2013**, *13*, 1780. [CrossRef] [PubMed]
33. Wong, I.; Ho, C.M. Surface molecular property modifications for poly(dimethylsiloxane) (PDMS) based microfluidic devices. *Microfluid. Nanofluid.* **2009**, *7*, 291–306. [CrossRef] [PubMed]
34. Goyanes, A.; Buanz, A.B.M.; Hatton, G.B.; Gaisford, S.; Basit, A.W. 3D printing of modified-release aminosalicylate (4-ASA and 5-ASA) tablets. *Eur. J. Pharm. Biopharm.* **2015**, *89*, 157–162. [CrossRef] [PubMed]
35. Leslie, D.C.; Waterhouse, A.; Berthet, J.B.; Valentin, T.M.; Watters, A.L.; Jain, A.; Kim, P.; Hatton, B.D.; Nedder, A.; Donovan, K.; et al. A bioinspired omniphobic surface coating on medical devices prevents thrombosis and biofouling. *Nat. Biotechnol.* **2014**, *32*, 1134–1140. [CrossRef] [PubMed]
36. Tu, Q.; Wang, J.C.; Liu, R.; He, J.; Zhang, Y.; Shen, S.; Xu, J.; Liu, J.; Yuan, M.S.; Wang, J. Antifouling properties of poly(dimethylsiloxane) surfaces modified with quaternized poly(dimethylaminoethyl methacrylate). *Colloids Surf. B Biointerfaces* **2013**, *102*, 361–370. [CrossRef] [PubMed]
37. Sunny, S.; Cheng, G.; Daniel, D.; Lo, P.; Ochoa, S.; Howell, C.; Vogel, N.; Majid, A.; Aizenberg, J. Transparent antifouling material for improved operative field visibility in endoscopy. *Proc. Natl. Acad. Sci. USA* **2016**, *113*, 201605272. [CrossRef] [PubMed]
38. Defense Sciences Office. DARPA DARPA-RA-17-01: Young Faculty Award. Available online: https://www.darpa.mil/attachments/DARPA-RA-17-01FAQ11-17-17.pdf (accessed on 27 November 2017).
39. Wan, A.M.D.; Devadas, D.; Young, E.W.K. Recycled polymethylmethacrylate (PMMA) microfluidic devices. *Sens. Actuators B Chem.* **2017**, *253*, 738–744. [CrossRef]
40. Bhattacharjee, N.; Urrios, A.; Kang, S.; Folch, A. The upcoming 3D-printing revolution in microfluidics. *Lab Chip* **2016**, *16*, 1720–1742. [CrossRef] [PubMed]
41. Li, X.; Cui, R.; Sun, L.; Aifantis, K.E.; Fan, Y.; Feng, Q.; Cui, F.; Watari, F. 3D-printed biopolymers for tissue engineering application. *Int. J. Polym. Sci.* **2014**, *2014*. [CrossRef]
42. Siddiqui, A.; Farhat, N.; Bucs, S.S.; Linares, R.V.; Picioreanu, C.; Kruithof, J.C.; Van Loosdrecht, M.C.M.; Kidwell, J.; Vrouwenvelder, J.S. Development and characterization of 3D-printed feed spacers for spiral wound membrane systems. *Water Res.* **2016**, *91*, 55–67. [CrossRef] [PubMed]
43. Waheed, S.; Cabot, J.M.; Macdonald, N.P.; Lewis, T.; Guijt, R.M.; Paull, B.; Breadmore, M.C. 3D printed microfluidic devices: Enablers and barriers. *Lab Chip* **2016**, *16*, 1993–2013. [CrossRef] [PubMed]
44. Singh, M.; Haverinen, H.M.; Dhagat, P.; Jabbour, G.E. Inkjet printing-process and its applications. *Adv. Mater.* **2010**, *22*, 673–685. [CrossRef] [PubMed]
45. Mazzoli, A. Selective laser sintering in biomedical engineering. *Med. Biol. Eng. Comput.* **2013**, *51*, 245–256. [CrossRef] [PubMed]
46. Xing, J.-F.; Zheng, M.-L.; Duan, X.-M. Two-photon polymerization microfabrication of hydrogels: an advanced 3D printing technology for tissue engineering and drug delivery. *Chem. Soc. Rev.* **2015**, *44*, 5031–5039. [CrossRef] [PubMed]
47. Maruo, S.; Nakamura, O.; Kawata, S. Three-dimensional microfabrication with two-photon-absorbed photopolymerization. *Opt. Lett.* **1997**, *22*, 132–134. [CrossRef] [PubMed]
48. Additively. Fused Deposition Modeling (FDM). Available online: https://www.additively.com/en/learn-about/fused-deposition-modeling (accessed on 27 November 2017).
49. Additively. Stereolithography (SL). Available online: https://www.additively.com/en/learn-about/stereolithography (accessed on 27 November 2017).
50. Additively. Photopolymer Jetting (PJ). Available online: https://www.additively.com/en/learn-about/photopolymer-jetting (accessed on 27 November 2017).
51. Stone, H.A.; Stroock, A.D.; Ajdari, A. Engineering Flows in Small Devices: Microfluidics Toward a Lab-on-a-Chip. *Annu. Rev. Fluid Mech.* **2004**, *36*, 381–411. [CrossRef]
52. Mata, A.; Fleischman, A.J.; Roy, S. Characterization of polydimethylsiloxane (PDMS) properties for biomedical micro/nanosystems. *Biomed. Microdevices* **2005**, *7*, 281–293. [CrossRef] [PubMed]
53. Qaderi, K. Polyethylene Glycol Diacrylate (PEGDA) Resin Development for 3D-Printed Microfluidic Devices. Master's Thesis, Brigham Young University, Provo, UT, USA, May 2015.
54. Rogers, C.I.; Qaderi, K.; Woolley, A.T.; Nordin, G.P. 3D printed microfluidic devices with integrated valves. *Biomicrofluidics* **2015**, *9*, 016501. [CrossRef] [PubMed]

55. Liu, J.; Lee, M.L. Permanent surface modification of polymeric capillary electrophoresis microchips for protein and peptide analysis. *Electrophoresis* **2006**, *27*, 3533–3546. [CrossRef] [PubMed]

56. Oskui, S.M.; Diamante, G.; Liao, C.; Shi, W.; Gan, J.; Schlenk, D.; Grover, W.H. Assessing and Reducing the Toxicity of 3D-Printed Parts. *Environ. Sci. Technol. Lett.* **2016**, *3*, 1–6. [CrossRef]

57. Johnston, I.D.; McCluskey, D.K.; Tan, C.K.L.; Tracey, M.C. Mechanical characterization of bulk Sylgard 184 for microfluidics and microengineering. *J. Micromech. Microeng.* **2014**, *24*, 35017. [CrossRef]

58. Tymrak, B.M.; Kreiger, M.; Pearce, J.M. Mechanical properties of components fabricated with open-source 3-D printers under realistic environmental conditions. *Mater. Des.* **2014**, *58*, 242–246. [CrossRef]

59. Ramot, Y.; Haim-Zada, M.; Domb, A.J.; Nyska, A. Biocompatibility and safety of PLA and its copolymers. *Adv. Drug Deliv. Rev.* **2016**, *107*, 153–162. [CrossRef] [PubMed]

60. Ma, Y.; Cao, X.; Feng, X.; Ma, Y.; Zou, H. Fabrication of super-hydrophobic film from PMMA with intrinsic water contact angle below 90°. *Polymer (Guildf.)* **2007**, *48*, 7455–7460. [CrossRef]

61. Brandhoff, L.; van den Driesche, S.; Lucklum, F.; Vellekoop, M.J. Creation of hydrophilic microfluidic devices for biomedical application through stereolithography. *SPIE Microtechnol.* **2015**, *9518*, 95180D. [CrossRef]

62. Formlabs Formlabs Clear Safety Data Sheet. Available online: https://formlabs.com/media/upload/Clear-SDS_u324bsC.pdf (accessed on 27 November 2017).

63. Takenaga, S.; Schneider, B.; Erbay, E.; Biselli, M.; Schnitzler, T.; Schöning, M.J.; Wagner, T. Fabrication of biocompatible lab-on-chip devices for biomedical applications by means of a 3D-printing process. *Phys. Status Solidi Appl. Mater. Sci.* **2015**, *212*, 1347–1352. [CrossRef]

64. Asiga PlasCLEAR Technical Datasheet. Available online: https://www.asiga.com/media/main/files/materials/PlasCLEAR_us_en.pdf (accessed on 27 November 2017).

65. Stratasys Stratasys PolyJet Materials Material Safety Data Sheets. Available online: http://www.stratasys.com/materials/material-safety-data-sheets/polyjet/transparent-materials (accessed on 27 November 2017).

66. Martino, C.; Berger, S.; Wootton, R.C.R.; deMello, A.J. A 3D-printed microcapillary assembly for facile double emulsion generation. *Lab Chip* **2014**, *14*, 4178–4182. [CrossRef] [PubMed]

67. 3D System Material Selection Guide for Stereolithography—SLA. Available online: https://www.3dsystems.com/sites/default/files/2017-06/3D-Systems_SLS_MaterialSelectionGuide_USEN_2017.06.28_WEB.pdf (accessed on 27 November 2017).

68. 3D Systems VisiJet S300 Specification Sheet. Available online: http://infocenter.3dsystems.com/projetmjp3600/user-guide/introduction/important-safety-information/visijet%C2%AE-s300-specification-sheet (accessed on 27 November 2017).

69. 3D Systems VisiJet MP200, VisiJet M3 StonePlast Safety Data Sheet. Available online: http://infocenter.3dsystems.com/materials/sites/default/files/sds-files/professional/24156-s12-02-asds_ghsenglishvisijet_mp200_and_m3_stoneplast.pdf (accessed on 27 November 2017).

70. Au, A.K.; Bhattacharjee, N.; Horowitz, L.F.; Chang, T.C.; Folch, A. 3D-Printed Microfluidic Automation. *Lab Chip* **2015**, *15*, 1934–1941. [CrossRef] [PubMed]

71. Wood, S.; Krishnamurthy, N.; Santini, T.; Raval, S.; Farhat, N.; Holmes, J.A.; Ibrahim, T.S. Design and fabrication of a realistic anthropomorphic heterogeneous head phantom for MR purposes. *PLoS ONE* **2017**, *12*, e0183168. [CrossRef] [PubMed]

72. Proto Labs Product Data: Somos WaterShed XC 11122. Available online: https://www.protolabs.com/media/1010884/somos-watershed-xc-11122.pdf (accessed on 27 November 2017).

73. MiiCraft. MiiCraft-Technical Data Sheet: BV-007 Clear Resin. Available online: http://www.miicraft.com/support/ (accessed on 27 November 2017).

74. DWS Lab Materials Range. Available online: http://www.dwslab.com/materials-range/?v=7516fd43adaa (accessed on 27 November 2017).

75. Choudhury, D.; Mo, X.; Iliescu, C.; Tan, L.L.; Tong, W.H.; Yu, H. Exploitation of physical and chemical constraints for three-dimensional microtissue construction in microfluidics. *Biomicrofluidics* **2011**, *5*. [CrossRef] [PubMed]

76. Waldbaur, A.; Rapp, H.; Länge, K.; Rapp, B.E. Let there be chip—Towards rapid prototyping of microfluidic devices: One-step manufacturing processes. *Anal. Methods* **2011**, *3*, 2681. [CrossRef]

77. O'Neill, P.F.; Ben Azouz, A.; Vázquez, M.; Liu, J.; Marczak, S.; Slouka, Z.; Chang, H.C.; Diamond, D.; Brabazon, D. Advances in three-dimensional rapid prototyping of microfluidic devices for biological applications. *Biomicrofluidics* **2014**, *8*, 52112. [CrossRef] [PubMed]

78. Spivey, E.C.; Xhemalce, B.; Shear, J.B.; Finkelstein, I.J. 3D-printed microfluidic microdissector for high-throughput studies of cellular aging. *Anal. Chem.* **2014**, *86*, 7406–7412. [CrossRef] [PubMed]

79. Villegas, M.; Cetinic, Z.; Shakeri, A.; Didar, T.F. Fabricating smooth PDMS microfluidic channels from low-resolution 3D printed molds using an omniphobic lubricant-infused coating. *Anal. Chim. Acta* **2018**, *1000*, 248–255. [CrossRef] [PubMed]

80. Hwang, Y.; Seo, D.; Roy, M.; Han, E.; Candler, R.N.; Seo, S. Capillary Flow in PDMS Cylindrical Microfluidic Channel Using 3-D Printed Mold. *J. Microelectromech. Syst.* **2016**, *25*, 238–240. [CrossRef]

81. Hwang, Y.; Paydar, O.H.; Candler, R.N. 3D printed molds for non-planar PDMS microfluidic channels. *Sensors Actuators, A Phys.* **2015**, *226*, 137–142. [CrossRef]

82. Saggiomo, V.; Velders, A.H. Simple 3D Printed Scaffold-Removal Method for the Fabrication of Intricate Microfluidic Devices. *Adv. Sci.* **2015**, *2*, 1–5. [CrossRef] [PubMed]

83. Gelber, M.K.; Bhargava, R. Monolithic multilayer microfluidics via sacrificial molding of 3D-printed isomalt. *Lab Chip* **2015**, *15*, 1736–1741. [CrossRef] [PubMed]

84. He, Y.; Qiu, J.; Fu, J.; Zhang, J.; Ren, Y.; Liu, A. Printing 3D microfluidic chips with a 3D sugar printer. *Microfluid. Nanofluid.* **2015**, *19*, 447–456. [CrossRef]

85. Bonyár, A.; Sántha, H.; Ring, B.; Varga, M.; Kovács, J.G.; Harsányi, G. 3D Rapid Prototyping Technology (RPT) as a powerful tool in microfluidic development. *Procedia Eng.* **2010**, *5*, 291–294. [CrossRef]

86. Shallan, A.I.; Smejkal, P.; Corban, M.; Guijt, R.M.; Breadmore, M.C. Cost-effective three-dimensional printing of visibly transparent microchips within minutes. *Anal. Chem.* **2014**, *86*, 3124–3130. [CrossRef] [PubMed]

87. Kitson, P.J.; Rosnes, M.H.; Sans, V.; Dragone, V.; Cronin, L. Configurable 3D-Printed millifluidic and microfluidic "lab on a chip" reactionware devices. *Lab Chip* **2012**, *12*, 3267. [CrossRef] [PubMed]

88. Donvito, L.; Galluccio, L.; Lombardo, A.; Morabito, G.; Nicolosi, A.; Reno, M. Experimental validation of a simple, low-cost, T-junction droplet generator fabricated through 3D printing. *J. Micromech. Microeng.* **2015**, *25*. [CrossRef]

89. Comina, G.; Suska, A.; Filippini, D. 3D printed unibody lab-on-a-chip: Features survey and check-valves integration. *Micromachines* **2015**, *6*, 437–451. [CrossRef]

90. Lee, W.; Kwon, D.; Choi, W.; Jung, G.Y.; Au, A.K.; Folch, A.; Jeon, S. 3D-Printed microfluidic device for the detection of pathogenic bacteria using size-based separation in helical channel with trapezoid cross-section. *Sci. Rep.* **2015**, *5*, 1–7. [CrossRef] [PubMed]

91. Anderson, K.B.; Lockwood, S.Y.; Martin, R.S.; Spence, D.M. A 3D printed fluidic device that enables integrated features. *Anal. Chem.* **2013**, *85*, 5622–5626. [CrossRef] [PubMed]

92. Erkal, J.L.; Selimovic, A.; Gross, B.C.; Lockwood, S.Y.; Walton, E.L.; McNamara, S.; Martin, R.S.; Spence, D.M. 3D printed microfluidic devices with integrated versatile and reusable electrodes. *Lab Chip* **2014**, *14*, 2023–2032. [CrossRef] [PubMed]

93. Chudobova, D.; Cihalova, K.; Skalickova, S.; Zitka, J.; Rodrigo, M.A.M.; Milosavljevic, V.; Hynek, D.; Kopel, P.; Vesely, R.; Adam, V.; et al. 3D-printed chip for detection of methicillin-resistant Staphylococcus aureus labeled with gold nanoparticles. *Electrophoresis* **2015**, *36*, 457–466. [CrossRef] [PubMed]

94. Krejcova, L.; Nejdl, L.; Rodrigo, M.A.M.; Zurek, M.; Matousek, M.; Hynek, D.; Zitka, O.; Kopel, P.; Adam, V.; Kizek, R. 3D printed chip for electrochemical detection of influenza virus labeled with CdS quantum dots. *Biosens. Bioelectron.* **2014**, *54*, 421–427. [CrossRef] [PubMed]

95. Sugioka, K.; Hanada, Y.; Midorikawa, K. 3D microstructuring of glass by femtosecond laser direct writing and application to biophotonic microchips. *Prog. Electromagn. Res. Lett.* **2008**, *1*, 942–946. [CrossRef]

96. Hanada, Y.; Sugioka, K.; Shihira-Ishikawa, I.; Kawano, H.; Miyawaki, A.; Midorikawa, K. 3D microfluidic chips with integrated functional microelements fabricated by a femtosecond laser for studying the gliding mechanism of cyanobacteria. *Lab Chip* **2011**, *11*, 2109. [CrossRef] [PubMed]

97. Huang, T.Q.; Qu, X.; Liu, J.; Chen, S. 3D printing of biomimetic microstructures for cancer cell migration. *Biomed. Microdevices* **2014**, *16*, 127–132. [CrossRef] [PubMed]

98. Neils, C.; Tyree, Z.; Finlayson, B.; Folch, A. Combinatorial mixing of microfluidic streams. *Lab Chip* **2004**, *4*, 342–350. [CrossRef] [PubMed]

99. Bhargava, K.C.; Thompson, B.; Malmstadt, N. Discrete elements for 3D microfluidics. *Proc. Natl. Acad. Sci. USA* **2014**, *111*, 15013–15018. [CrossRef] [PubMed]

100. Robinson, C.D.; Auchtung, J.M.; Collins, J.; Britton, R.A. Epidemic Clostridium difficile strains demonstrate increased competitive fitness compared to nonepidemic isolates. *Infect. Immun.* **2014**, *82*, 2815–2825. [CrossRef] [PubMed]

101. Lind, J.U.; Busbee, T.A.; Valentine, A.D.; Pasqualini, F.S.; Yuan, H.; Yadid, M.; Park, S.J.; Kotikian, A.; Nesmith, A.P.; Campbell, P.H.; et al. Instrumented cardiac microphysiological devices via multimaterial three-dimensional printing. *Nat. Mater.* **2017**, *16*, 303–308. [CrossRef] [PubMed]

102. Guan, A.; Wang, Y.; Phillips, K.S.; Li, Z. A contact-lens-on-a-chip companion diagnostic tool for personalized medicine. *Lab Chip* **2016**, *16*, 1152–1156. [CrossRef] [PubMed]

103. Chapman, R.G.; Ostuni, E.; Liang, M.N.; Meluleni, G.; Kim, E.; Yan, L.; Pier, G.; Warren, H.S.; Whitesides, G.M. Polymeric thin films that resist the adsorption of proteins and the adhesion of bacteria. *Langmuir* **2001**, *17*, 1225–1233. [CrossRef]

104. Zhang, H.; Hao, R.; Ren, X.; Yu, L.; Yang, H.; Yu, H. PEG/lecithin–liquid-crystalline composite hydrogels for quasi-zero-order combined release of hydrophilic and lipophilic drugs. *RSC Adv.* **2013**, *3*, 22927. [CrossRef]

105. Goddard, J.M.; Hotchkiss, J.H. Polymer surface modification for the attachment of bioactive compounds. *Prog. Polym. Sci.* **2007**, *32*, 698–725. [CrossRef]

106. Zheng, J.; Li, L.; Tsao, H.-K.; Sheng, Y.-J.; Chen, S.; Jiang, S. Strong Repulsive Forces between Protein and Oligo (Ethylene Glycol) Self-Assembled Monolayers: A Molecular Simulation Study. *Biophys. J.* **2005**, *89*, 158–166. [CrossRef] [PubMed]

107. Herrwerth, S.; Eck, W.; Reinhardt, S.; Grunze, M. Factors that determine the protein resistance of oligoether self-assembled monolayers-Internal hydrophilicity, terminal hydrophilicity, and lateral packing density. *J. Am. Chem. Soc.* **2003**, *125*, 9359–9366. [CrossRef] [PubMed]

108. Vermette, P.; Meagher, L. Interactions of phospholipid- and poly(ethylene glycol)-modified surfaces with biological systems: Relation to physico-chemical properties and mechanisms. *Colloids Surf. B Biointerfaces* **2003**, *28*, 153–198. [CrossRef]

109. Morra, M. On the molecular basis of fouling resistance. *J. Biomater. Sci. Polym. Ed.* **2000**, *11*, 547–569. [CrossRef] [PubMed]

110. Zhang, Z.; Wang, J.; Tu, Q.; Nie, N.; Sha, J.; Liu, W.; Liu, R.; Zhang, Y.; Wang, J. Surface modification of PDMS by surface-initiated atom transfer radical polymerization of water-soluble dendronized PEG methacrylate. *Colloids Surf. B Biointerfaces* **2011**, *88*, 85–92. [CrossRef] [PubMed]

111. Bernards, M.T.; Cheng, G.; Zhang, Z.; Chen, S. Nonfouling Polymer Brushes via Surface-Initiated, Two-Component Atom Transfer Radical Polymerization—Macromolecules (ACS Publications). *Macromolecules* **2008**, *41*, 4216–4219. [CrossRef]

112. Bernards, M.; He, Y. Polyampholyte polymers as a versatile zwitterionic biomaterial platform. *J. Biomater. Sci. Polym. Ed.* **2014**, *25*, 1479–1488. [CrossRef] [PubMed]

113. Anthony Yesudass, S.; Mohanty, S.; Nayak, S.K.; Rath, C.C. Zwitterionic–polyurethane coatings for non-specific marine bacterial inhibition: A nontoxic approach for marine application. *Eur. Polym. J.* **2017**, *96*, 304–315. [CrossRef]

114. Li, G.; Xue, H.; Gao, C.; Zhang, F.; Jiang, S. Nonfouling polyampholytes from an ion-pair comonomer with biomimetic adhesive groups. *Macromolecules* **2010**, *43*, 14–16. [CrossRef] [PubMed]

115. Parker, A.P.; Reynolds, P.A.; Lewis, A.L.; Kirkwood, L.; Hughes, L.G. Investigation into potential mechanisms promoting biocompatibility of polymeric biomaterials containing the phosphorylcholine moiety: A physicochemical and biological study. *Colloids Surf. B Biointerfaces* **2005**, *46*, 204–217. [CrossRef] [PubMed]

116. Murphy, E.F.; Keddie, J.L.; Lu, J.R.; Brewer, J.; Russell, J. The reduced adsorption of lysozyme at the phosphorylcholine incorporated polymer/aqueous solution interface studied by spectroscopic ellipsometry. *Biomaterials* **1999**, *20*, 1501–1511. [CrossRef]

117. Yu, L.; Li, C.M.; Liu, Y.; Gao, J.; Wang, W.; Gan, Y. Flow-through functionalized PDMS microfluidic channels with dextran derivative for ELISAs. *Lab Chip* **2009**, *9*, 1243. [CrossRef] [PubMed]

118. Hu, S.G.; Jou, C.H.; Yang, M.C. Protein adsorption, fibroblast activity and antibacterial properties of poly(3-hydroxybutyric acid-co-3-hydroxyvaleric acid) grafted with chitosan and chitooligosaccharide after immobilized with hyaluronic acid. *Biomaterials* **2003**, *24*, 2685–2693. [CrossRef]

119. McArthur, S.L.; McLean, K.M.; Kingshott, P.; St John, H.A.W.; Chatelier, R.C.; Griesser, H.J. Effect of polysaccharide structure on protein adsorption. *Colloids Surf. B Biointerfaces* **2000**, *17*, 37–48. [CrossRef]

120. Martwiset, S.; Koh, A.E.; Chen, W. Nonfouling characteristics of dextran-containing surfaces. *Langmuir* **2006**, *22*, 8192–8196. [CrossRef] [PubMed]

121. Zhao, C.; Li, L.; Wang, Q.; Yu, Q.; Zheng, J. Effect of film thickness on the antifouling performance of poly(hydroxy-functional methacrylates) grafted surfaces. *Langmuir* **2011**, *27*, 4906–4913. [CrossRef] [PubMed]

122. Wu, D.; Luo, Y.; Zhou, X.; Dai, Z.; Lin, B. Multilayer poly(vinyl acohol)-adsorbed coating on poly(dimethylsiloxane) microfluidic chips for biopolymer separation. *Electrophoresis* **2005**, *26*, 211–218. [CrossRef] [PubMed]

123. Huang, X.; Doneski, L.J.; Wirth, M.J. Surface-Confined Living Radical Polymerization for Coatings in Capillary Electrophoresis. *Anal. Chem.* **1998**, *70*, 4023–4029. [CrossRef] [PubMed]

124. Huang, X.; Wirth, M.J. Surface-initiated radical polymerization on porous silica. *Anal. Chem.* **1997**, *69*, 4577–4580. [CrossRef]

125. Yu, H.J.; Luo, Z.H. Novel superhydrophobic silica/poly(siloxane-fluoroacrylate) hybrid nanoparticles prepared via two-step surface-initiated ATRP: Synthesis, characterization, and wettability. *J. Polym. Sci. Part A Polym. Chem.* **2010**, *48*, 5570–5580. [CrossRef]

126. Wang, Y.; Betts, D.E.; Finlay, J.A.; Brewer, L.; Callow, M.E.; Callow, J.A.; Wendt, D.E.; Desimone, J.M. Photocurable amphiphilic perfluoropolyether/poly(ethylene glycol) networks for fouling-release coatings. *Macromolecules* **2011**, *44*, 878–885. [CrossRef]

127. Sundaram, H.S.; Cho, Y.; Dimitriou, M.D.; Finlay, J.A.; Cone, G.; Williams, S.; Handlin, D.; Gatto, J.; Callow, M.E.; Callow, J.A.; et al. Fluorinated amphiphilic polymers and their blends for fouling-release applications: The benefits of a triblock copolymer surface. *ACS Appl. Mater. Interfaces* **2011**, *3*, 3366–3374. [CrossRef] [PubMed]

128. Martinelli, E.; Agostini, S.; Galli, G.; Chiellini, E.; Glisenti, A.; Pettitt, M.E.; Callow, M.E.; Callow, J.A.; Graf, K.; Bartels, F.W. Nanostructured films of amphiphilic fluorinated block copolymers for fouling release application. *Langmuir* **2008**, *24*, 13138–13147. [CrossRef] [PubMed]

129. Wyss Institute. SLIPS (Slippery Liquid-Infused Porous Surfaces). Available online: https://wyss.harvard.edu/technology/slips-slippery-liquid-infused-porous-surfaces (accessed on 27 November 2017).

130. Wong, T.-S.; Kang, S.H.; Tang, S.K.Y.; Smythe, E.J.; Hatton, B.D.; Grinthal, A.; Aizenberg, J. Bioinspired self-repairing slippery surfaces with pressure-stable omniphobicity. *Nature* **2011**, *477*, 443–447. [CrossRef] [PubMed]

131. Epstein, A.K.; Wong, T.-S.; Belisle, R.A.; Boggs, E.M.; Aizenberg, J. Liquid-infused structured surfaces with exceptional anti-biofouling performance. *Proc. Natl. Acad. Sci. USA* **2012**, *109*, 13182–13187. [CrossRef] [PubMed]

132. Vogel, N.; Belisle, R.A.; Hatton, B.; Wong, T.S.; Aizenberg, J. Transparency and damage tolerance of patternable omniphobic lubricated surfaces based on inverse colloidal monolayers. *Nat. Commun.* **2013**, *4*, 1–10. [CrossRef] [PubMed]

133. Juthani, N.; Howell, C.; Ledoux, H.; Sotiri, I.; Kelso, S.; Kovalenko, Y.; Tajik, A.; Vu, T.L.; Lin, J.J.; Sutton, A.; et al. Infused polymers for cell sheet release. *Sci. Rep.* **2016**, *6*, 1–9. [CrossRef] [PubMed]

134. Kota, A.K.; Kwon, G.; Tuteja, A. The design and applications of superomniphobic surfaces. *NPG Asia Mater.* **2014**, *6*, e109. [CrossRef]

135. Li, X.-M.; Reinhoudt, D.; Crego-Calama, M. What do we need for a superhydrophobic surface? A review on the recent progress in the preparation of superhydrophobic surfaces. *Chem. Soc. Rev.* **2007**, *36*, 1350. [CrossRef] [PubMed]

136. Grinthal, A.; Aizenberg, J. Mobile interfaces: Liquids as a perfect structural material for multifunctional, antifouling surfaces. *Chem. Mater.* **2014**, *26*, 698–708. [CrossRef]

137. Kim, P.; Kreder, M.J.; Alvarenga, J.; Aizenberg, J. Hierarchical or not? Effect of the length scale and hierarchy of the surface roughness on omniphobicity of lubricant-infused substrates. *Nano Lett.* **2013**, *13*, 1793–1799. [CrossRef] [PubMed]

138. Maccallum, N.; Howell, C.; Kim, P.; Sun, D.; Friedlander, R.; Ranisau, J.; Ahanotu, O.; Lin, J.J.; Vena, A.; Hatton, B.; et al. Liquid-Infused Silicone As a Biofouling-Free Medical Material. *ACS Biomater. Sci. Eng.* **2015**, *1*, 43–51. [CrossRef]

139. Howell, C.; Vu, T.L.; Johnson, C.P.; Hou, X.; Ahanotu, O.; Alvarenga, J.; Leslie, D.C.; Uzun, O.; Waterhouse, A.; Kim, P.; et al. Stability of surface-immobilized lubricant interfaces under flow. *Chem. Mater.* **2015**, *27*, 1792–1800. [CrossRef]

140. Keller, D.; Besch, W. Plasma-Induced Surface Functionalization of Polymeric Biomaterials in Ammonia Plasma. *Contrib. Plasma Phys.* **2001**, *41*, 562–572. [CrossRef]

141. Badv, M.; Jaffer, I.H.; Weitz, J.I.; Didar, T.F. An omniphobic lubricant-infused coating produced by chemical vapor deposition of hydrophobic organosilanes attenuates clotting on catheter surfaces. *Sci. Rep.* **2017**, *7*, 1–10. [CrossRef] [PubMed]

142. Milton Harris, J.; Chess, R.B. Effect of pegylation on pharmaceuticals. *Nat. Rev. Drug Discov.* **2003**, *2*, 214–221. [CrossRef] [PubMed]

143. Sotiri, I.; Overton, J.C.; Waterhouse, A.; Howell, C. Immobilized liquid layers: A new approach to anti-adhesion surfaces for medical applications. *Exp. Biol. Med.* **2016**, *241*, 909–918. [CrossRef] [PubMed]

144. Zhang, F.; Sautter, K.; Larsen, A.M.; Findley, D.A.; Davis, R.C.; Samha, H.; Linford, M.R. Chemical vapor deposition of three aminosilanes on silicon dioxide: Surface characterization, stability, effects of silane concentration, and cyanine dye adsorption. *Langmuir* **2010**, *26*, 14648–14654. [CrossRef] [PubMed]

145. Liu, Z.Z.; Wang, Q.; Liu, X.; Bao, J.Q. Effects of amino-terminated self-assembled monolayers on nucleation and growth of chemical vapor-deposited copper films. *Thin Solid Films* **2008**, *517*, 635–640. [CrossRef]

146. Baker, J.; Stephenson, T.; Dard, S.; Cote, P. Characterisation of fouling of nanofiltration membranes used to treat surface waters. *Environ. Technol.* **1995**, *16*, 977–985. [CrossRef]

147. Van Paassen, J.A.M.; Kruithof, J.C.; Bakker, S.M.; Kegel, F.S. Integrated multi-objective membrane systems for surface water treatment: Pre-treatment of nanofiltration by riverbank filtration and conventional ground water treatment. *Desalination* **1998**, *118*, 239–248. [CrossRef]

148. Cornelissen, E.R.; Vrouwenvelder, J.S.; Heijman, S.G.J.; Viallefont, X.D.; Van Der Kooij, D.; Wessels, L.P. Periodic air/water cleaning for control of biofouling in spiral wound membrane elements. *J. Membr. Sci.* **2007**, *287*, 94–101. [CrossRef]

149. Majamaa, K.; Aerts, P.E.M.; Groot, C.; Paping, L.L.M.J.; van den Broek, W.; van Agtmaal, S. Industrial water reuse with integrated membrane system increases the sustainability of the chemical manufacturing. *Desalin. Water Treat.* **2010**, *18*, 17–23. [CrossRef]

150. Miller, D.J.; Araújo, P.A.; Correia, P.B.; Ramsey, M.M.; Kruithof, J.C.; van Loosdrecht, M.C.M.; Freeman, B.D.; Paul, D.R.; Whiteley, M.; Vrouwenvelder, J.S. Short-term adhesion and long-term biofouling testing of polydopamine and poly(ethylene glycol) surface modifications of membranes and feed spacers for biofouling control. *Water Res.* **2012**, *46*, 3737–3753. [CrossRef] [PubMed]

151. Ronen, A.; Resnick, A.; Lerman, S.; Eisen, M.S.; Dosoretz, C.G. Biofouling suppression of modified feed spacers: Localized and long-distance antibacterial activity. *Desalination* **2015**, *393*, 159–165. [CrossRef]

152. Ronen, A.; Lerman, S.; Ramon, G.Z.; Dosoretz, C.G. Experimental characterization and numerical simulation of the anti-biofuling activity of nanosilver-modified feed spacers in membrane filtration. *J. Membr. Sci.* **2015**, *475*, 320–329. [CrossRef]

153. Wibisono, Y.; Yandi, W.; Golabi, M.; Nugraha, R.; Cornelissen, E.R.; Kemperman, A.J.B.; Ederth, T.; Nijmeijer, K. Hydrogel-coated feed spacers in two-phase flow cleaning in spiral wound membrane elements: A novel platform for eco-friendly biofouling mitigation. *Water Res.* **2015**, *71*, 171–186. [CrossRef] [PubMed]

154. Araújo, P.A.; Kruithof, J.C.; Van Loosdrecht, M.C.M.; Vrouwenvelder, J.S. The potential of standard and modified feed spacers for biofouling control. *J. Membr. Sci.* **2012**, *403–404*, 58–70. [CrossRef]

155. McCullough, E.J.; Yadavalli, V.K. Surface modification of fused deposition modeling ABS to enable rapid prototyping of biomedical microdevices. *J. Mater. Process. Technol.* **2013**, *213*, 947–954. [CrossRef]

156. Upcraft, S.; Fletcher, R. The rapid prototyping technologies. *Assem. Autom.* **2003**, *23*, 318–330. [CrossRef]

micromachines

MDPI

Article

3D-Printed Capillary Circuits for Calibration-Free Viscosity Measurement of Newtonian and Non-Newtonian Fluids

Sein Oh and Sungyoung Choi *

Department of Biomedical Engineering, Kyung Hee University, 1732 Deogyeong-daero, Giheung-gu, Yongin-si, Gyeonggi-do 17104, Korea; hottaek_90@khu.ac.kr
* Correspondence: s.choi@khu.ac.kr; Tel.: +82-31-201-3496

Received: 25 May 2018; Accepted: 16 June 2018; Published: 21 June 2018

Abstract: Measuring viscosity is important for the quality assurance of liquid products, as well as for monitoring the viscosity of clinical fluids as a potential hemodynamic biomarker. However, conventional viscometers and their microfluidic counterparts typically rely on bulky and expensive equipment, and lack the ability for rapid and field-deployable viscosity analysis. To address these challenges, we describe 3D-printed capillary circuits (3D-CCs) for equipment- and calibration-free viscosity measurement of Newtonian and non-Newtonian fluids. A syringe, modified with an air chamber serving as a pressure buffer, generates and maintains a set pressure to drive the pressure-driven flows of test fluids through the 3D-CCs. The graduated fluidic chambers of the 3D-CCs serve as a flow meter, enabling simple measurement of the flow rates of the test fluids flowing through the 3D-CCs, which is readable with the naked eye. The viscosities of the test fluids can be simply calculated from the measured flow rates under a set pressure condition without the need for peripheral equipment and calibration. We demonstrate the multiplexing capability of the 3D-CC platform by simultaneously measuring different Newtonian-fluid samples. Further, we demonstrate that the shear-rate dependence of the viscosity of a non-Newtonian fluid can be analyzed simultaneously under various shear-rate conditions with the 3D-CC platform.

Keywords: viscometer; 3D printing; capillary circuit; microfluidics; Newtonian fluid; non-Newtonian fluid

1. Introduction

Measuring viscosity is important for assessing the quality of liquid products [1–5], optimizing the performance of microfluidic devices [6,7], and monitoring the viscosity of clinical fluids as a hemodynamic biomarker [8]. Testing the viscosity of liquid products is an indispensable process to maximize production efficiency and to screen formulation candidates across diverse industrial fields [1–5]. Whole blood viscosity has a significant correlation with Alzheimer's disease and can be used as a potential biomarker for the monitoring and control of the disease [8]. Hyperviscosity syndrome is a combination of clinical symptoms (e.g., mucosal hemorrhage, visual abnormalities, or neurological disorders) caused by increased blood viscosity. It can be diagnosed by measuring whole blood viscosity or plasma viscosity, and treated by plasmapheresis and blood transfusion [9].

Two types of viscometers are commonly used for such applications and categorized into cone-plate viscometers [10] and capillary viscometers [11]. Cone-plate viscometers utilize a rotating cone that applies a shearing force and measures the resistance of a test fluid between the cone and a stationary plate [10]. Although cone-plate viscometers allow viscosity measurement over a wide range of viscosities and shear rates, their use at the sampling point for rapid and on-site viscosity measurement is typically limited by low measurement throughput and the requirement of bulky and costly equipment. Capillary viscometers are based on the Hagen-Poiseuille law and determine the viscosity of a test fluid from a measured flow rate under a given pressure condition or a measured pressure under a given flow-rate condition [11]. While relatively small and simple to operate, capillary viscometers require expensive, high-precision sensors and actuators for accurate viscosity measurement, thereby, increasing the cost of analysis.

Recently, miniaturized viscometers have been developed by adopting the advantages of microfluidic technologies, including low sample consumption, cost-effectiveness, small device footprint, and ease-of-use [12], and can be categorized as being based on a single microchannel [13,14], a falling ball [15], droplet microfluidics [16,17], a microfluidic comparator [18–20], or an optical tweezer [21]. The microchannel approach utilizes the relationship between the flow rate and viscosity of a test fluid flowing through a microchannel under a given capillary pressure [13,14]. Falling-ball viscometers measure the settling velocity of a ball falling in a test fluid that is correlated to the viscosity of the fluid of interest [15]. In droplet-based viscometers, micro-droplets are generated through a flow-focusing device and the length of the droplets, which is a function of rheological parameters, determines the viscosity of an aqueous or non-aqueous fluid [16,17]. The comparator approach uses a parallel microchannel in which the co-flowing laminar streams of two fluids, namely, a reference fluid with a known viscosity and a test fluid with an unknown viscosity, can form different interfacial positions or equilibrium states depending on their viscosities [18–20]. Thereby, the stream-width ratio or the flow-rate ratio of the two fluids determines the relative viscosity of the test fluid. An optical tweezer has been used to measure viscosity by monitoring the trajectory of an optically-trapped bead [21]. Although this method enables accurate viscosity measurement with low sample consumption in a few microliters per measurement, it typically requires complex and costly optical setup and lacks the ability to perform multiplexed viscosity measurement. The above approaches have been proven effective to analyze various kinds of fluids, such as whole blood [15,18,19], blood plasma [13], and polymer products [14], but there are limitations that impede the wide-spread use of these approaches in industrial and clinical applications. The microchannel approach requires cumbersome microscopic flow observation for viscosity measurement. The falling ball approach lacks sufficient controllability to generate a range of shear rates for the analysis of a non-Newtonian fluid. The comparator and droplet-based viscometers rely on bulky and high-precision pumps for the stable and accurate generation of co-flowing laminar streams and micro-droplets, respectively, which increases the size and cost of the overall system. To address these challenges, our group has recently developed 3D-printed capillary circuits (3D-CCs) that consist of parallel capillary channels for simple viscosity measurement [22,23]. The 3D-CCs, though enabling equipment-free and multiplexed viscosity measurement, is highly dependent on the accuracy of a reference-fluid viscosity, requiring its additional calibration for accurate viscosity measurement.

Here, we report new 3D-CCs that enable equipment- and calibration-free viscosity measurement of Newtonian and non-Newtonian fluids (Figure 1). The 3D-CC platform enables simple flow-rate measurement using the graduated fluidic chambers of the 3D-CCs and portable generation of a set pressure using a modified syringe, called a 'smart pipette', that is powered by hand. Based on the Hagen-Poiseuille law, the viscosities of test fluids flowing through the capillaries of the 3D-CCs can be calculated from the measured flow rates under a set pressure condition. The number of test fluid samples to be simultaneously analyzed can be augmented simply by adding more fluidic channels. We demonstrated the multiplexed capability of the 3D-CC platform by simultaneously measuring four Newtonian fluid samples within two minutes and comparing the measurement results with a conventional cone-plate viscometer. We also demonstrated that the 3D-CC platform enables simple analysis of a non-Newtonian fluid under multiple shear-rate conditions, proving the potential of the 3D-CC platform for rapid and on-site viscosity analysis of Newtonian and non-Newtonian fluids.

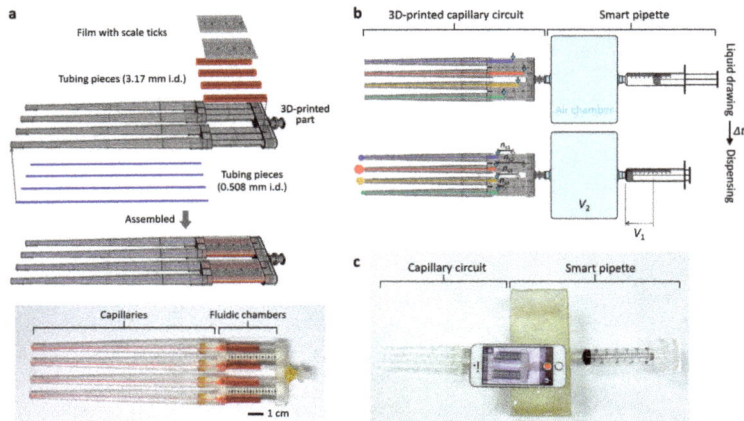

Figure 1. Working principle of the 3D-printed capillary circuit (3D-CC) platform for equipment- and calibration-free viscosity measurement. (**a**) (Top) Fabrication process of the 3D-CC having four identical fluidic channels for the multiplexed analysis of Newtonian fluids. (Bottom) Optical image of the fabricated 3D-CC filled with red ink for visualization; (**b**) schematic of the operation process: (top) Scale reading after withdrawing test liquids and (bottom) expelling the liquids by fully depressing the plunger. The volumetric flow rate (Q) of each test fluid can be measured by the number of scale ticks (n_s) counted for a given time. Knowing the Q, capillary dimensions, and drop in pressure through each capillary, the viscosities of the test liquids can be simultaneously determined without cumbersome calibration and expensive equipment; (**c**) experimental setup for multiplexed viscosity measurement using the 3D-CC platform composed of the 3D-CC and smart pipette.

2. Materials and Methods

2.1. Device Fabrication

The 3D-CCs were fabricated by assembling a 3D-printed housing and cut tubing pieces to be used as the capillaries and fluidic chambers of the 3D-CCs (Figure 1). The 3D-printed parts were fabricated using a stereolithography-based 3D printer (DWS Systems, Thiene, Italy). The details of fabrication using stereolithography can be found in previous literature [24,25]. Tygon® tubings, with an inner diameter of 0.508 and 3.17 mm (Cole-Parmer, Vernon Hills, IL, USA), were cut into pieces to serve as the capillaries and fluidic chambers of the 3D-CCs, respectively. The air chamber, with an inner volume of 1.27 L, was fabricated using a 3D printer. Luer-Lock fittings were attached to the ends of the chamber for reversible and airtight interconnection between a 3D-CC and a syringe (Figure 1c). A film mask with scale ticks was fabricated, cut into pieces, and attached to the assembled 3D-CC for volume indication (Figure 1).

A microfluidic viscometer for comparison with the 3D-CC for analysis of a non-Newtonian fluid was fabricated by photolithography and polydimethylsiloxane (PDMS) replica molding processes as previously reported (Figure S1) [22]. The comparator is composed of an inlet device and an outlet device. The inlet device has two inlets, a junction channel, and two outlets, and the outlet device has two inlets and one outlet. The devices were connected through cut tubing pieces, having an inner diameter of 0.508 mm and a length of 20 cm. The microfluidic comparator is based on the comparison of two different fluid streams (i.e., a test fluid and a reference fluid) in the junction channel. The ratio of their viscosities can be determined by the ratio of their flow rates at the equilibrium state, where the interfacial position of the fluids is stationary in the junction channel. This approach requires additional calibration for the reference-fluid viscosity to calculate the absolute viscosity of the test fluid.

2.2. Sample Preparation and Analysis

A glycerol solution was purchased from Sigma-Aldrich (St. Louis, MO, USA) and diluted in distilled water at various concentrations (50 to 64.5 wt %) for the preparation of Newtonian fluids of different viscosities. The viscosities of the glycerol-water mixtures were determined using a cone-plate viscometer (Brookfield AMETEK Inc., Middleborough, MA, USA). A xanthan gum solution was obtained from Sigma-Aldrich and diluted in distilled water at a concentration of 0.1 wt % for analysis of a non-Newtonian fluid. All experiments with the 3D-CCs were recorded using a smartphone and changes in liquid level were measured during 15 s. The volumetric flow rate in each channel of the 3D-CCs was calculated by multiplying the cross-sectional area of the fluidic chamber by the number of measurement ticks counted per unit time. The microfluidic comparator was operated using two syringe pumps (KD Scientific, Holliston, MA, USA), and the flow rates of the test and reference fluids were manually adjusted to determine the equilibrium state.

2.3. Computational Fluid Dynamics Simulation

The pressure drop profile along the channel of the 3D-CC for multiplexed analysis of Newtonian fluids was analyzed using COMSOL Multiphysics (COMSOL, Burlington, MA, USA) to verify the design criterion for the pressure-drop ratio between the capillary and fluidic chamber. The wall shear rate and pressure drop along the length of each capillary of the 3D-CC for analysis of a non-Newtonian fluid were also analyzed using COMSOL Multiphysics to determine capillary lengths to achieve desired shear-rate conditions. Three-dimensional finite element models were created in the same dimensions as the 3D-CCs and solved using the incompressible Naiver-Stokes equation. The inlet boundary condition was set to a constant pressure, while the outlet boundary condition was set to atmospheric pressure. All other boundary conditions were set with no-slip boundary conditions.

3. Results and Discussion

3.1. Measurement Principle

The 3D-CC platform is composed of a 3D-CC and a smart pipette to measure the volumetric flow rates (Q) of test fluids discharging through the channels of the 3D-CC and to pump the fluids at a fixed pressure drop, respectively (Figure 1). The channels of the 3D-CC are arranged in a successive structure of a capillary and a fluidic chamber. The scales marked on the fluidic chamber enable accurate measurement of the liquid volume discharging through each channel over a given time (Δt), thereby, determining Q. The smart pipette has an air chamber serving as a pressure buffer, thereby, generating a fixed pressure condition [26–29]. The Hagen-Poiseuille law is the basis of the 3D-CC platform to measure viscosity without a reference fluid [30]:

$$Q = \frac{\pi r^4 \Delta P}{8 \mu L} \tag{1}$$

where ΔP is the pressure drop through each capillary of the 3D-CC, r is the capillary radius, μ is the fluid viscosity, and L is the capillary length. Test fluids of an unknown viscosity are filled and dispensed through the channels of the 3D-CC under a fixed pressure-drop condition (Figure 1b). The Q of each test fluid can be measured by the number of scale ticks (n_s) for Δt. Knowing the Q, capillary dimensions and drop in pressure through each capillary, the viscosities of test fluids can be simply calculated using Equation (1).

There are two design parameters that affect viscosity measurement accuracy. The first parameter is the volume setting of the smart pipette. Depressing the syringe plunger in a set volume (V_1) compresses the air inside the air chamber and generates a desired pressure (P). Since the product of pressure and volume is a constant at a constant temperature based on Boyle's law, P can be defined as $P = (V_1/V_2)P_{atm}$, where V_2 is the volume of compressed air. However, a continual expansion in

V_2 occurs due to the discharging volume of liquid, V_3, thereby, affecting P. Thus, V_1 and V_2 should be sufficiently larger than V_3 to maintain P. The smart pipette was, thus, designed with a large air chamber volume of 1.27 L, at which V_3 becomes negligible. At $(V_1)_{min}$ = 20 mL, V_2 = 1.27 L and $(V_3)_{max}$ = 0.5 mL, the smart pipette can, theoretically, maintain P within 99.96%. A pressure sensor can be further integrated into the smart pipette for accurate pressure measurement.

The resistance ratio between the capillary and fluidic chamber of the 3D-CC is another design parameter for accurate viscosity measurement. Under a fixed P, the applied pressure drops along the capillary and fluidic chamber. In the serial connection, the upstream fluidic chamber resistance (R_f) can affect the downstream ΔP, i.e., the pressure drop through the capillary. During liquid dispensation, R_f decreases due to a decrease in the length of the liquid filling in each fluidic chamber. The decreasing R_f can result in changes in ΔP, leading to an inaccurate viscosity measurement. To maintain the pressure drop through the capillary at P regardless of R_f, it is necessary to have a capillary, $R_c > 410 R_f$, as previously reported [22], where R_c is the capillary resistance. Under this condition, it can be assumed that P drops only in the capillaries of the 3D-CC. Q can be measured by multiplying the cross-sectional area of the 3D-CC chamber and the length difference of the liquid meniscuses moved during liquid dispensation. As a result, Equation (1) can be rewritten as:

$$\mu = \frac{\pi r^4}{8LQ}P = \frac{\pi r^4 \Delta t}{8L\left(\pi r_{fl}^2 dn_s\right)}\left(\frac{V_1}{V_2}\right)P_{atm} \tag{2}$$

where r_{fl} is the radius of the fluidic chamber and d is the distance between adjacent scale ticks. Therefore, simply pushing the plunger of the smart pipette by V_1 and counting n_s during Δt enables accurate viscosity measurement by calculating Equation (2), with a set P, known geometric dimensions, and measured n_s. The concept of connecting multiple capillaries and generating fluid flows through the capillaries using the portable pressure generator enables multiplexed viscosity analysis of different fluid samples, equipment-free viscosity measurement and viscosity analysis of a non-Newtonian fluid over a range of shear rates, and provides ease of use similar to a conventional pipette, as well as a new perspective for 3D-printed fluidic sensors.

3.2. Multiplexed Analysis of Newtonian Fluids

We first tested the ability of the 3D-CC platform for multiplexed analysis of Newtonian fluids. The 3D-CC for multiplexed analysis of Newtonian fluids was fabricated by incorporating 12 cm-long Tygon® tubing pieces (inner diameter: 0.508 mm) as a capillary, 3.9 cm-long Tygon® tubing pieces (inner diameter: 3.17 mm) as a fluidic chamber, and a scale-patterned film into a 3D-printed part (Figure 1). Under this condition, the ratio of R_c and R_f was greater than 4000 and the pressure drop along the fluidic chamber was negligible, satisfying the above design criteria (Figure 2). Glycerol-in-water mixtures in different weight ratios were used as a model system for Newtonian fluids of different viscosities. Their viscosities of the glycerol solutions were measured using a commercial cone-plate viscometer and compared with the results obtained from the 3D-CC. Viscosity measurements were performed by a simple two-step process: Withdrawing and dispensing test fluids through the channels of the 3D-CC. At P = 2.4 kPa and Δt = 15 s, n_s highly depended on the viscosity of a fluid discharging through each channel of the 3D-CC (Figure 3a). As shown in Figure 3b, the viscosity data obtained using the 3D-CC in the range of V_1 (30 to 50 mL), which corresponds to the range of P (2.4 to 4.0 kPa), had good agreement with those measured using the cone-plate viscometer. We note that the coefficient of variation (CV) for V_1 = 20 mL and μ_c = 14.4 cP was 9.1%, which was significantly higher than the averaged CV of 3.5% for other V_1 and μ_c conditions (range; 0–6.7%) (Figure 3b), where μ_c is the viscosity measured using the cone-plate viscometer. This is likely due to the low count of n_s at V_1 = 20 mL and μ_c = 14.4 cP, which was 6.7 ± 0.6. If n_s is too low, then the measured μ may not be representative of the actual μ due to an increased effect of stochastic variables. It was, thus, necessary to use $n_s \geq 9$ to ensure reliable viscosity measurement. In addition, the viscosity value (15.6 ± 1.4 cP) measured using the 3D-CC at $V_1 = 20$ mL and $\mu_c = 14.4$ cP was higher than μ_c (14.4 $\pm$ 0.1 cP) (Figure 3b). This might be due to the low shear rate condition (108.7 s^{-1}) for viscosity measurement under which fluid-wall interactions and gravitational effects can become dominant. Such additional effects make the test fluid appear to have a high viscosity. As shown in Figure 4, V_1 or the corresponding shear rate condition is an important operational parameter for accurate viscosity measurement at fixed capillary geometry and V_2. The shear rate condition for 3D-CC operation should be higher than 108.7 s^{-1} for accurate viscosity measurement.

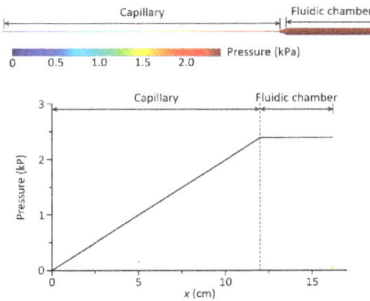

Figure 2. Computational fluid dynamics simulation showing the pressure profile along the fluidic channel of the 3D-CC for multiplexed analysis at $\mu = 11.0$ cP and $P = 2.4$ kPa. Compared to the capillary, the drop in pressure along the fluidic chamber is negligible.

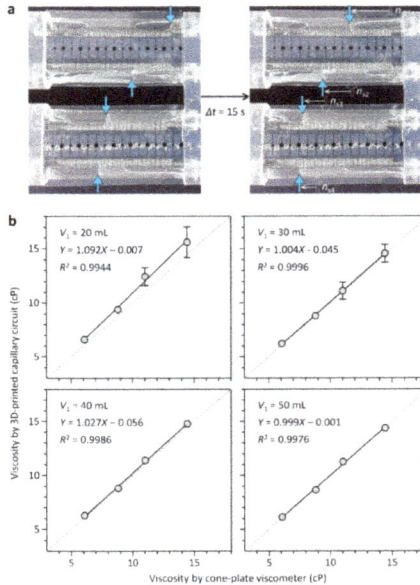

Figure 3. Multiplexed analysis of Newtonian fluids using the 3D-CC, having four identical fluidic channels. (**a**) Glycerol-water mixtures with different viscosities were analyzed using the 3D-CC platform at $V_1 = 30$ mL. Their viscosities were 6.1, 8.8, 11.0, and 14.4 cP (from top to bottom) as determined by the cone-plate viscometer. The blue arrows indicate meniscuses; (**b**) effect of V_1 on viscosity measurement. Solid lines denote linear regressions, while dotted lines represent a unity slope. Error bars: s.d. ($n = 3$).

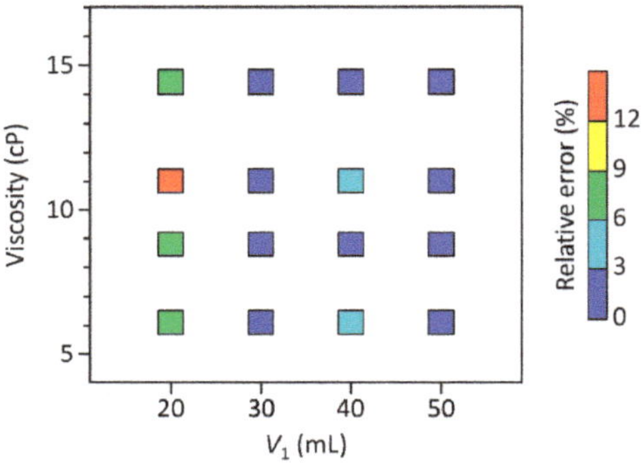

Figure 4. Relative errors of the 3D CC-based viscosity measurement results in Figure 3b as a function of V_1 and the absolute viscosities measured using the cone-plate viscometer. Optimal operation conditions for accurate viscosity measurement can be determined from the color map for further experiments.

3.3. Analysis of a Non-Newtonian Fluid versus Shear Rate

We next tested the capability of the 3D-CC platform to analyze a non-Newtonian fluid, xanthan gum, at different shear-rate conditions. The 3D-CC for analysis of a non-Newtonian fluid was fabricated using Tygon® tubing pieces (inner diameter: 0.508 mm) in different lengths for the generation of different shear-rate conditions (Figure 5a). To generate a range of shear rates (150 to 900 s^{-1}), the viscosities of a 0.1 wt % xanthan gum solution were first measured at γ = 178, 353, 653, and 930 s^{-1} using a conventional microfluidic comparator (Figure S1); where γ is the wall shear rate. With the measured viscosities, we numerically calculated ΔP per unit capillary length (1 cm) required for the operation of the 3D-CC (Figure 6a). The capillary lengths of the 3D-CC were then determined using Equation (1) at ΔP = 2.8 kPa and r = 0.254 mm, which were 8.4, 10.4, 14.6, and 20.6 cm for γ = 930, 653, 353, and 178 s^{-1}, respectively (Figure 6b). For convenience of fabrication, L was set to 8.5, 10.9, 15.5, and 20.0 cm, and r_{fl} of the 3D-CC was set to 1.59 mm. The dimensional conditions satisfy the design criteria described above, while the resistance ratio between the capillary and fluidic chamber was greater than 4000. Depressing the plunger of the smart pipette, set at V_1 = 35 mL, simultaneously generated four-different shear-rate conditions according to capillary length (Figure 5b), which corresponded to 864, 541, 257, and 159 s^{-1} for L = 8.5, 10.9, 15.5, and 20.0 cm, respectively. For all conditions, n_s at Δt = 15 s was greater than 9, satisfying the design criteria described above. As shown in Figure 5c, the viscosity results measured using the 3D-CC were closely matched with those obtained using the microfluidic comparator, demonstrating the effectiveness of the 3D-CC platform for simultaneously testing multiple shear-rate conditions. Compared with current state-of-the-art viscometers [13–22], the proposed 3D-CC platform allows rapid, simple, and intuitive operation without bulky, complicated, high-precision equipment and additional calibration. Thus, non-experts can easily carry out the necessary procedures in a similar manner to a conventional micropipette. In addition, the scalability of the 3D-CC enables further enhancement in measurement throughput by simply adding more channels. Although xanthan gum, as a model system of a non-Newtonian fluid, was tested for proof-of-principle demonstration, the 3D-CC platform can be widely applicable for quality-control testing of liquid products and for clinical analysis of blood viscosity as a potential hemodynamic biomarker.

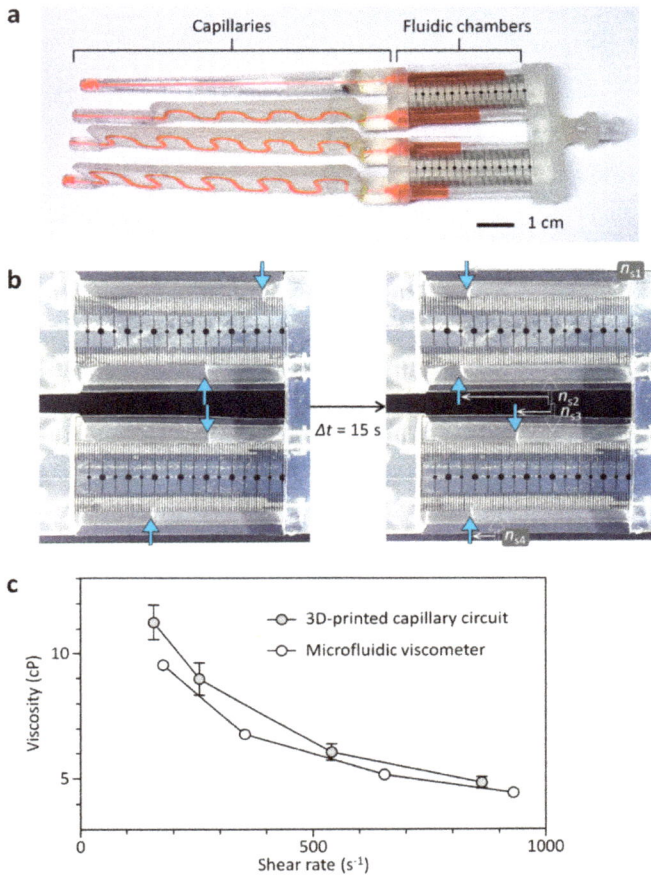

Figure 5. 3D-CC for analysis of a non-Newtonian fluid under multiple shear-rate conditions. (**a**) Optical image of the fabricated 3D-CC having four-different fluidic channels in length. The lengths were 8.5, 10.9, 15.5, and 20.0 cm (from top to bottom). The channels of the 3D-CC are filled with red ink for visualization; (**b**) a xanthan gum solution, which served as a non-Newtonian fluid, was analyzed using the 3D-CC at different shear-rate conditions of 864, 541, 257, and 159 s^{-1} (from top to bottom); (**c**) analysis results for the xanthan gum solution using the 3D-CC and microfluidic viscometer. Error bars: s.d. ($n = 3$).

Micromachines **2018**, *9*, 314

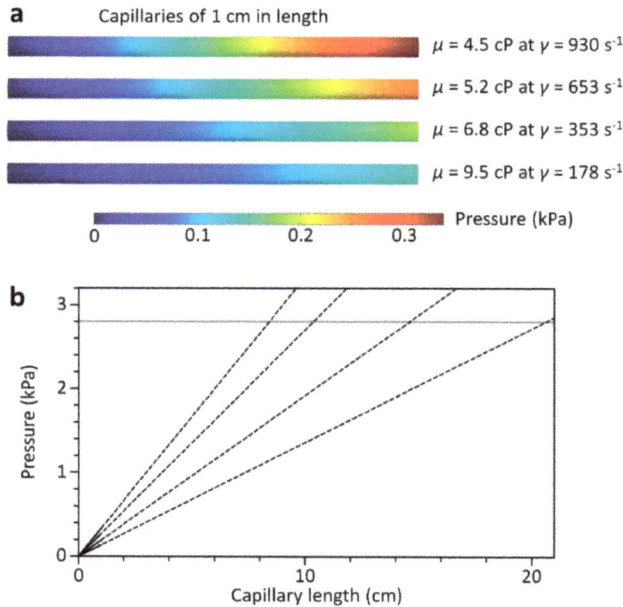

Figure 6. Computational fluid dynamics simulation for determining the capillary lengths of the 3D-CC for analysis of the xanthan gum solution under multiple shear-rate conditions. (**a**) Simulated pressure profiles along the capillaries (0.508 mm in inner diameter) at different μ and γ to calculate the pressure drop per unit length at each condition. The viscosities were obtained using the microfluidic comparator and varied according to the corresponding γ; (**b**) the different capillary lengths were determined from the relationship between ΔP and L, the Hagen-Poiseuille law, at the different μ and γ conditions. The dotted lines denote the relationships between ΔP and L for μ = 4.5, 5.2, 6.8, and 9.5 cP from left to right. At ΔP = 2.8 kPa, the capillary lengths were, respectively, determined as 8.4, 10.4, 14.6, and 20.6 cm that can, subsequently, generate the corresponding γ conditions for the 3D-CC.

4. Conclusions

We present a new 3D-CC platform for equipment- and calibration-free analysis of Newtonian and non-Newtonian fluids. The viscometer platform allows hand-powered generation of pressure-driven flows and visual (naked-eye) determination of flow rates in parallel capillary networks. We demonstrated multiplexed viscosity measurement of Newtonian fluids of different viscosities and viscosity measurement of a non-Newtonian fluid at various shear-rate conditions. Given these novel features and the importance of viscosity measurement as a quality control means of liquid products and a diagnostic tool for evaluating hemorheological disorders, we believe that the proposed platform represents a promising approach for viscosity measurement in various industrial and clinical applications.

Supplementary Materials: The following are available online at http://www.mdpi.com/2072-666X/9/7/314/s1, Figure S1: Schematic showing a microfluidic viscometer to measure a test fluid of unknown viscosity.

Author Contributions: S.C. conceptualized and directed the research project. S.C. and S.O. prepared the manuscript. S.O. performed the experiments and data analysis with S.C. All authors discussed the results and commented on the manuscript.

Funding: This research was supported by a grant from the Pioneer Research Center Program through the National Research Foundation (NRF) of Korea funded by the Ministry of Science, ICT & Future Planning (MSIP) (2013M3C1A3064777), a NRF grant funded by the Korea government (MSIP) (2015R1C1A1A01053990), and a NRF grant funded by the Korea government (MSIP) (2016R1A5A1010148).

Conflicts of Interest: The authors declare no conflict of interest.

References

1. Fügel, R.; Carle, R.; Schieber, A. Quality and authenticity control of fruit purées, fruit preparations and jams—A review. *Trends Food Sci. Technol.* **2005**, *16*, 433–441. [CrossRef]
2. Hoekman, S.K.; Broch, A.; Robbins, C.; Ceniceros, E.; Natarajan, M. Review of biodiesel composition, properties, and specifications. *Renew. Sustain. Energy Rev.* **2012**, *16*, 143–169. [CrossRef]
3. Lebedeva, N.; Malkova, E.; Vyugin, A.; Koifman, O.; Gubarev, Y. Spectral and hydrodynamic studies of complex formation of tetraalkoxy substituted zinc(II)phthalocyanines with defatted and nondefatted bovine serum albumin. *Biochip J.* **2015**, *10*, 1–8. [CrossRef]
4. Jimtaisong, A.; Saewan, N. Utilization of carboxymethyl chitosan in cosmetics. *Int. J. Cosmet. Sci.* **2014**, *36*, 12–21. [CrossRef] [PubMed]
5. Tao, W.; Ma, Y.; Chen, Y.; Li, L.; Wang, M. The influence of adhesive viscosity and elastic modulus on laser spot weld bonding process. *Int. J. Adhes. Adhes.* **2014**, *51*, 111–116. [CrossRef]
6. Chiarello, E.; Derzsi, L.; Pierno, M.; Mistura, G.; Piccin, E. Generation of Oil Droplets in a Non-Newtonian Liquid Using a Microfluidic T-Junction. *Micromachines* **2015**, *6*, 1825–1835. [CrossRef]
7. Sklodowska, K.; Debski, P.R.; Michalski, J.A.; Korczyk, P.M.; Dolata, M.; Zajac, M.; Jakiela, S. Simultaneous Measurement of Viscosity and Optical Density of Bacterial Growth and Death in a Microdroplet. *Micromachines* **2018**, *9*, 251. [CrossRef]
8. Smith, M.M.; Chen, P.C.Y.; Li, C.-S.; Ramanujam, S.; Cheung, A.T.W. Whole blood viscosity and microvascular abnormalities in Alzheimer's Disease. *Clin. Hemorheol. Microcirc.* **2009**, *41*, 229–239. [CrossRef] [PubMed]
9. Gertz, M.A.; Kyle, R.A. Hyperviscosity syndrome. *J. Intensive Care Med.* **1995**, *10*, 128–141. [CrossRef] [PubMed]
10. McKennell, R. Cone-plate viscometer. *Anal. Chem.* **1956**, *28*, 1710–1714. [CrossRef]
11. Cooke, B.M.; Stuart, J. Automated measurement of plasma viscosity by capillary viscometer. *J. Clin. Pathol.* **1988**, *41*, 1213–1216. [CrossRef] [PubMed]
12. Gupta, S.; Wang, W.S.; Vanapalli, S.A. Microfluidic viscometers for shear rheology of complex fluids and biofluids. *Biomicrofluidics* **2016**, *10*, 043402. [CrossRef] [PubMed]
13. Srivastava, N.; Davenport, R.D.; Burns, M.A. Nanoliter viscometer for analyzing blood plasma and other liquid samples. *Anal. Chem.* **2005**, *77*, 383–392. [CrossRef] [PubMed]
14. Srivastava, N.; Burns, M.A. Analysis of non-Newtonian liquids using a microfluidic capillary viscometer. *Anal. Chem.* **2006**, *78*, 1690–1696. [CrossRef] [PubMed]
15. Kim, W.-J.; Kim, S.; Huh, C.; Kim, B.K.; Kim, Y.J. A novel hand-held viscometer applicable for point-of-care. *Sens. Actuators B* **2016**, *234*, 239–246. [CrossRef]
16. Liu, M.; Xie, S.; Ge, J.; Xu, Z.; Wu, Z.; Ru, C.; Luo, J.; Sun, Y. Microfluidic Assessment of Frying Oil Degradation. *Sci. Rep.* **2016**, *6*, 27970. [CrossRef] [PubMed]
17. Li, Y.; Ward, K.R.; Burns, M.A. Viscosity Measurements Using Microfluidic Droplet Length. *Anal. Chem.* **2017**, *89*, 3996–4006. [CrossRef] [PubMed]
18. Kang, Y.J.; Yang, S. Integrated microfluidic viscometer equipped with fluid temperature controller for measurement of viscosity in complex fluids. *Microfluid. Nanofluid.* **2013**, *14*, 657–668. [CrossRef]
19. Kang, Y.J.; Ryu, J.; Lee, S.-J. Label-free viscosity measurement of complex fluids using reversal flow switching manipulation in a microfluidic channel. *Biomicrofluidics* **2013**, *7*, 44106. [CrossRef] [PubMed]
20. Choi, S.; Park, J.K. Microfluidic rheometer for characterization of protein unfolding and aggregation in microflows. *Small* **2010**, *6*, 1306–1310. [CrossRef] [PubMed]
21. Tassieri, M.; Del Giudice, F.; Robertson, E.J.; Jain, N.; Fries, B.; Wilson, R.; Glidle, A.; Greco, F.; Netti, P.A.; Maffettone, P.L.; et al. Microrheology with optical tweezers: Measuring the relative viscosity of solutions 'at a glance'. *Sci. Rep.* **2015**, *5*, 8831. [CrossRef] [PubMed]
22. Oh, S.; Kim, B.; Lee, J.K.; Choi, S. 3D-printed capillary circuits for rapid, low-cost, portable analysis of blood viscosity. *Sens. Actuators B* **2018**, *259*, 106–113. [CrossRef]
23. Oh, S.; Kim, B.; Choi, S. A 3D-Printed Multichannel Viscometer for High-Throughput Analysis of Frying Oil Quality. *Sensors* **2018**, *18*, 1625. [CrossRef] [PubMed]

24. Mao, M.; He, J.; Li, X.; Zhang, B.; Lei, Q.; Liu, Y.; Li, D. The Emerging Frontiers and Applications of High-Resolution 3D Printing. *Micromachines* **2017**, *8*, 113. [CrossRef]

25. Waheed, S.; Cabot, J.M.; Macdonald, N.P.; Lewis, T.; Guijt, R.M.; Paull, B.; Breadmore, M.C. 3D printed microfluidic devices: Enablers and barriers. *Lab Chip* **2016**, *16*, 1993–2013. [CrossRef] [PubMed]

26. Kim, B.; Choi, S. Smart Pipette and Microfluidic Pipette Tip for Blood Plasma Separation. *Small* **2016**, *12*, 190–197. [CrossRef] [PubMed]

27. Kim, B.; Hahn, Y.K.; You, D.; Oh, S.; Choi, S. A smart multi-pipette for hand-held operation of microfluidic devices. *Analyst* **2016**, *141*, 5753–5758. [CrossRef] [PubMed]

28. Kim, B.; Oh, S.; You, D.; Choi, S. Microfluidic Pipette Tip for High-Purity and High-Throughput Blood Plasma Separation from Whole Blood. *Anal. Chem.* **2017**, *89*, 1439–1444. [CrossRef] [PubMed]

29. You, D.; Oh, S.; Kim, B.; Hahn, Y.K.; Choi, S. Rapid preparation and single-cell analysis of concentrated blood smears using a high-throughput blood cell separator and a microfabricated grid film. *J. Chromatogr. A* **2017**, *1507*, 141–148. [CrossRef] [PubMed]

30. Truskey, G.A.; Yuan, F.; Katz, D.F. *Transport Phenomena in Biological Systems*, 1st ed.; Pearson Education: Upper Saddle River, NJ, USA, 2004.

micromachines

MDPI

Article

3D Printed Microfluidic Features Using Dose Control in X, Y, and Z Dimensions

Michael J. Beauchamp [1], Hua Gong [2], Adam T. Woolley [1,*] and Gregory P. Nordin [2,*]

[1] Department of Chemistry and Biochemistry, Brigham Young University, Provo, UT 84602, USA; mikejbeau@byu.edu

[2] Department of Electrical and Computer Engineering, Brigham Young University, Provo, UT 84602, USA; gonghuabupt@gmail.com

* Correspondence: atw@byu.edu (A.T.W.); nordin@byu.edu (G.P.N.); Tel.: +1-801-422-1701 (A.T.W.); +1-801-422-1863 (G.P.N.)

Received: 6 June 2018; Accepted: 26 June 2018; Published: 28 June 2018

Abstract: Interest has grown in recent years to leverage the possibilities offered by three-dimensional (3D) printing, such as rapid iterative changes; the ability to more fully use 3D device volume; and ease of fabrication, especially as it relates to the creation of complex microfluidic devices. A major shortcoming of most commercially available 3D printers is that their resolution is not sufficient to produce features that are truly microfluidic ($<100 \times 100 \ \mu m^2$). Here, we test a custom 3D printer for making ~30 μm scale positive and negative surface features, as well as positive and negative features within internal voids (i.e., microfluidic channels). We found that optical dosage control is essential for creating the smallest microfluidic features (~30 μm wide for ridges, ~20 μm wide for trenches), and that this resolution was achieved for a number of different exposure approaches. Additionally, we printed various microfluidic particle traps, showed capture of 25 μm diameter polymer beads, and iteratively improved the trap design. The rapid feedback allowed by 3D printing, as well as the ability to carefully control optical exposure conditions, should lead to new innovations in the types and sizes of devices that can be created for microfluidics.

Keywords: 3D printing; microfluidics; particle traps; stereolithography

1. Introduction

3D printing is a valuable technique for custom and rapid design change and optimization in fabrication of millifluidic devices [1]. Miniature device applications stand to benefit from the advantages offered from 3D printing, such as the ability to create and test devices with rapid feedback, allowing changes to be quickly tested. Device optimization based on empirical results could save time and money compared to traditional device fabrication techniques that involve conventional machining or micromachining.

A number of groups have recently sought to use 3D printing to produce fluidic devices for various applications. Devices for nitrite [2] or anemia [3] detection, measuring endocrine secretion [4], sorting bacteria [5], and cell culture [6,7], have all been shown to name a few. Although these are promising assays, a key issue from these works is the size of printed features. Most commercially available printers and resins are only able to achieve feature sizes down to 250 μm, with typical features around 500 μm, which are not suitable for many microfluidics applications. Additionally, these commercial printers lack flexibility in terms of resin development, individual layer custom exposure time control, or multiple exposures per layer. For 3D printing of fluidic devices with low surface roughness, stereolithography (SLA) printers are the best suited [8]. The material left in the channels after printing is a liquid and thus is much easier to clear than solid sacrificial materials formed with either polyjet or fused deposition modeling printers [9–11]. Reviews of 3D printing over the past several years have offered helpful insights regarding types of printers and their applications. Many different outlooks are given for future directions in 3D printing of fluidic devices, such as resin improvements, material removal techniques, throughput, and printer resolution [1,12–16]. Several groups have undertaken work to investigate and compare the resolution of various 3D printers in an effort to make fluidic devices [8,11,17–19]; however, much of this work is still well above 100 μm needed for most microfluidic applications.

Our group has developed an SLA 3D printer, as well as a custom resin formulated specifically for creating truly microfluidic structures with this printer [20], and we have made small (18 × 20 μm) microfluidic channels [21], as well as fluid control systems involving pumps and valves [22]. To make the smallest channels, an edge compensation technique was employed which overexposed the pixels at the channel edge to make it narrower [21]. However, we have not yet examined how this edge compensation approach affects features in the channels.

In this work, we investigate precise control over printing exposure areas and dosage conditions to create microscale substructures within microfluidic features. First, we look at positive and negative features on the exterior of prints to see what size features can be printed with various exposure times and with exposure edge compensation. Next, we evaluate positive and negative features in interior void areas to see the impact of exposure times and edge compensation. Finally, we create microfluidic particle traps to demonstrate how the ability to control specific dosing parameters allows improved function.

2. Materials and Methods

2.1. Material Sources

Acetone, 2-propanol (IPA), and 25 μm polystyrene microspheres were purchased from Thermo-Fisher Scientific (Salt Lake City, UT, USA). Triethoxysilyl propylmethacrylate, hydroxypropyl cellulose (HPC) and polyethylene glycol diacrylate (PEGDA, 258 Da MW) were purchased from Sigma-Aldrich (Milwaukee, WI, USA). Toluene and glass microscope slides (3″ × 1″ × 1.2 mm) were purchased from Avantor (Center Valley, PA, USA). 2-Nitrodiphenylsulfide (NPS) was purchased from TCI America (Portland, OR, USA). Phenylbis (2,4,6-trimethylbenzoyl)phosphine oxide (Irgacure 819) was provided by BASF (Midland, MI, USA). All chemicals were used as received.

2.2. Glass Slide Preparation

Glass microscope slides were scored on one side using a laser cutter (Universal Laser Systems, Scottsdale, AZ, USA). The settings for cutting were 50% power, 10% speed, and 165 points per inch. The glass slides were then broken along the scored mark, washed with acetone and IPA, and dried with air. A fresh preparation of 10% triethoxysilyl propylmethacrylate in toluene was made. Glass slides were submerged in this silane solution in a shallow covered dish for a minimum of two hours, after which they were rinsed with IPA and dried with air. For longer term storage, slides were kept in a container under toluene.

2.3. Resin Preparation

The resin was prepared by mixing 2% NPS with 1% Irgacure 819 in 97% PEGDA. The resin was sonicated until all solid components dissolved and was stored in an amber bottle wrapped in aluminum foil to protect it from light. Details regarding resin formulation can be found in reference [21] for the choice of photoinitiator and absorber matched to the LED of the printer.

2.4. Device Designs

Designs for 3D printed parts were made using open source OpenSCAD software (openscad.org). Schematics of the resolution prints can be seen in Figure 1. For the exterior ridges (Figure 1a) and trenches (Figure 1b), the features are 100 μm tall or deep, and the widths are from 1 to 10 pixels (7.6 to 76 μm), with a spacing between individual ridges or trenches of 100 pixels.

For the interior resolution features (Figure 1c), the height of the feature area is 100 µm. The internal ridges are all 5 pixels wide, 250 pixels long, and have heights from 1 to 10 layers (10 to 100 µm). For the trench sections, the trenches are all 100 µm deep and vary in width from 1 to 10 pixels. The pillars are designed with diameters ranging from 1 to 10 pixels, and all of the pillars in a given row are identical. In between each internal feature (or set of features with the pillars) is a support beam to help hold up the microchannel ceiling. These ceiling supports are all 5 pixels wide and go from the floor of the feature areas to the ceiling.

The trapping devices consist of 6 straight channels 30 pixels wide and 8 layers tall with fluidic reservoirs at both ends (Figure 1d). The traps consist of two L-shaped pieces facing each other that are 8 pixels long, 4 pixels wide, and spaced 2 pixels apart (Figure 1e). The traps are spaced 20 pixels apart down the length of the channel. Each print contains 6 different channels for testing a variety of trap layouts. Three different configurations of the traps were tested, one with traps only down the center of the channel, one with traps staggered along the edges, and one with traps staggered in the middle of the channel and along the sides (see Figure 1d, inserts).

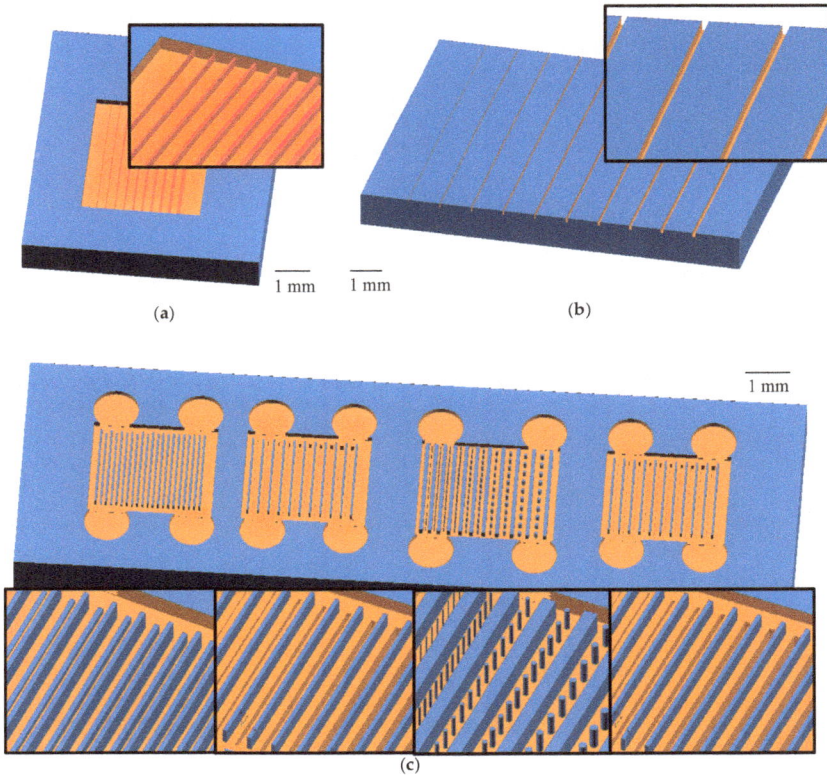

(a)   (b)

(c)

Figure 1. *Cont.*

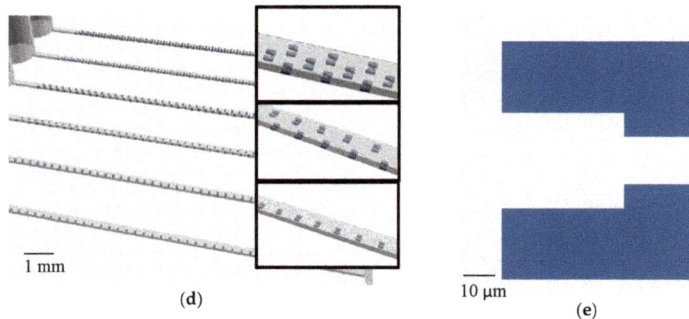

Figure 1. OpenSCAD designs of prints for exterior and interior resolution features. (**a**) Ridge device. The ridges are shown in red and there is a support box around the ridges. The ridges are 100 μm tall and have widths of 1–10 pixels (7.6–76 μm) from left to right. The cutout shows a zoom view; (**b**) Trench device with the trenches shown in orange. The trenches are 100 μm deep and have widths of 1–10 pixels (7.6–76 μm) from left to right. The cutout shows a zoom view; (**c**) Interior features device. The top layers of the device have been removed in the schematic to show the features. From left to right the regions are ridges, trenches with exposure compensation, pillars, and trenches without exposure compensation. Each set of features has a series of ceiling support ridges running the length of the feature area. Below each feature void area is a zoom view; (**d**) Bead trap device design showing 6 channels and traps within channels. Inserts show a zoom view of the three different trap layouts; (**e**) Schematic of bead trap. The large gap is designed to allow the beads to enter and the smaller gap allows fluid to pass through the trap.

2.5. 3D Printing Parameters

The 3D printer used for this work is the same as described in reference [21]. This printer operates with a nominally 385 nm light source and 7.6 μm pixel size in the image plane. The build layer height for all prints was 10 μm, the image plane irradiance was 21.2 mW·cm^{-2}, and the exposure time was chosen to be either 500, 750, 1000, or 1500 ms to ensure thorough attachment of the print to the glass slide the first four layers were overexposed for 20, 10, 5, and 1 s, respectively. For prints in which the normal layer exposure time exceeded 1 s, only the first three layers were exposed in this manner. After printing, the remaining liquid resin in the print was flushed out three times with IPA using vacuum. Finally, the device was cured under an 11 mW 430 nm LED (ThorLabs, Newton, NJ, USA) for 10 min before use. This LED allows the photoinitiator to further cure the print at a wavelength at which the UV absorber (NPS) does not absorb the light.

2.6. Edge Compensation Technique

An edge compensation technique similar to the one in reference [21] was used where indicated for both interior and exterior trenches (negative features). This technique exposes the two pixels forming the edge of the trench for double the normal exposure time. The purpose of this technique is to cause a wider trench design to be narrower when printed. For example, a 3D printed trench that is designed to be 4 pixels wide without compensation has the same width as a 6 pixel wide design formed with compensation.

2.7. Measurement of Print Featuures

Exterior feature heights, depths, and widths were measured using a Zeta 20 optical profilometer (Zeta Instruments, San Jose, CA, USA). The width was measured as the full width at half height or depth. SEM imaging was done using an ESEM XL30 (FEI, ThermoFisher Scientific, Waltham, MA, USA). Samples were prepared by cutting them open with a razor blade and sputtering with 80:20

Au:Pd to allow the side profile to be observed. Images were processed using ImageJ (NIH) to measure the widths and heights of interior pillars, ridges, and trenches.

For trapping experiments, imaging was done using a Zeiss AXIO Observer A1 inverted microscope (Thornwood, NY, USA) using a 10× objective connected to a Photometrics coolSNAP HQ2 CCD camera (Tucson, AZ, USA). The exposure time for the CCD was 10 ms. The images were recorded and processed using ImageJ. Measurement of print features was performed three times ($n = 3$), and the standard deviation is given for each measurement.

For our printer and this resin, surface roughness has previously been characterized using optical profilometry with prints at various exposure times [22]. For all exposure times tested (600–1200 ms), the RMS surface roughness was less than 100 nm, typically 55–60 ± 15 nm. While some surfaces may appear to show pixelation (Figure 4 and Supplementary Materials) as they did in this previous work, it is expected that the surface roughness should still be the same.

2.8. Trapping Opperation

The bead solution for trapping was made by suspending the beads in deionized water at a concentration of 1 mg/mL with 0.5% HPC to prevent aggregation. 1.5 µL of bead solution was pipetted into the left reservoir as oriented in Figure 1d and drawn through the channel with vacuum over ~7 s, which resulted in a flow rate of 13 µL/min. This was repeated three times so a total of 4.5 µL of bead solution was pulled through the channel, after which CCD images were taken.

3. Results and Discussion

3.1. Exterior Features

3.1.1. Ridges

Initial testing focused on the exterior resolution features of our 3D printer by testing a set of ridges and trenches on the surface of prints. The purpose of these features is to evaluate how positive features turn out on the surface of print. For the surface ridges, the design shown in Figure 1a was created; the design was printed three times with exposure times of 500, 1000, or 1500 ms for each build layer. We found that the ends of the ridges became warped when they were not anchored, so a support box was placed around the ridges. A photograph though the microscope can be seen in Figure 2a, showing example ridges that are 3 and 4 pixels wide for 1500 ms exposure. The complete set of images can be found in the Supplementary Materials (Figures S1–S3). The heights and widths of these features were then measured with an optical profilometer. Figure 3a shows the measured ridge width plotted against the designed width for ridges that reached >90% of the full height. This shows that increasing light dosage from 500 ms (blue line in Figure 3a) to 1500 ms (red line in Figure 3a) allows smaller ridges to successfully reach full height and be closer to their designed width. Additionally, a ridge that is designed to a certain width will print smaller than expected if the exposure time is insufficient. The minimum width ridge that could be successfully printed was 30 ± 1 µm, which was with an exposure time of 1500 ms. These positive features benefit from increased light exposure, indicating the need to be able to give sufficient exposure to positive features.

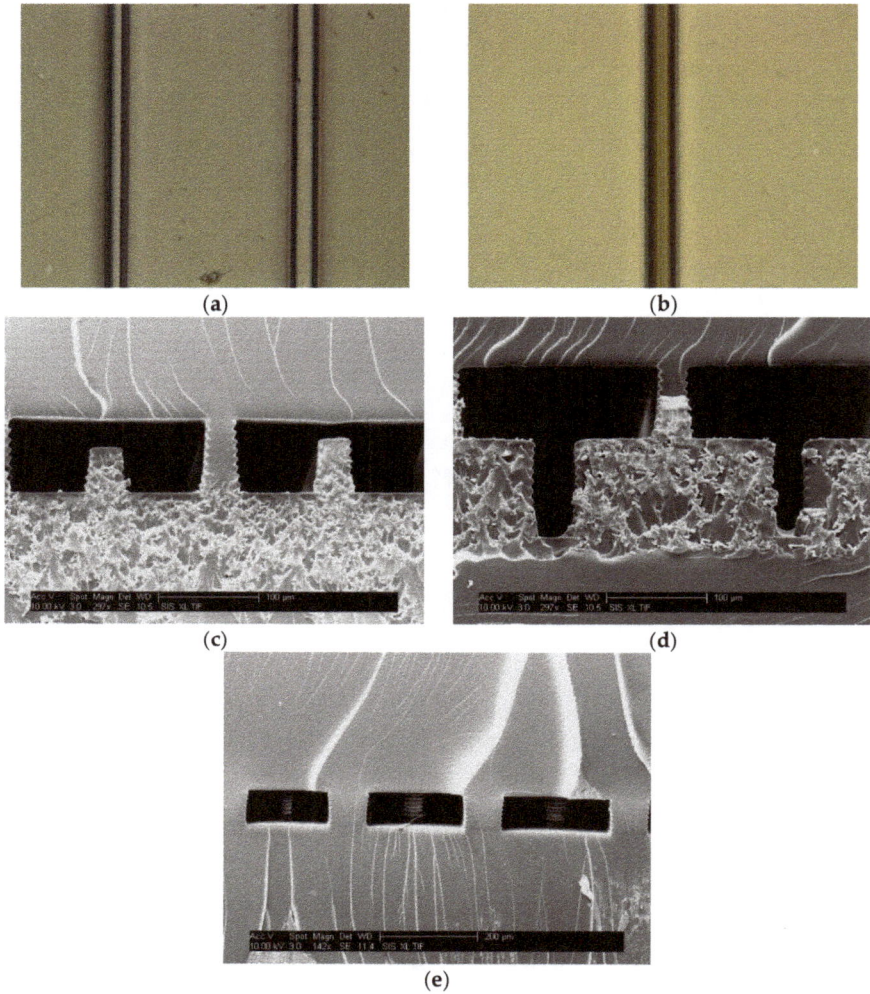

Figure 2. Images of 3D printed features. (**a**) Top view photograph of 1500 ms exposure ridges designed 3 and 4 pixels (23 and 30 µm) wide. The ridges measured 25 ± 1 and 29 ± 1 µm; (**b**) Top view photograph of a 500 ms exposure (without compensation) trench designed 4 pixels (30 µm) wide, which measures 21 ± 0.5 µm; (**c**) SEM images of 1000 ms exposure interior ridges designed 5 and 6 layers tall. The ridges measured 46 ± 1 and 55 ± 1 µm tall, respectively; a support pillar is in the middle of the image; (**d**) SEM image of interior trenches at 1000 ms exposure without compensation designed 5 and 6 pixels (38 and 46 µm) wide, which measured 22 ± 0.7 and 34 ± 2 µm wide; (**e**) SEM image of interior pillar structures at 1500 ms exposure designed to be 5–7 pixels (38–53 µm) in diameter.

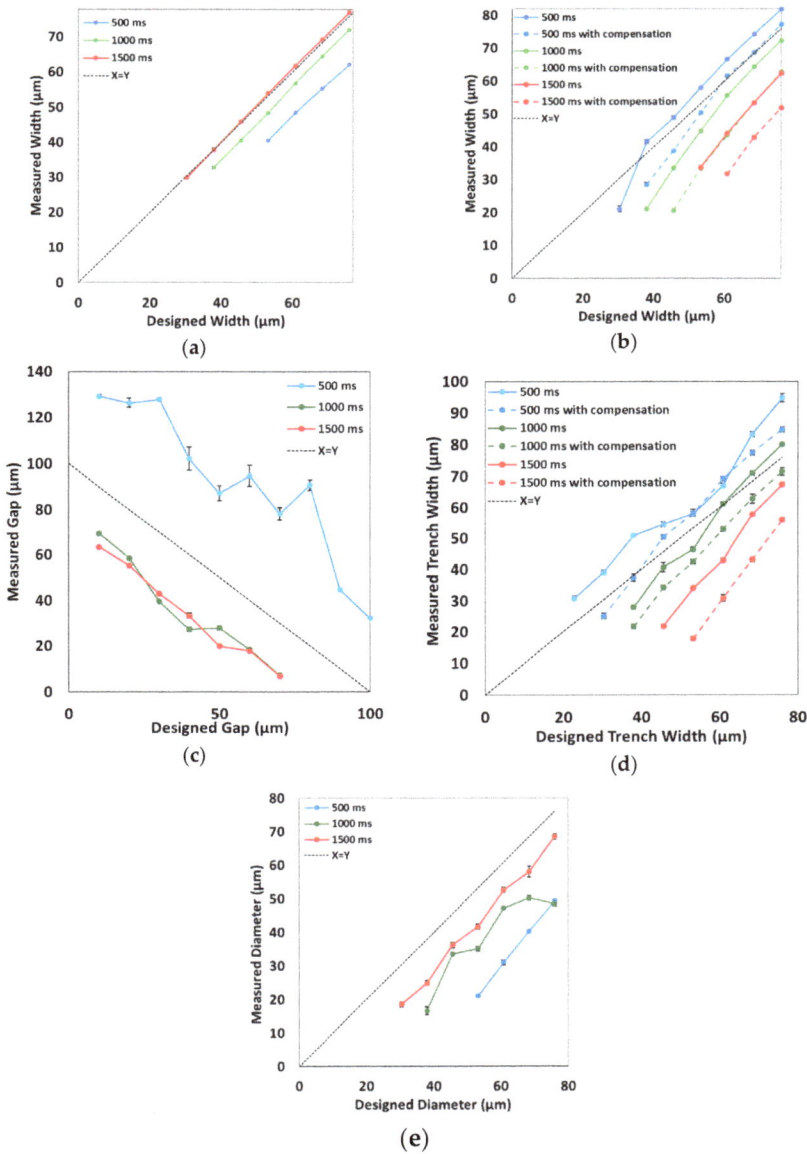

Figure 3. Designed feature width versus measured width for different exposure conditions. Only features that reached >90 μm in height (ridges) or depth (trenches) were included. For the pillars, only those that were attached floor to ceiling were included. Standard deviations are indicated for each included point (*n* = 3). (**a**) Exterior ridges; (**b**) Exterior trenches; (**c**) Interior ridges; (**d**) Interior trenches; (**e**) Interior pillars.

3.1.2. Trenches

For the exterior negative features, we created the design shown in Figure 1b. This design was printed with and without the exposure compensation pattern at 500, 1000, and 1500 ms for a total of six prints. An example microscope image showing a trench 4 pixels wide with 500 ms exposure can be seen in Figure 2b with the full data in Figures S4–S9. The heights and depths of these trenches were measured with optical profilometry. As seen in Figure 3b, measured width was plotted against designed width; only the features that achieved >90% of the designed feature depth were included. The first observation is that the compensation pattern caused the trenches to turn out narrower than the uncompensated devices due to the additional exposure at the trench edge. The second effect that can be seen is that increasing the layer exposure time for the uncompensated case results in narrower trenches, as expected. Finally we note that, to achieve a minimum trench width at full depth, there are three different possibilities: 500 ms exposure without compensation, 1000 ms exposure without compensation, or 1000 ms exposure with compensation. All three of these approaches produced a trench 100 μm deep and about 20 ± 0.5 μm wide; however, the designed widths were all different (4, 5, and 6 pixels, respectively). These results show that having precise control over both exposure and printing parameters allows for careful control of final feature sizes for exterior trenches.

3.2. Interior Features

For interior feature resolution, we undertook a similar study of positive and negative features in a confined space according to our design in Figure 1c. When prototyping this design, we found that a large void space without a support for the ceiling resulted in irregular top layers and erratic feature measurements. Thus, an alternating pattern of feature and ceiling support pieces exists in each of the interior feature areas. Each print contains two sets of trenches, either with or without exposure compensation, such that both could be tested in a single print. Additionally, Figure 2c–e shows characteristic SLA printing artifacts including uneven sides showing where each layer of the print sits on the next one, as well as a somewhat trapezoidal shape for the interior ridges and ceiling support pieces. Smaller layer thicknesses (finer z resolution) could help mitigate the layering, while the trapezoidal shape may be a result of material shrinkage.

3.2.1. Ridges

We again used ridges as positive features to determine the types of structures that could be placed in an interior void in a 3D printed part. Five pixels width was chosen, because it formed reliably for surface ridges. An example SEM image of two ridges at 1000 ms exposure is shown in Figure 2c; the interior ridge 5 layers tall is shown on the left, with the ceiling support in the middle, and the ridge 6 layers tall is on the right. For interior ridges, we measured the gap between the top of the ridge and the ceiling of the void area, which gave information about interior z resolution. Gap distance (excluding any ridges that attached to the ceiling) is plotted for these interior ridges as shown in Figure 3c. The 500 ms exposure time resulted in a void region that was taller than the designed 100 μm, due to insufficient adhesion between layers, as can be observed in the Supplementary Materials (Figure S10). The ceiling support pieces appear to have broken off, resulting in voids taller than the designed size. For 1000 and 1500 ms, the void height is smaller than designed, likely due to exposure of the top layer of the chamber polymerizing significantly more than to 10 μm of resin, thereby making the first ceiling layer substantially thicker than designed, resulting in reduced overall chamber height and hence reduced gap size (Figures S11–S12). This is consistent with our previous work, in which we analyzed the layer exposure profile as a function of z [20]. The height of the void area was about 70–75 μm instead of 100 μm, and thus any ridges designed to be >7 layers tall were attached to the ceiling. For the ridges that were not attached, however, there was a linear relationship between the designed and measured gap distance. From this data, gaps between the feature and ceiling area as small as 7 ± 0.7 μm can be produced with either 1000 or 1500 ms exposures. As long as the print receives sufficient light exposure (>500 ms for this formulation), the gap height between features and ridges is independent of the exposure time for 1000 and 1500 ms exposures, as seen in Figure 3c.

3.2.2. Trenches

For the interior trenches, a similar approach to the exterior trenches was used. The widths of printed trenches were measured with SEM imaging, and an example image is shown in Figure 2d for 1000 ms exposure (without compensation) of trenches that were designed to be 5 and 6 pixels (38 and 46 μm) wide. In this image the trench 6 pixels wide is on the left, and the trench 5 pixels wide is on the right, with the ceiling support pillar in the middle. The full set of interior trench images can been found in the Supplementary Materials Figures S13–S18. The measured width was compared to the designed width in Figure 3d, including only those trenches that reached >90% of full depth. The 500 ms trenches (both with and without compensation) turn out wider than the 1500 ms trenches, indicating that lower exposure times work better for forming wider trenches. Similar to the effect observed with the exterior trenches, the use of a compensation pattern leads to trenches smaller than they would have otherwise printed, as the compensated trenches are all narrower than the uncompensated ones. Similar to the exterior trenches, the smallest trenches that successfully formed were about 20 μm wide, though they were printed with different exposure times: 5 pixels wide for 1000 ms with compensation (21 ± 0.8 μm), 6 pixels wide for 1500 ms without compensation (21 ± 0.7 μm), and 7 pixels wide for 1500 ms with compensation (18 ± 0.7 μm). This again demonstrates the concept that there are several different ways to obtain a minimum feature size if the right exposure conditions are determined, especially when there is full control over the exposure properties.

3.2.3. Pillars

A final type of resolution feature we investigated was cylindrical pillars in the interior of a void area. SEM images were taken of the pillars, and an example can be seen in Figure 2e showing 1500 ms exposure pillars that are 5–7 pixels wide (38–53 μm), with more data in Figures S19–S21 of the Supplementary Materials. Figure 3e shows a plot comparing the designed pillar diameter and the measured diameter, including only those pillars which were fully formed. From this graph, it can be seen that the pillars all printed narrower than their designed width and followed the trend that longer exposure times led to wider pillars. The smallest pillars that were successfully printed were about 14 ± 1 μm in diameter, printed with 1000 ms exposure for a 5 pixel designed width. The dependency of pillar diameter on light exposure is another example of how control over dosing parameters is essential to achieve desired feature dimensions during SLA 3D printing.

3.2.4. Trapping Devices

To demonstrate the utility of tight control of both positive and negative interior features we created trapping devices to catch particles as they flowed through a channel, as a first step toward trapping of cells. The beads selected were 25 μm in diameter, approximately the same size as the smallest trench that was successfully printed. Because our 3D printing designs can be easily edited to optimize configurations and print parameters, we attempted a number of different trap placements (see Figure 1d inserts) to determine the best trapping efficiency. Figure 4 shows images of channels after trapping experiments were carried out. We found that traps exclusively in the center of the channel (Figure 4a) did not trap as efficiently as traps along the sides (Figure 4b), which in turn were not as efficient as traps that alternated between the channel sides and center (Figure 4c). This was largely due to the flow of the beads around (instead of through) the traps. Additionally, we found that traps that were 3 layers tall did not trap as efficiently as traps that were the full height of the channel; when the traps are shorter many beads simply pass over the traps.

Finally, it is critical to have traps that have the correct dimensions. The degree of trap openness could be controlled through the exposure time used for the print. If the exposure time was too short (<600 ms), the traps would be partially formed, and the beads would pass through the trap without getting caught. Conversely, if the exposure time was too high (>1000 ms), the traps would end up overexposed, and there would be no flow through the traps, which would result in no beads being trapped and, often, small bubbles that stuck instead. The exposure time of ~750 ms was nearly ideal for forming traps that worked well for beads. Figure 4 shows this effect, with underexposed in (d), optimal exposure in (e), and overexposed in (f).

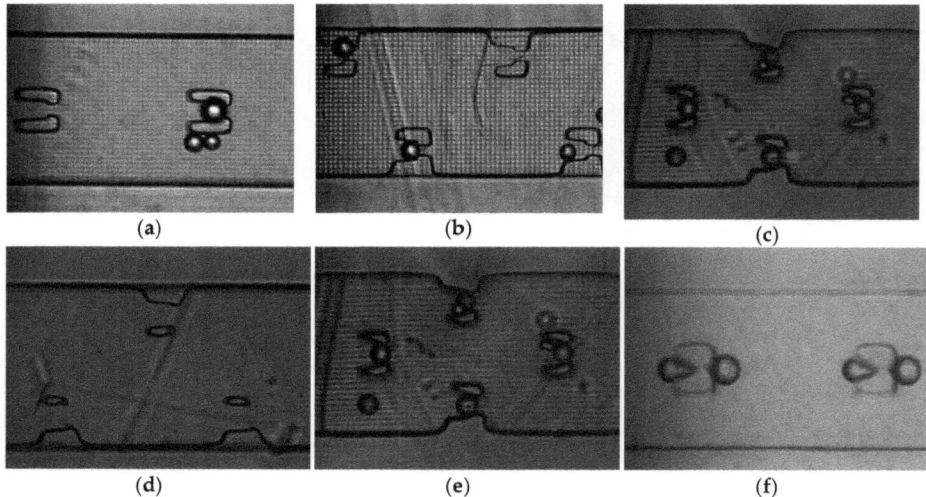

Figure 4. CCD images showing the effect of trap placement and different exposure times on trap shape. (**a**) Channel with traps exclusively in the center; (**b**) Channel with traps staggered along the sides; (**c**) Channel with traps staggered along the sides and in the middle of the channel; (**d**) Prints exposed 500 ms showing partially formed traps with no bead capture; (**e**) Prints exposed for 750 ms with beads trapped well; (**f**) Prints exposed at 1000 ms showing overexposed traps. Bubbles are stuck at the front and back of the traps hindering bead capture.

4. Conclusions

In this work, we have characterized the 3D printing of sub-100 μm external and internal positive and negative resolution features. We have shown the importance of controlling light dosage, as well as the benefits of multiple different exposure patterns within one layer of a print. Finally, we created a particle trapping device and leveraged the rapid iterative design capabilities of 3D printing to improve trap placement and efficiency.

These developments demonstrate the need for careful control of dosing parameters in making complex 3D printed microfluidic devices. More customization of printer control, resin development, and higher resolution projectors should lead to smaller, more intricate 3D printed microfluidic systems. Improved microfluidics of this type could provide even smaller traps, potentially allowing their use in cell capture or isolation experiments.

Supplementary Materials: The following are available online at http://www.mdpi.com/2072-666X/9/7/326/s1: Figure S1: exterior features, ridges, 500 ms exposure; Figure S2: exterior features, ridges, 1000 ms exposure; Figure S3: exterior features, ridges, 1500 ms exposure; Figure S4: exterior features, trenches with edge compensation, 500 ms; Figure S5: exterior features, trenches with edge compensation, 1000 ms; Figure S6: exterior features, trenches with edge compensation, 1500 ms; Figure S7: exterior features, trenches without edge compensation, 500 ms; Figure S8: exterior features, trenches without edge compensation, 1000 ms; Figure S9: exterior features, trenches without edge compensation, 1500 ms; Figure S10: interior features, ridges, 500 ms; Figure S11: interior features, ridges, 1000 ms; Figure S12: interior features, ridges, 1500 ms; Figure S13: interior features, trenches with edge compensation, 500 ms; Figure S14: interior features, trenches with edge compensation, 1000 ms; Figure S15: interior features, trenches with edge compensation, 1500 ms; Figure S16: interior features, trenches without edge compensation, 500 ms; Figure S17: interior features, trenches without edge compensation,

1000 ms; Figure S18: interior features, trenches without edge compensation, 1500 ms; Figure S19: interior features, pillars, 500 ms; Figure S20: interior features, pillars, 1000 ms; Figure S21: interior features, pillars, 1500 ms.

Author Contributions: Conceptualization: A.T.W., G.P.N. and M.J.B.; methodology: A.T.W., G.P.N., H.G. and M.J.B.; software: H.G.; validation: A.T.W. and M.J.B.; formal analysis: A.T.W. and M.J.B.; investigation: M.J.B.; data curation: M.J.B.; writing—original draft preparation: M.J.B.; writing—review & editing: A.T.W., G.P.N. and M.J.B.; visualization: M.J.B.; supervision: A.T.W. and G.P.N.; funding acquisition: A.T.W. and G.P.N.

Funding: This research was funded by The United States National Institutes of Health grant numbers R01EB006124 and R15GM123405-01A1.

Conflicts of Interest: The authors declare no conflict of interest. The founding sponsors had no role in the design of the study; in the collection, analyses, or interpretation of data; in the writing of the manuscript; or in the decision to publish the results.

References

1. Au, A.K.; Huynh, W.; Horowitz, L.F.; Folch, A. 3D-printed microfluidics. *Angew. Chem. Int. Ed.* **2016**, *55*, 3862–3881. [CrossRef] [PubMed]
2. Shallan, A.I.; Smejkal, P.; Corban, M.; Guijt, R.M.; Breadmore, M.C. Cost effective 3D printing of visible transparent microchips within minutes. *Anal. Chem.* **2014**, *86*, 3124–3130. [CrossRef] [PubMed]
3. Plevniak, K.; Campbell, M.; Myers, T.; Hodges, A.; He, M. 3D Printed Auto-Mixing chip enables rapid smartphone analysis of anemia. *Biomicrofluidics* **2016**, *10*, 054113. [CrossRef] [PubMed]
4. Brooks, J.C.; Fort, K.I.; Holder, D.H.; Holtan, M.D.; Easley, C.J. Macro to micro interfacing to microfluidic channels using 3d printed templates: Application to time resolved secretion sampling of endocrine tissue. *Analyst* **2016**, *141*, 5714–5721. [CrossRef] [PubMed]
5. Lee, W.; Kwon, D.; Choi, W.; Jung, G.Y.; Au, A.K.; Folch, A.; Jeon, S. 3D-printed microfluidic device for the detection of pathogenic bacteria using size-based separation in helical channel with trapezoid cross-section. *Sci. Rep.* **2015**, *5*, 7717. [CrossRef] [PubMed]
6. Takenaga, S.; Schneider, B.; Erbay, E.; Biselli, M.; Schnitzler, T.; Schöning, M.J.; Wagner, T. Fabrication of biocompatible lab-on-chip devices for biomedical applications by means of a 3D-printing process. *Phys. Status Solidi A* **2015**, *212*, 1347–1352. [CrossRef]
7. Urrios, A.; Parra-Cabrera, C.; Bhattacharjee, N.; Gonzalez-Suarez, A.M.; Rigat-Brugarolas, L.G.; Nallapatti, U.; Samitier, J.; DeForest, C.A.; Posas, F.; Garcia-Cordero, J.L.; et al. 3D printing of transparent bio-microfluidic devices in PEGDA. *Lab Chip* **2016**, *16*, 2287–2294. [CrossRef] [PubMed]
8. Macdonald, N.P.; Cabot, J.M.; Smejkal, P.; Guit, R.M.; Paull, B.; Breadmore, M.C. Comparing microfluidic performance of three-dimensional (3D) printing platforms. *Anal. Chem.* **2017**, *89*, 3858–3866. [CrossRef] [PubMed]
9. Sochol, R.D.; Sweet, E.; Glick, C.C.; Venkatesh, S.; Avetisyan, A.; Ekman, K.G.; Raulinaitis, A.; Tsai, A.; Wienkers, A.; Korner, K.; et al. 3D printed microfluidic circuitry via multijet based additive manufacturing. *Lab Chip* **2016**, *16*, 668–678. [CrossRef] [PubMed]
10. Lee, K.G.; Park, K.J.; Seok, S.; Shin, S.; Kim, D.H.; Park, J.Y.; Heo, Y.S.; Lee, S.J.; Lee, T.J. 3D printed modules for integrated microfluidic devices. *RSC Adv.* **2014**, *4*, 32876–32880. [CrossRef]
11. Lee, J.M.; Zhang, M.; Yeong, W.Y. Characterization and evaluation of 3D printed microfluidic chip for cell processing. *Microfluid. Nanofluid.* **2016**, *20*, 1–15. [CrossRef]
12. He, Y.; Wu, Y.; Fu, J.; Gao, Q.; Qiu, J. Developments of 3D printing Microfluidics and Applications in Chemistry and Biology: A Review. *Electroanalysis* **2016**, *28*, 1–22. [CrossRef]
13. Yazdi, A.A.; Popma, A.; Wong, W.; Nguyen, T.; Pan, Y.; Xu, J. 3D Printing: An emerging tool for novel microfluidics and lab-on-a-chip applications. *Microfluid. Nanofluid.* **2016**, *20*, 1–18. [CrossRef]
14. Chen, C.; Mehl, B.T.; Munshi, A.S.; Townsend, A.D.; Spence, D.M.; Martin, R.S. 3D-printed microfluidic devices: Fabrication, advantages and limitations-a mini review. *Anal. Methods* **2016**, *8*, 6005–6012. [CrossRef] [PubMed]
15. Waheed, S.; Cabot, J.M.; Macdonald, N.P.; Lewis, T.; Guijt, R.M.; Paull, B.; Breadmore, M.C. 3D printed microfluidic devices: Enablers and barriers. *Lab Chip* **2016**, *16*, 1993–2013. [CrossRef] [PubMed]
16. Beauchamp, M.J.; Nordin, G.P.; Woolley, A.T. Moving from millifluidic to truly microfluidic sub-100 μm cross-section 3D printed devices. *Anal. Bioanal. Chem.* **2017**, *409*, 4311–4318. [CrossRef] [PubMed]

17. Ukita, Y.; Takamura, Y.; Utsumi, Y. Direct digital manufacturing of autonomous centrifugal microfluidic device. *Jpn. J. Appl. Phys.* **2016**, *55*, 06GN02. [CrossRef]
18. Walczak, R.; Adamski, K. Inkjet 3D printing of microfluidic structures-on the selction of the printer towards printing your own microfluidic chips. *J. Micromech. Microeng.* **2015**, *25*, 085013. [CrossRef]
19. Fuad, N.M.; Carve, M.; Kaslin, J.; Wlodkowic, D. Characterization of 3D-printed moulds for soft lithography of millifluidic devices. *Micromachines* **2018**, *9*, 116. [CrossRef]
20. Gong, H.; Beauchamp, M.J.; Perry, S.; Woolley, A.T.; Nordin, G.P. Optical approach to resin formulation for 3D printed microfluidics. *RSC Adv.* **2015**, *5*, 106621–106632. [CrossRef] [PubMed]
21. Gong, H.; Bickham, B.P.; Woolley, A.T.; Nordin, G.P. Custom 3D printer and resin for 18 μm × 20 μm microfluidic flow channels. *Lab Chip* **2017**, *17*, 2899–2909. [CrossRef] [PubMed]
22. Gong, H.; Woolley, A.T.; Nordin, G.P. 3D printed high density, reversible, chip-to-chip microfluidic interconnects. *Lab Chip* **2018**, *4*, 639–647. [CrossRef] [PubMed]

micromachines

MDPI

Review

3D-Printed Biosensor Arrays for Medical Diagnostics

Mohamed Sharafeldin [1,2]**, Abby Jones** [1] **and James F. Rusling** [1,3,4,5,*]

[1] Department of Chemistry (U-3060), University of Connecticut, 55 North Eagleville Road, Storrs, CT 06269, USA; mohamed.sharafeldin@uconn.edu (M.S.); abby.jones@uconn.edu (A.J.)
[2] Analytical Chemistry Department, Faculty of Pharmacy, Zagazig University, Zagazig 44519, Sharkia, Egypt
[3] Institute of Materials Science, University of Connecticut, 97 North Eagleville Road, Storrs, CT 06269, USA
[4] Department of Surgery and Neag Cancer Center, UConn Health, Farmington, CT 06032, USA
[5] School of Chemistry, National University of Ireland, Galway, University Road, Galway, Ireland
* Correspondence: james.rusling@uconn.edu; Tel: +1-860-486-4909

Received: 14 June 2018; Accepted: 2 August 2018; Published: 7 August 2018

Abstract: While the technology is relatively new, low-cost 3D printing has impacted many aspects of human life. 3D printers are being used as manufacturing tools for a wide variety of devices in a spectrum of applications ranging from diagnosis to implants to external prostheses. The ease of use, availability of 3D-design software and low cost has made 3D printing an accessible manufacturing and fabrication tool in many bioanalytical research laboratories. 3D printers can print materials with varying density, optical character, strength and chemical properties that provide the user with a vast array of strategic options. In this review, we focus on applications in biomedical diagnostics and how this revolutionary technique is facilitating the development of low-cost, sensitive, and often geometrically complex tools. 3D printing in the fabrication of microfluidics, supporting equipment, and optical and electronic components of diagnostic devices is presented. Emerging diagnostics systems using 3D bioprinting as a tool to incorporate living cells or biomaterials into 3D printing is also reviewed.

Keywords: 3D printing; diagnostics; optics; bioprinting; electronics; microfluidics

1. Introduction

Charles W. Hull in 1986 was the first to report stereolithography [1] as a tool to fabricate 3D structures. Since then, 3D printing has evolved into a multifunctional fabrication tool that offers unique advantages for biomedical applications including diagnostics [2], scaffolds for 3D implants [3], prosthesis [4] and tissue engineering [5]. In recent years, the ability to convert computer-assisted design (CAD) files into 3D-printed pieces, also known as additive manufacturing, has sparked significant progress in the field of diagnostics [6]. 3D printing has been utilized in a wide spectrum of applications with excellent design and performance. As an additive manufacturing technique, production costs are lower compared to traditional subtractive manufacturing techniques like milling or ablation due to reduction of the labor and material cost. In addition, versatile 3D printers can be used to produce different devices and parts without the need for pre-fabrication changes normally required in subtractive manufacturing techniques [7,8]. These criteria make 3D printing a valuable tool in prototyping, testing and production of tools and equipment for analytical and diagnostic laboratories. In principle, CAD files of previously reported devices can be downloaded and printed in any laboratory. In this way, advanced diagnostic tools can be directly utilized by researchers without the need for purchase from a commercial vendor. This approach has the potential to bring advanced diagnostic tools more rapidly to the research lab than ever before.

In this review, we focus on applications of 3D printing techniques in medical diagnostics. We discuss different 3D printing techniques and how these techniques impact many design aspects including resolution, cost and fabrication of complex diagnostic devices in a continuous process [9–11]. 3D-printed microfluidic devices have been used to fabricate semi and fully automated diagnostic approaches for diseases like cancer [12,13], infectious diseases [14–16], and xenobiotic genotoxicity [17]. 3D printing can also make tailored supporting devices that improve performance of existing diagnostics like spectrophotometers [18] and Polymerase Chain Reaction (PCR) devices [14,19] and is used to assist with smartphone integration for remote sensing [20,21]. The ability to print materials with special properties allows for the creation of new equipment that can dramatically reduce the cost of diagnostic devices like Surface Plasmon Resonance (SPR) [22]. All these applications use 3D printing for cost-effective multifunctional production to integrate several functions in one device [23].

Fabrication of diagnostic devices with embedded electronics and circuits have also been accomplished by 3D printing. The ability to print different materials simultaneously permits the fabrication of electrodes that can be incorporated into the insulator plastic matrices allowing for subsequent electrochemical detection of metals [24–26], organic compounds [27,28] and biologically active molecules [29]. 3D printing avoids disadvantages associated with screen printing like the need for masking and drying steps, while exhibiting better resolution and faster fabrication [30].

3D bioprinting is another emerging modification to traditional 3D printing where cells, enzymes or proteins may be encapsulated or loaded into printable bio-ink solutions [31]. A major focus of this technique is to provide cell growth medium for tissue and organ repair and regeneration, but it has also been explored as a tool for diagnostic applications [32]. Bioprinting offers an opportunity to fabricate 3D-printed implantable sensors that are biocompatible, geometrically complex, and cheap. With 3D printing, there is a limited need for specialized training, and devices can be tailored to the users' requirements [33,34]. In this review, the most common techniques for 3D-printed diagnostics are briefly described with several examples of diagnostic platforms incorporating microfluidics, device supports, optical components, electronics and biomaterials.

2. Additive Manufacturing Techniques

2.1. Fused Deposition Modeling (FDM)

This technique utilizes thermoplastic polymeric materials extruded to print objects layer-by-layer from a heated nozzle onto a surface or platform where it is cooled to below its thermoplastic temperature (Figure 1). Several materials have been utilized in this printing technique, including acrylonitrile butadiene styrene (ABS), polycarbonate (PC), PC-ABS blend, and polylactic acid (PLA) [35]. Single-, double- and triple-print-head machines are available for FDM, making it a good choice for simultaneous multi-material 3D printing [36]. The ability to incorporate conductive materials like pyrolytic graphite, graphene, carbon nanotubes and metal nanoparticles into the thermoplastic matrix enables FDM printing of conductive inks to fabricate electrodes and circuits [37–40]. FDM is good for rapid prototyping and fabrication of holders and supporting devices, but still suffers from several limitations, including mechanical strength, roughness and shape integrity of the final product. Microfluidic devices printed using FDM can show leakage and shape deformation if printing parameters and the thermoplastic polymer are not carefully tuned [37]. FDM has been successfully used to print 3D scaffolds that can be seeded with living cells without loss of cell viability [41,42] and to print bio-friendly polymer materials [43,44].

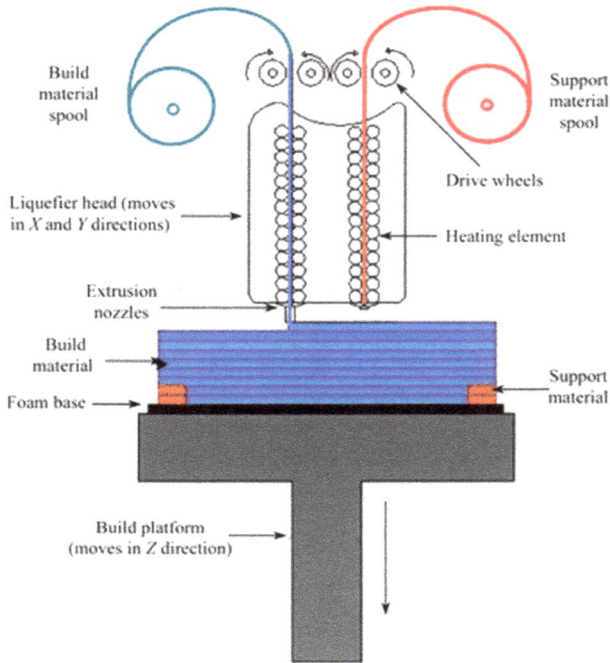

Figure 1. Schematic representation of a dual-head fused deposition modeling 3D printer. Thermoplastic polymer is extruded from a heated nozzle into a printing platform, where it is cooled to below its thermoplastic temperature. Reproduced with permission from [35]. Copyright (2015) Springer.

2.2. Direct Ink Writing

Similar to fused deposition modeling, Direct Ink Writing (DIW) relies on extrusion of ink through a fine deposition nozzle to form a 3D structure in layer-by-layer approach [45]. Two different strategies are utilized in this technique, based on the ink type. First is the extrusion of low-viscosity ink that undergoes gelation via a chemical, photochemical or noncovalent process [46]. The second strategy is the use of a shear thinning hydrogel ink that possesses a viscoelastic response toward applied pressure. Hydrogels like sodium alginate and gelatin are commonly used [47]. In addition to hydrogel inks, epoxy-based direct writing was developed by Compton and Lewis [48], where epoxy ink with significant shear thinning is extruded through the printing nozzle. Once extruded, the ink has sufficient shear to maintain its printed filamentary shape. Direct ink writing is utilized in the 3D fabrication of injectable therapeutics [49], cell-laden scaffolds [50,51], degradable biomaterials [52] and stretchable complex cellularized structures [53]. The utilization of DIW in bioprinting offers a tool to develop multifunctional diagnostic devices with high resolution, which may improve assay sensitivities.

2.3. Stereolithography

Stereolithography, or digital light processing, employs a photocurable polymeric resin which, when exposed to light, cures into a solid. Initially, curing was only possible with UV light, but polymers cured with visible wavelengths have recently been introduced. Highly focused lasers or LED beams with high intensity are used and the spot size of the light beam determines printing resolution [54]. Each layer of the object is printed as a point-by-point 2D cross section cured by the scanning focused beam onto a printing platform immersed in a photocurable tank that holds the liquid resin [5]

(Figure 2A). Recently, projection-based stereolithography has been introduced with a promise to decrease print time while maintaining almost the same resolution as line-based stereolithography. Projection-based stereolithography replaces point-by-point curing with entire-layer curing under one single UV or visible light exposure [55,56] (Figure 2B). Stereolithography resin materials have been extensively studied to produce devices with different properties, including transparency, color, flexibility and thermal stability [57]. Stereolithography has also been used for printing cells using biocompatible resins maintaining >90% cell viability after printing [58].

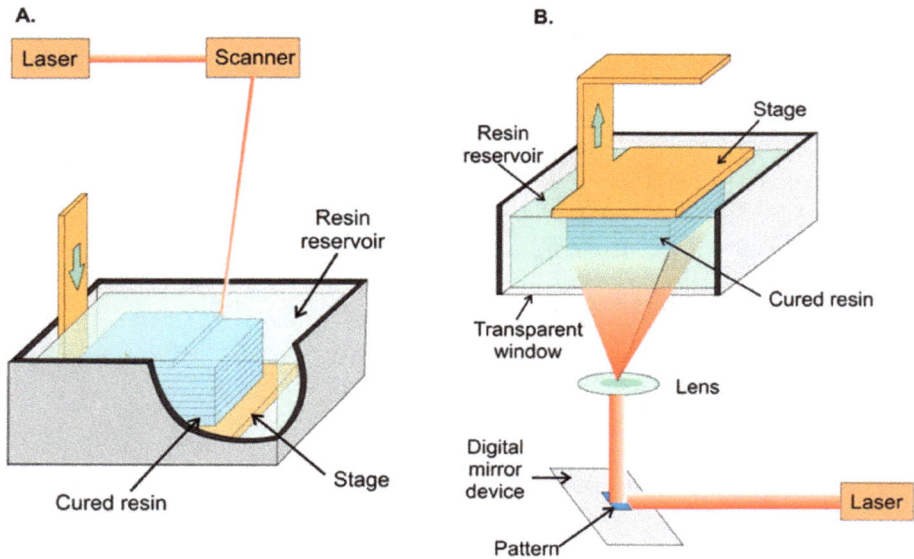

Figure 2. Schematic representation of stereolithographic 3D printing. (**A**) Scanning laser stereolithography, where the focused laser beam scans point-by-point to cure a layer of resin on top of a previously fabricated layer. (**B**) Projection-based stereolithography, where an entire layer is printed in a single step by projecting the entire layer on top of the previous layer. In both strategies, a printing platform is immersed in a tank filled with liquid photocurable resin. Reproduced with permission from [54]. Copyright (2014) American Chemical Society.

2.4. Photopolymer Inkjet Printing (Multi-Jet Modeling—MJM)

This technique utilizes multi-head printers with print heads similar to inkjet printers. The print head extrudes layers of photocurable resin or molten wax, usually with a second head printing support material to maintain the shape of the design until cured (Figure 3). After printing, the object is cured by UV irradiation or heat and support material can be removed by heating or dissolving in a specific solvent [59]. Researchers have been able to utilize this printing technique to print metal nanoparticles for printed electronics [60], preceramic polymers for 3D-printed ceramics [61] and even metallic electrodes on flexible substrates [62]. The ability to print multiple materials with varying chemical and physical properties simultaneously makes MJM a good candidate for diagnostic device fabrication. Microfluidic channels integrated with either electrodes for electrochemical signal detection [63] or porous membranes that can be seeded with viable cells for drug permeability and toxicity studies have been printed using this technique [64]. Most printing resins and materials are proprietary, which makes the cost of using MJM relatively higher than other 3D-printing techniques [65].

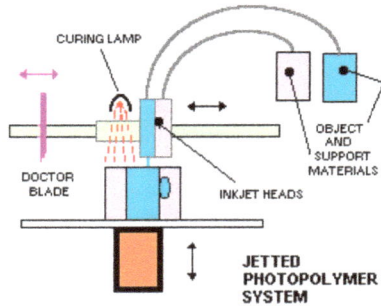

Figure 3. Schematic representation of multi-jet printing technique, a photocurable resin is printed simultaneously with a support material that can be removed after curing. Up to 10 printing heads can be used. Reproduced with permission from [59]. Copyright (2014) American Chemical Society.

2.5. Selective Laser Sintering (SLS)

A focused Infra-Red (IR) laser beam supplies enough localized energy to sinter a fine powdered polymer into layers of solid. The IR laser scans through the surface of powder in the shape of each layer of the sliced 3D design (Figure 4). Due to the high energy required to sinter powders, high-energy CO2/Nd:YAG laser sources are typically used [66]. SLS can be divided into two distinctive subcategories based on the printing temperature. The first is solid-state sintering, where binding occurs at a temperature lower than the melting temperature and is usually used with polymers like polycarbonate. The second is full melting SLS and is used for metals and ceramics where sintering requires a high temperature above the melting temperature [67]. Printing resolution is affected by powder particle size and can be controlled by the scan speed and intensity of the laser beam, which also affects the density and strength of the printed parts [68]. SLS has utilized several printing substrates, including natural and synthesized polymers like cellulose and polycarbonate, making it compatible with bioprinting for tissue engineering and cartilage repair [69]. Other printing substrates include metals, ceramics and polymer/ceramic composites. It is important to note that the printing resolution of polymers is much lower compared to that of metals or ceramics [70]. Due to the high-energy laser source required and the substrate specifications, SLS is currently considered to be the most expensive 3D printing technique [71]. Recently, Formlabs introduced Fuse1, a desktop SLS printer that provides end-users with a more affordable option [72].

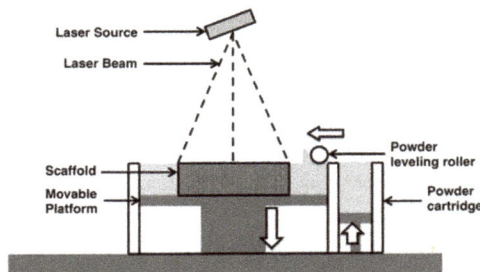

Figure 4. Schematic representation of Selective Laser Sintering, a rolling ball pushes powdered substrate to the surface of the printing platform. A high-energy focused laser beam scans the surface where it sinters the powder particles into a solid layer. The printing chamber is sealed under vacuum or inert gas atmosphere. Reproduced with permission from [70]. Copyright (2015) Springer Nature available under Creative Commons Attribution.

2.6. Direct Laser Writing (DLW) 3D Lithography

Direct laser writing is an emerging technique for 3D printing of high-resolution structures utilizing a highly pulsed femtosecond laser beam to cure a photosensitive resin material [73]. Two-photon absorption and polymerization facilitates fast fabrication of 3D scaffolds with high resolution [74]. This short pulsed laser is suitable for encapsulating living cells and biomaterials in 3D structures as it does not generate localized overheating or UV toxicity. Due to versatility in substrate materials and the ability to print high-resolution 3D scaffolds, DLW has been used in piezoelectric scaffolds for in vitro cell stimulation [74], cartilage tissue engineering [75] and cells and whole organisms containing 3D structures [76–78]. Due to the high resolution achieved by DLW, it has been utilized in fabrication of microvalve assembly [79], custom microstructures [80,81] and complex microfluidic constructs [82,83].

2.7. Summary of 3D Printing Techniques

In order to select the appropriate 3D printing technique, the user must have in mind the properties required for the printed piece. Several criteria, such as flexibility, resolution, complexity, transparency, thermal and chemical stability, are crucial in determining the best technique. Table 1 summarizes the 3D printing techniques discussed in this section.

Table 1. Summary of 3D printing techniques. FDM: fused deposition modeling, PLA: polylactic acid, ABS: acrylonitrile butadiene styrene, PC: polycarbonate, DIW: direct ink writing, SLA: stereolithography, MJM: multijet modeling, SLS: selective laser sintering, DLW: direct laser writing.

3D Printing Technique	Principle	Materials	Pros	Cons	Commercially Available Printers
FDM [35,36]	Filament Extrusion	PLA, ABS, PC, Acrylates	Inexpensive, Fast, Multiple Materials	Low resolution, Roughness, Leakage	Makerbot, Ultimaker Prusa
DIW [46–48]	Ink Extrusion	Alginates, Gelatin, Hyaluronates, Epoxy resin	Biocompatible, High Resolution	Extensive Optimization Required	3D-Bioplotter, BioAssemblyBot
SLA [54,55]	Light Assisted Polymerization	Acrylates	Good Resolution, Flexibility	Single Material	Form2, FabPro, Nobel
MJM [59,64,65]	Printable Photocurable Resin on Support	Multiple Materials	Multiple Materials, High Resolution	Expensive	ProJet, Multijet
SLS [67,68]	IR Beam to Sinter Powdered Polymer	Ceramics, Metals, Polymers	Good Resolution, Variety of Substartes	Expensive, Special handling	Fuse 1, Sintratec
DLW [73,74]	Two-Photonn Absorption and Polymerization	Polymeric Resin	Exceptional Resolution, Biocompatible	Bulky Instrument	Femtowriter, Tungsten-LAM

3. Applications of 3D Printing in Diagnostics

3D printing has improved biomedical diagnostics in many ways, specifically with advantages in ease of onsite design and fabrication, providing researchers with the means to develop or modify devices and equipment. Here we concentrated on the main areas in which biomedical diagnostic research has been focused recently.

3.1. 3D-Printed Microfluidics

The most representative use of 3D printing technology in diagnostics is the design and development of microfluidic devices. The ability to fine-tune geometrically complex structures at the micrometer level is an attractive feature 3D printing can offer while maintaining low-cost and time-efficient processing. Several applications that have used 3D-printed microfluidic devices are discussed.

3.1.1. Sample Pretreatment

Sample pretreatment is an essential step in many diagnostics, as it helps reduce the complexity of the matrix and improve the sensitivity of the assay. 3D-printed microfluidics facilitates the integration of sample pretreatment compartments into real applications including sample injection valves [84,85], preconcentration [86] and sample reactors [87]. Rafeie et al. utilized 3D printing to fabricate an ultrafast microfluidic blood plasma separator, an essential sample pretreatment step in most assays requiring blood samples. They were able to fabricate a spiral microfluidic device (Figure 5A) where cells would flow close to the inner wall of the channel and concentrate in a narrow band near the outlet allowing the separation of cell/platelet free plasma. [88]. Lee et al. separated pathogenic bacteria, *E. coli*, from milk using a 3D-printed helical channel [89]. They flowed magnetic nanoclusters through the helical microfluidic channel (Figure 5B) where free magnetic nanoclusters were separated from bacteria-bound clusters. Yan et al. proposed a portable hand operated microfluidic device that can specifically separate platelets from peripheral blood mononuclear cells [90]. Their device is composed of a microfluidic channel equipped with a groove (Figure 5C) that effectively sorts platelets from blood samples with 100% purity where the user pumps the fluid manually with a hand-held syringe. While fluctuation in the flow rate did not affect the platelet purity, the percent recovery of blood mononuclear cells varied. A microfluidic pre-concentrator for detection of *E. coli* was also proposed by Park et al. [91]. Magnetic nanoparticles labeled with *E. coli*-specific antibodies were allowed to capture bacteria from blood samples. The microfluidic device was equipped with a magnet to separate (Figure 5D) magnetic nanoparticles from the blood matrix which then transferred with buffer for adenosine triphosphate (ATP) luminescence analysis. Although these devices are interesting applications for 3D printing in sample pretreatment, they still require a manual transfer of the treated samples for detection. This manual transfer can negatively affect the assays sensitivity and reproducibility required for a good diagnostic approach.

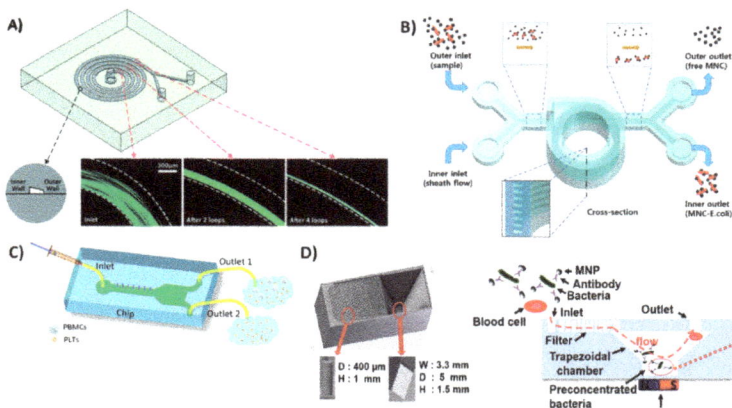

Figure 5. 3D-printed devices for sample pretreatment. (**A**) Spiral microfluidic device to separate blood cells and platelets from plasma, as the cells and platelets tend to flow in a narrowing band near the inner wall of the spiral channel. Reproduced with permission from [88]. Copyright (2016) Royal Society of Chemistry. (**B**) Helical microfluidic device to separate magnetic nanoclusters coupled to *E. coli* from free magnetic nanoclusters. Reproduced with permission from [89]. Copyright (2015) Springer Nature, available under Creative Commons Attribution. (**C**) A hand-driven microfluidic channel with a groove-like structure to separate platelets from blood mononuclear cells. Reproduced with permission from [90] Copyright (2018) Springer Nature. (**D**) Trapezoidal filter equipped with a microfluidic channel for the preconcentration of *E. coli* captured on magnetic beads. Reproduced with permission from [91] Copyright (2017) Elsevier.

3.1.2. Microfluidic Flow Devices

Microfluidic devices offer the most promising approach for miniature fluidic devices due to their ability to handle small sample volumes and assay reagents in a controlled manner. 3D printing has pushed prototyping and development of microfluidics forward by supporting fast and easy design with lower production costs compared to traditional microfabrication techniques. 3D printing also offers an efficient tool to generate geometrically complex microfluidic devices with the aid of 3D design software, thus eliminating the hassle associated with traditional manufacturing tools. Utilizing these advantages, Oh et al. designed and fabricated a 3D-printed blood viscosity analysis capillary circuit [92]. They designed a hand-held device that can be operated and read manually that measures blood viscosity using the same principle as commercial viscometers which are very expensive and complex (Figure 6A). Surprisingly, their device did not utilize the resolution advances of 3D printing, but instead they added Tygon tubing, with inner diameter of 0.508 mm, to build a capillary circuit inside a 3D-printed channel. Santangelo et al. proposed a highly sensitive 3D-printed continuous-flow microfluidic device for quantification of adenosine triphosphate (ATP) molecules (Figure 6B). The device comprised two main functions: mixing of the ATP sample with the luminescence reagent mixture (Luciferin/Luciferase mixture) and a detection chamber that brings the produced luminescence close to a silicon photomultiplier detector [93]. Tang et al. utilized 3D printing to fabricate a unibody ELISA-inspired chemiluminescence assay to detect and quantify prostate specific antigen (PSA) and platelet factor-4 (PF-4) as cancer biomarker proteins (Figure 6C) [94]. They proposed a design that can reduce the assay time to 30 min while approaching an ultra-low sensitivity. Their design is divided into three connected compartments: first, a mixing chamber to accelerate the interaction between reagents; second, a compartment of sample and reagent reservoirs; and third, a transparent detection compartment. The ability to 3D print transparent objects allowed them to directly detect the chemiluminescent signal in their device using a CCD camera without the need for complex processing. Recently, a Lego-like modular microfluidic capillary-driven 3D-printed flow device was introduced by Nie et al. [95]. This approach proposed a strategy to build microfluidic devices tailored to different applications. Flow in such devices is driven by capillary forces, with improved flow rate programmability and biocompatibility. They were able to design different modules assembled in various designs and utilized them in diverse applications, like degradable bone scaffolds and cell culture. Kadimisetty et al. proposed a 3D-printed microfluidic unit that manually controls the flow of sample and assay reagents for electrochemiluminescent detection of PSA, PF-4 and prostate specific membrane antigen (PSMA) in human serum [12]. The printed device had a slot to incorporate a screen-printed carbon electrode labeled with detection antibodies for each of the selected protein biomarkers (Figure 6D). Similar 3D-printed microfluidic flow devices were fabricated and utilized for flow chemical analysis [96], evaluation of blood components [97], electrochemiluminescence DNA studies [98] and salivary cortisol detection [99]. In these discussed examples, 3D printing was the key for better diagnostic performance by providing low-cost incorporation of multiple fluidic functions easily and without the need for laborious manufacturing procedures.

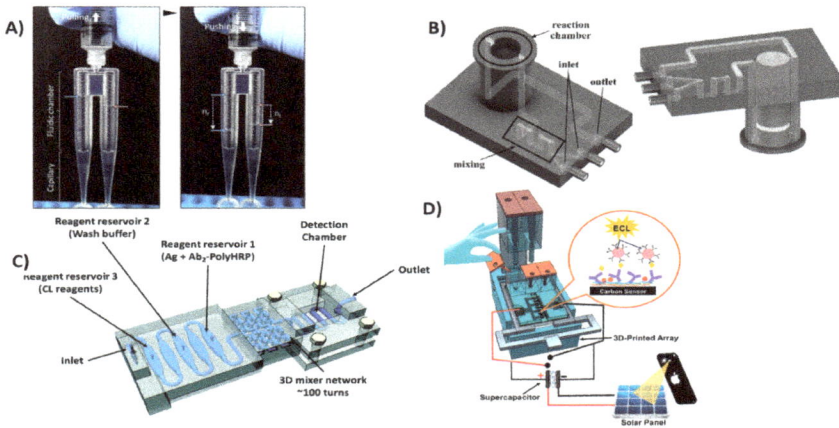

Figure 6. 3D-printed microfluidic devices for flow control. (**A**) Viscometer like 3D-printed syringe attachment for blood viscosity measurement. Reproduced with permission from [92]. Copyright (2018) Elsevier. (**B**) Mixing and detection microfluidic device for luminescent detection of ATP. Reproduced with permission from [93]. Copyright (2018) Elsevier. (**C**) Unibody 3D-printed microfluidic chip for detection of PSA and PF-4. Reproduced with permission from [94]. Copyright (2017) Royal Society of Chemistry. (**D**) Manually controlled flow regulatory system for electrochemiluminescence detection of PSA, PSMA and PF-4. Reproduced with permission from [12]. Copyright (2016) Elsevier.

3.1.3. Microfluidic Mixers

Efficient mixing can be used to improve diagnostic tests by enhancing interaction kinetics between reactants. Microfluidics can be configured for efficient mixing to enhance chaotic convection in solutions, increasing the frequency of interactions between solution components [100]. 3D-printed microfluidic mixers have been successfully used to improve passive mixing enhancing mixing efficiency that improve diagnostics sensitivity [101]. Devices equipped with 3D-printed mixers have been used in amperometric quantitation of hydrogen peroxide [102] and DNA assembly [103]. 3D-printed mixers have been successfully integrated with optical spectroscopic probes including UV/Vis, infrared and fluorescence probes [104]. Plevniak et al. proposed a 3D-printed microfluidic mixer for diagnosis of anemia (Figure 7A). In their work, they were able to integrate the device with smartphone-aided colorimetric signal detection to overcome the distance barrier for efficient screening [105]. The device can analyze a finger prick of blood (~5 μL) driven by capillary force into the mixing chamber where it is mixed with an oxidizing agent in less than 1 sec with cost 50 cents/chip. Another mixing device was introduced by Mattio et al. [106], where a complex valve design was fabricated using 3D printing (Figure 7B). The device has eight inlets for sample and reagents connected to a valve where samples and reagents are mixed and transferred to detector. The device was used to quantify Lead and Cadmium in water samples extracted from soil. One inlet was used for nitric acid required for column conditioning, other inlets were used for fluorescence reagent, Rhod-5N™, and co-reagents potassium iodide, N,N,N′,N′-tetrakis-(2-Pyridylmethyl)ethylenediamine (TPEN) and ammonium oxalate.

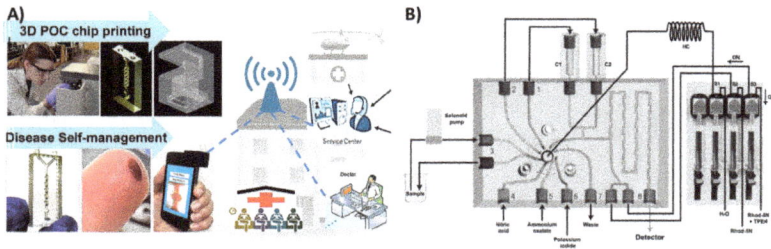

Figure 7. 3D-printed microfluidic mixers. (**A**) Microfluidic mixer for tele-diagnosis of anemia. Less than one second of mixing required for the blood sample with the oxidizing agent; generated colorimetric signal detected with a smartphone. Reproduced with permission from [105]. Copyright (2016) AIP Publishing. (**B**) A lab on valve complex 3D-printed microfluidic chip for quantification of lead and cadmium in water samples. Reproduced with permission from [106]. Copyright (2018) Elsevier.

3.1.4. Multifunctional Microfluidics

In the previous examples, 3D printing was utilized to fabricate microfluidics that served only one purpose. Several researchers have proposed multifunctional 3D-printed microfluidic devices capable of performing several tasks simultaneously. Kadimisetty et al. introduced a microfluidic device that can analyze extracts from e-cigarette vapors [17]. The device is equipped with sample and reagent reservoirs, in addition to an electrochemiluminescence signal detection compartment (Figure 8A). Another multifunctional microfluidic device was also introduced recently by Kadimisetty et al. [9], where they were able to extract, concentrate and isothermally amplify nucleic acids in different bodily fluids as an approach for microfluidic point of care diagnostics (Figure 8B). The microfluidic device is integrated with a membrane to isolate nucleic acids, then placed in a chamber where loop mediated isothermal amplification is induced. Finally, the signal is either detected colorimetrically by a mobile phone or by fluorescence with a portable USB fluorescence microscope. This demonstrates the promising utility of 3D-printed microfluidic devices in Point-of-care (POC) applications. Other multifunctional microfluidic devices have been proposed to detect Human immunodeficiency virus (HIV) antibodies [107], zika virus [108] and glucose [109].

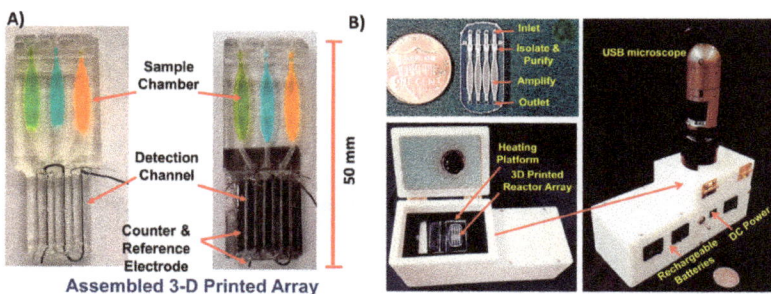

Figure 8. Multifunctional 3D-printed microfluidics. (**A**) A 3D-printed chip to detect genotoxicity of metabolites from e-cigarette extracts. The device has a sample and reagent reservoir compartment and a detection compartment equipped with platinum counter electrode and Ag/AgCl reference electrode. Reproduced with permission from [17]. Copyright (2017) American Chemical Society. (**B**) A 3D-printed microfluidic array for isolation of nucleic acids equipped with a separation membrane and heating compartment to amplify nucleic acids using loop mediated isothermal amplification that can be attached to a USB microscope for fluorescence detection. Reproduced with permission from [9]. Copyright (2018) Elsevier.

3.2. 3D-Printed Sensing Electronics

A number of researchers, especially those employing electrochemistry, are interested in 3D printing for its ability to design and fabricate sensing electronics. Supported by the versatility of printable materials, 3D printers have the ability to produce well defined shapes without masking required in traditional screen printing or photolithography This enables 3D printing to fabricate integrated electrode biosensors and electronic sensors that can be utilized as personal diagnostics devices and POC sensors. Li et al. used a home-made 3D printer to print a conductive polymer in polydimethylsiloxane (PDMS) or EcoflexTM to fabricate stretchable electrode sensors [110] (Figure 9A). Using this 3D printer, they achieved a resolution of 400 μm with an electrode height of 1 mm and detected sodium chloride electrochemically with a 1 μM detection limit and good sensitivity and reproducibility. Another approach for 3D printing electrodes using fused deposition modeling was proposed by Palenzuela et al. [111]. A commercially available graphene/polylactic acid filament was used to print electrodes of distinctive shapes designed on CAD software (Figure 9B). The printed electrodes were characterized using different redox probes and utilized to detect picric acid and ascorbic acid in solution. In order to fabricate more complex electronics, Leigh et al. used a triple-head fused deposition modeling printer to impede conductive filament within a nonconductive ABS or PLA matrix [38]. Using this approach, they were able to fabricate a variety of complex functional objects like a 3D flex sensor, capacitive buttons and a smart vessel (Figure 9C). The ability of 3D printing to develop electronic biosensing devices was demonstrated in the fabrication of strain sensors in biological systems [112,113] and skin-like sensors using thermo-responsive hydrogels [114]. Most of these applications are directed towards the fabrication of electronic skin, a promising diagnostic tool that composed of flexible and stretchable sensor that can perform several health monitoring functions like temperature, glucose, sodium chloride and pressure sensing [115].

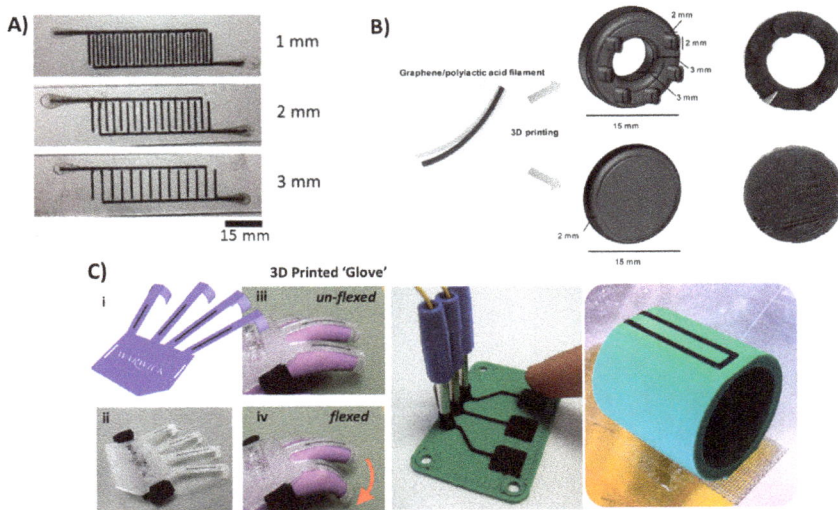

Figure 9. 3D-printed electronics. (**A**) 3D-printed tactile electrode sensor. Conductive PDMS doped with carbon nanotubes were printed on PDMS or EcoflexTM to fabricate flexible electrode sensors. Reproduced with permission from [110]. Copyright (2018) IOP Publishing. (**B**) A 3D-printed graphene/polylactic acid electrode with ring or disc shape. Reproduced with permission from [111]. Copyright (2018) American Chemical Society. (**C**) A 3D-printed conductive carbon black electrode in different objects from left to right: flexible glove sensor, capacitive buttons and smart vessel. Reproduced from [38]. Copyright (2012) PLOS available under Creative Commons Attribution.

3.3. 3D-Printed Supporting Devices

Versatility, ease of design and modification in a fast and economic manner made 3D printing the method of choice to develop the supporting equipment and pieces required for diagnostics. Shanmugam et al. used 3D printing to fabricate a custom designed mobile phone microscopy support unit. This unit perfectly aligns the sample compartment with simple optics and a mobile phone camera (Figure 10A) [116]. They also proposed a holder that can incorporate a microfluidic chamber for analyzing flowing samples rather than stationary samples (Figure 10B). Using such equipment, they were able to perform screening of soil-transmitted parasitic worms in resource-limited areas. Another supporting device for a paper-based electrochemical sensor was proposed by Scordo et al. [117]. A reagent-free sensor was proposed to test butyrylcholinesterase activity by detecting thiocholine. A 3D-printed support equipped with a sample application hole was used to provide the supporting strength and insulation required for electric connections (Figure 10C). Several 3D-printed supports for mobile phone-assisted diagnostics have been developed [118–120], in addition to equipment pieces that can lower the cost of current diagnostic strategies [19,121,122] without compromising performance.

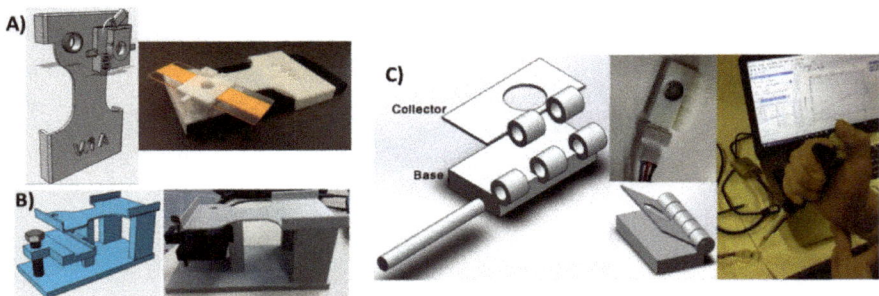

Figure 10. 3D-printed support devices. (**A**) Soil analysis system with 3D-printed mobile phone holder equipped with a glass slide holder where samples were fixed in a lens in between the mobile camera and the sample holder. This jig has a replaceable filter just above the lens for fluorescence imaging. Reproduced from [115]. Copyright (2018) PLOS available under Creative Commons Attribution. (**B**) Alternate soil analysis system with the same support components, but modified to hold a microfluidic chip for flowing samples. Reproduced from [115]. Copyright (2018) PLOS available under Creative Commons Attribution. (**C**) Support device with sample application hole for paper-based electrochemical detection of butyrylcholinesterase activity. Reproduced with permission from [116]. Copyright (2018) Elsevier.

3.4. 3D-Printed Optics

Despite the current limitations of the 3D printing of fully transparent surfaces without defects that could affect light reflection and transmission, researchers have tried to print functional optical components to reduce the cost and improve the performance of diagnostic devices. An interesting example from Hinamn et al. describes a 3D-printed prism that can be used for plasmonic sensing [22] (Figure 11A). In order to prove functionality, they deposited a layer of gold on one side of the prism and used it to detect cholera toxins. They also printed prisms with different geometries and used them to monitor nanoparticle growth (Figure 11B). Other researchers used two-photon polymerization 3D printing to fabricate high-resolution micro-optic components of optic fiber ends [123] and other micro-optics [124]. 3D-printed optical tweezers for sample trapping [125] were also developed to aid chemical and spectroscopic sample analysis. Some other 3D-printed fine optics that have not been used yet in diagnostics have also been proposed [126,127]. Integrating compact, lost-cost, effective optical components is a crucial development for POC diagnostics. These examples are good candidates for integration in POC optical diagnostic systems for medical testing [128].

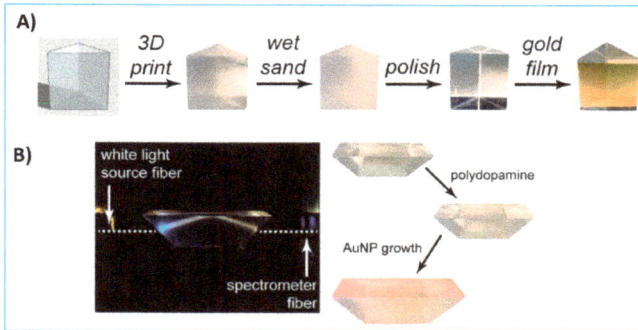

Figure 11. 3D-printed optics. (**A**) 3D-printed prism polished with simple benchtop polishing decorated with a layer of gold and used for plasmonic sensing of cholera toxins. Reproduced with permission from [22]. Copyright (2017) American Chemical Society. (**B**) 3D-printed prism with a different geometry than (A) used to monitor nanoparticle growth. Reproduced with permission from [22]. Copyright (2017) American Chemical Society.

3.5. 3D Bioprinting

The ability to use biocompatible 3D printing substrates allowed the incorporation of biomaterials in 3D-printed scaffolds. This facilitated the further investigation of multifunctional 3D-printed devices that could express biomimetic activity in diagnostic applications. A bioinspired microfluidic chip that can be attached to a whole organ was proposed by Singh et al. [129]. This microfluidic chip was fabricated based on structured light scanning of a whole organ followed by stereolithographic 3D printing using the scanned conformation. The as-printed device was attached to porcine kidney for biomarker extraction and profiling (Figure 12A) without the need for tissue removal. This approach enables the study of metabolic activities in a living whole organ, paving the way for further investigation into drug toxicity screening and biomarker discovery.

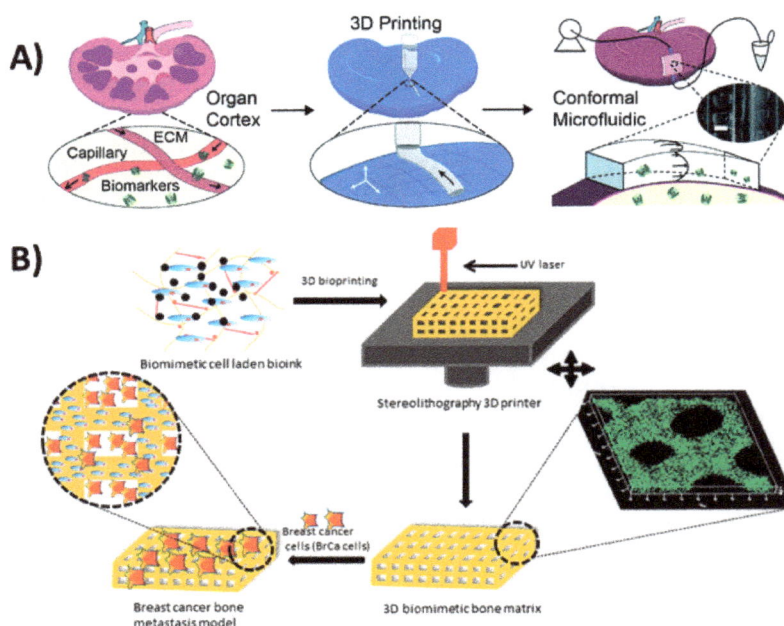

Figure 12. 3D bioprinting. (**A**) 3D-printed perfusion chip for extraction of metabolites and biomarkers from whole organs. Reproduced with permission from [129]. Copyright (2017) Royal Society of Chemistry (**B**) 3D-printed bone-like scaffold carrying bone stromal cells to study their interactions with breast cancer cells. Reproduced with permission from [34]. Copyright (2016) American Chemical Society.

A cell-laden bone matrix was proposed by Zhou et al. [34] to study breast cancer metastasis. They printed a gelatin-based methacrylate hydrogel with incorporated bone stromal cells to study their interactions with breast cancer cells (Figure 12B). An in vivo alkaline phosphatase testing platform was introduced by Park et al. [32] using 3D-printed biocompatible calcium-deficient hydroxyapatite. Although 3D bioprinting developments are mainly used in tissue constructs for therapeutics [130], they offer great advancement opportunities in the field of diagnostics. These developments include the immobilization of aptamers on silicon nitride surfaces [131], functionalizing gold electrodes with bacterial reaction centers [132] and embedding bacteria in 3D constructs [133], all which can be useful in diagnostic applications.

4. Conclusions and Outlook

Evolving applications and developments suggest that 3D printing will be a major player in fabricating readily available, cheap, miniaturized, multifunctional and sensitive diagnostic devices. Researchers from different backgrounds have developed diagnostic assays using this versatile technology. 3D printing has been used as a tool for device prototyping and development with photolithography most commonly used because of the availability of materials exhibiting different properties and high resolution. However, its applications now go well beyond prototyping into real-world device fabrication technology. That is, the fully optimized device becomes the final diagnostic tool to be used in hospitals and clinics. In addition, 3D printing is pushing biomedical diagnostic research towards multifunctional devices that can perform several functions, like protein and metabolite extraction and detection using optical and electrochemical signal detection.

Micromachines **2018**, *9*, 394

3D printing technology still needs improvement in order to enhance current diagnostic abilities. First, simultaneous printing of multiple materials with high resolution and good compatibility is essential, especially for functional materials like conductive inks and biomimetic substrates. Printing multiple materials with different physical properties would greatly improve the capabilities of 3D printers to produce more complex functional architectures. The ability to print active biomaterials like enzymes and proteins in 3D formats without compromising their basic activity is also an important requirement for better diagnostic devices.

Given the progressive nature of 3D printing, more complex microfluidic architectures can be expected in the near future. Recent research has focused on the development of microfluidic pumps [134], automated flow control valves [135], atomic force microscopes [136] and sophisticated scanning electron microscope sample holders [137]. These are examples of very complex architectures that cannot be readily approached in the averaged bioanalytical laboratory without 3D printers. This illustrates again the significance of incorporating 3D printing in bioanalytical and diagnostic testing research providing a platform for achieving what was believed to be imaginary in the pre-3D printing era.

Funding: The authors thank the University of Connecticut for an Academic Plan Project Grant and NIH and for Grants No. ES03154 from the National institute of Environmental Health Sciences (NIEHS), and No. EB016707 from the National Institute of Biomedical Imaging and Bioengineering (NIBIB), for financial support in preparing this article.

Conflicts of Interest: The authors declare no conflict of interest.

References

1. Hull, C. Apparatus for Production of Three-Dimensional Objects by Stereolithography. U.S. Patent No 4,575,330, 11 March 1986.
2. Gross, B.; Lockwood, S.Y.; Spence, D.M. Recent Advances in Analytical Chemistry by 3D Printing. *Anal. Chem.* **2017**, *89*, 57–70. [CrossRef] [PubMed]
3. Chia, H.N.; Wu, B.M. Recent advances in 3D printing of biomaterials. *J. Biol. Eng.* **2015**, *9*, 4. [CrossRef] [PubMed]
4. Diment, L.E.; Thompson, M.S.; Bergmann, J.H. Three-dimensional printed upper-limb prostheses lack randomised controlled trials: A systematic review. *Prosthet. Orthot. Int.* **2018**, *42*, 7–13. [CrossRef] [PubMed]
5. Bose, S.; Vahabzadeh, S.; Bandyopadhyay, A. Bone tissue engineering using 3D printing. *Mater. Today* **2013**, *16*, 496–504. [CrossRef]
6. Waheed, S.; Cabot, J.M.; Macdonald, N.P.; Lewis, T.; Guijt, R.M.; Paull, B.; Breadmore, M.C. 3D printed microfluidic devices: Enablers and barriers. *Lab Chip* **2016**, *16*, 1993–2013. [CrossRef] [PubMed]
7. Why 3D Printing Could Be a Manufacturing and Logistics Game Changer. Available online: https://www.manufacturing.net/blog/2013/10/why-3d-printing-could-be-manufacturing-and-logistics-game-changer (accessed on 30 April 2018).
8. Au, A.K.; Lee, W.; Folch, A. Mail-order microfluidics: Evaluation of stereolithography for the production of microfluidic devices. *Lab Chip* **2014**, *14*, 1294–1301. [CrossRef] [PubMed]
9. Kadimisetty, K.; Song, J.; Doto, A.M.; Hwang, Y.; Peng, J.; Mauk, M.G.; Bushman, F.D.; Gross, R.; Jarvis, J.N.; Liu, C. Fully 3D printed integrated reactor array for point-of-care molecular diagnostics. *Biosens. Bioelectron.* **2018**, *109*, 156–163. [CrossRef] [PubMed]
10. Yazdi, A.A.; Popma, A.; Wong, W.; Nguyen, T.; Pan, Y.; Xu, J. 3D printing: An emerging tool for novel microfluidics and lab-on-a-chip applications. *Microfluid. Nanofluid.* **2016**, *20*, 50. [CrossRef]
11. Zhang, Z.; Xu, J.; Hong, B.; Chen, X. The effects of 3D channel geometry on CTC passing pressure—Towards deformability-based cancer cell separation. *Lab Chip* **2014**, *14*, 2576–2584. [CrossRef] [PubMed]
12. Kadimisetty, K.; Mosa, I.M.; Malla, S.; Satterwhite-Warden, J.E.; Kuhns, T.M.; Faria, R.C.; Lee, N.H.; Rusling, J.F. 3D-printed supercapacitor-powered electrochemiluminescent protein immunoarray. *Biosens. Bioelectron.* **2016**, *77*, 188–193. [CrossRef] [PubMed]

13. Damiati, S.; Peacock, M.; Leonhardt, S.; Damiati, L.; Baghdadi, M.A.; Becker, H.; Kodzius, R.; Schuster, B. Embedded Disposable Functionalized Electrochemical Biosensor with a 3D-Printed Flow Cell for Detection of Hepatic Oval Cells (HOCs). *Genes* **2018**, *9*, 89. [CrossRef] [PubMed]

14. Mulberry, G.; White, K.A.; Vaidya, M.; Sugaya, K.; Kim, B.N. 3D printing and milling a real-time PCR device for infectious disease diagnostics. *PLoS ONE* **2017**, *12*, e0179133. [CrossRef] [PubMed]

15. Singh, H.; Shimojima, M.; Shiratori, T.; An, L.V.; Sugamata, M.; Yang, M. Application of 3D Printing Technology in Increasing the Diagnostic Performance of Enzyme-Linked Immunosorbent Assay (ELISA) for Infectious Diseases. *Sensors* **2015**, *15*, 16503–16515. [CrossRef] [PubMed]

16. Chudobova, D.; Cihalova, K.; Skalickova, S.; Zitka, J.; Rodrigo, M.A.M.; Milosavljevic, V.; Hynek, D.; Kopel, P.; Vesely, R.; Adam, V.; et al. 3D-printed chip for detection of methicillin-resistant Staphylococcus aureus labeled with gold nanoparticles. *Electrophoresis* **2015**, *36*, 457–466. [CrossRef] [PubMed]

17. Kadimisetty, K.; Malla, S.; Rusling, J.F. Automated 3-D Printed Arrays to Evaluate Genotoxic Chemistry: E-Cigarettes and Water Samples. *ACS Sens.* **2017**, *2*, 670–678. [CrossRef] [PubMed]

18. Schäfer, M.; Bräuler, V.; Ulber, R. Bio-sensing of metal ions by a novel 3D-printable smartphone spectrometer. *Sens. Actuators B Chem.* **2018**, *255*, 1902–1910. [CrossRef]

19. Mendoza-Gallegos, R.A.; Rios, A.; Garcia-Cordero, J.L. An Affordable and Portable Thermocycler for Real-Time PCR Made of 3D-Printed Parts and Off-the-Shelf Electronics. *Anal. Chem.* **2018**, *90*, 5563–5568. [CrossRef] [PubMed]

20. Wang, L.-J.; Sun, R.; Vasile, T.; Chang, Y.-C.; Li, L. High-Throughput Optical Sensing Immunoassays on Smartphone. *Anal. Chem.* **2016**, *88*, 8302–8308. [CrossRef] [PubMed]

21. Chen, Y.; Fu, Q.; Li, D.; Xie, J.; Ke, D.; Song, Q.; Tang, Y.; Wang, H. A smartphone colorimetric reader integrated with an ambient light sensor and a 3D printed attachment for on-site detection of zearalenone. *Anal. Bioanal. Chem.* **2017**, *409*, 6567–6574. [CrossRef] [PubMed]

22. Hinman, S.S.; McKeating, K.S.; Cheng, Q. Plasmonic Sensing with 3D Printed Optics. *Anal. Chem.* **2017**, *89*, 12626–12630. [CrossRef] [PubMed]

23. Rusling, J.F. Developing Microfluidic Sensing Devices Using 3D Printing. *ACS Sens.* **2018**, *3*, 522–526. [CrossRef] [PubMed]

24. Lee, K.Y.; Ambrosi, A.; Pumera, M. 3D-printed Metal Electrodes for Heavy Metals Detection by Anodic Stripping Voltammetry. *Electroanalysis* **2017**, *29*, 2444–2453. [CrossRef]

25. Rymansaib, Z.; Iravani, P.; Emslie, E.; Medvidović-Kosanović, M.; Sak-Bosnar, M.; Verdejo, R.; Marken, F. All-Polystyrene 3D-Printed Electrochemical Device with Embedded Carbon Nanofiber-Graphite-Polystyrene Composite Conductor. *Electroanalysis* **2016**, *28*, 1517–1523. [CrossRef]

26. Honeychurch, K.C.; Rymansaib, Z.; Iravani, P. Anodic stripping voltammetric determination of zinc at a 3-D printed carbon nanofiber–graphite–polystyrene electrode using a carbon pseudo-reference electrode. *Sens. Actuators B Chem.* **2018**, *267*, 476–482. [CrossRef]

27. Cheng, T.S.; Nasir, M.Z.M.; Ambrosi, A.; Pumera, M. 3D-printed metal electrodes for electrochemical detection of phenols. *Appl. Mater. Today* **2017**, *9*, 212–219. [CrossRef]

28. Tan, C.; Nasir, M.Z.M.; Ambrosi, A.; Pumera, M. 3D printed electrodes for detection of nitroaromatic explosives and nerve agents. *Anal. Chem.* **2017**, *89*, 8995–9001. [CrossRef] [PubMed]

29. Liyarita, B.R.; Ambrosi, A.; Pumera, M. 3D-printed electrodes for sensing of biologically active molecules. *Electroanalysis* **2018**, *30*, 1319–1326. [CrossRef]

30. Lewis, J.A.; Ahn, B.Y. Device fabrication: Three-dimensional printed electronics. *Nature* **2015**, *518*, 42–43. [CrossRef] [PubMed]

31. Derakhshanfar, S.; Mbeleck, R.; Xu, K.; Zhang, X.; Zhong, W.; Xing, M. 3D bioprinting for biomedical devices and tissue engineering: A review of recent trends and advances. *Bioact. Mater.* **2018**, *3*, 144–156. [CrossRef] [PubMed]

32. Park, C.S.; Ha, T.H.; Kim, M.; Raja, N.; Yun, H.; Sung, M.J.; Kwon, O.S.; Yoon, H.; Lee, C.-S. Fast and sensitive near-infrared fluorescent probes for ALP detection and 3D printed calcium phosphate scaffold imaging in vivo. *Biosens. Bioelectron.* **2018**, *105*, 151–158. [CrossRef] [PubMed]

33. Knowlton, S.; Onal, S.; Yu, C.H.; Zhao, J.J.; Tasoglu, S. Bioprinting for cancer research. *Trends Biotechnol.* **2015**, *33*, 504–513. [CrossRef] [PubMed]

34. Zhou, X.; Zhu, W.; Nowicki, M.; Miao, S.; Cui, H.; Holmes, B.; Glazer, R.I.; Zhang, L.G. 3D bioprinting a cell-laden bone matrix for breast cancer metastasis study. *ACS Appl. Mater. Interfaces* **2016**, *8*, 30017–30026. [CrossRef] [PubMed]

35. Mohamed, O.A.; Masood, S.H.; Bhowmik, J.L. Optimization of fused deposition modeling process parameters: A review of current research and future prospects. *Adv. Manuf.* **2015**, *3*, 42–53. [CrossRef]

36. Dul, S.; Fambri, L.; Pegoretti, A. Fused deposition modelling with ABS–graphene nanocomposites. *Compos. Part Appl. Sci. Manuf.* **2016**, *85*, 181–191. [CrossRef]

37. Wang, X.; Jiang, M.; Zhou, Z.; Gou, J.; Hui, D. 3D printing of polymer matrix composites: A review and prospective. *Compos. Part B Eng.* **2017**, *110*, 442–458. [CrossRef]

38. Leigh, S.J.; Bradley, R.J.; Pursell, C.P.; Billson, D.R.; Hutchins, D.A. A Simple, Low-Cost Conductive Composite Material for 3D Printing of Electronic Sensors. *PLoS ONE* **2012**, *7*, e49365. [CrossRef] [PubMed]

39. Cruz, M.A.; Ye, S.; Kim, M.J.; Reyes, C.; Yang, F.; Flowers, P.F.; Wiley, B.J. Multigram synthesis of cu-ag core–shell nanowires enables the production of a highly conductive polymer filament for 3D printing electronics. *Part. Part. Syst. Charact.* **2018**, *25*, 1700385. [CrossRef]

40. Lee, J.-Y.; An, J.; Chua, C.K. Fundamentals and applications of 3D printing for novel materials. *Appl. Mater. Today* **2017**, *7*, 120–133. [CrossRef]

41. Rimington, R.P.; Capel, A.J.; Christie, S.D.; Lewis, M.P. Biocompatible 3D printed polymers via fused deposition modelling direct C 2 C 12 cellular phenotype in vitro. *Lab Chip* **2017**, *17*, 2982–2993. [CrossRef] [PubMed]

42. Rosenzweig, D.H.; Carelli, E.; Steffen, T.; Jarzem, P.; Haglund, L. 3D-Printed ABS and PLA Scaffolds for Cartilage and Nucleus Pulposus Tissue Regeneration. *Int. J. Mol. Sci.* **2015**, *16*, 15118–15135. [CrossRef] [PubMed]

43. Skliutas, E.; Kasetaite, S.; Jonušauskas, L.; Ostrauskaite, J.; Malinauskas, M. Photosensitive naturally derived resins toward optical 3-D printing. *Opt. Eng.* **2018**, *57*, 041412. [CrossRef]

44. Voet, V.S.D.; Strating, T.; Schnelting, G.H.M.; Dijkstra, P.; Tietema, M.; Xu, J.; Woortman, A.J.J.; Loos, K.; Jager, J.; Folkersma, R. Biobased acrylate photocurable resin formulation for stereolithography 3D printing. *ACS Omega* **2018**, *3*, 1403–1408. [CrossRef]

45. Malek, S.; Raney, J.R.; Lewis, J.A.; Gibson, L.J. Lightweight 3D cellular composites inspired by balsa. *Bioinspir. Biomim.* **2017**, *12*, 026014. [CrossRef] [PubMed]

46. Smith, P.T.; Basu, A.; Saha, A.; Nelson, A. Chemical modification and printability of shear-thinning hydrogel inks for direct-write 3D printing. *Polymer* **2018**, in press. [CrossRef]

47. He, Y.; Yang, F.; Zhao, H.; Gao, Q.; Xia, B.; Fu, J. Research on the printability of hydrogels in 3D bioprinting. *Sci. Rep.* **2016**, *6*, 29977. [CrossRef] [PubMed]

48. Compton, B.G.; Lewis, J.A. 3D-printing of lightweight cellular composites. *Adv. Mater.* **2014**, *26*, 5930–5935. [CrossRef] [PubMed]

49. Loebel, C.; Rodell, C.B.; Chen, M.H.; Burdick, J.A. Shear-thinning and self-healing hydrogels as injectable therapeutics and for 3D-printing. *Nat. Protoc.* **2017**, *12*, 1521–1541. [CrossRef] [PubMed]

50. Fedorovich, N.E.; Schuurman, W.; Wijnberg, H.M.; Prins, H.-J.; van Weeren, P.R.; Malda, J.; Alblas, J.; Dhert, W.J.A. Biofabrication of osteochondral tissue equivalents by printing topologically defined, cell-laden hydrogel scaffolds. *Tissue Eng. Part C Methods* **2011**, *18*, 33–44. [CrossRef] [PubMed]

51. Billiet, T.; Gevaert, E.; De Schryver, T.; Cornelissen, M.; Dubruel, P. The 3D printing of gelatin methacrylamide cell-laden tissue-engineered constructs with high cell viability. *Biomaterials* **2014**, *35*, 49–62. [CrossRef] [PubMed]

52. Ifkovits, J.L.; Burdick, J.A. Review: Photopolymerizable and degradable biomaterials for tissue engineering applications. *Tissue Eng.* **2007**, *13*, 2369–2385. [CrossRef] [PubMed]

53. Hong, S.; Sycks, D.; Chan, H.F.; Lin, S.; Lopez, G.P.; Guilak, F.; Leong, K.W.; Zhao, X. 3D printing of highly stretchable and tough hydrogels into complex, cellularized structures. *Adv. Mater.* **2015**, *27*, 4035–4040. [CrossRef] [PubMed]

54. Gross, B.C.; Erkal, J.L.; Lockwood, S.Y.; Chen, C.; Spence, D.M. Evaluation of 3D Printing and Its Potential Impact on Biotechnology and the Chemical Sciences. *Anal. Chem.* **2014**, *86*, 3240–3253. [CrossRef] [PubMed]

55. Gauvin, R.; Chen, Y.-C.; Lee, J.W.; Soman, P.; Zorlutuna, P.; Nichol, J.W.; Bae, H.; Chen, S.; Khademhosseini, A. Microfabrication of complex porous tissue engineering scaffolds using 3D projection stereolithography. *Biomaterials* **2012**, *33*, 3824–3834. [CrossRef] [PubMed]

56. Zhang, A.P.; Qu, X.; Soman, P.; Hribar, K.C.; Lee, J.W.; Chen, S.; He, S. Rapid Fabrication of Complex 3D Extracellular Microenvironments by Dynamic Optical Projection Stereolithography. *Adv. Mater.* **2012**, *24*, 4266–4270. [CrossRef] [PubMed]

57. Macdonald, N.P.; Cabot, J.M.; Smejkal, P.; Guijt, R.M.; Paull, B.; Breadmore, M.C. Comparing Microfluidic Performance of Three-Dimensional (3D) Printing Platforms. *Anal. Chem.* **2017**, *89*, 3858–3866. [CrossRef] [PubMed]

58. Lin, H.; Zhang, D.; Alexander, P.G.; Yang, G.; Tan, J.; Cheng, A.W.-M.; Tuan, R.S. Application of visible light-based projection stereolithography for live cell-scaffold fabrication with designed architecture. *Biomaterials* **2013**, *34*, 331–339. [CrossRef] [PubMed]

59. Hofmann, M. 3D printing gets a boost and opportunities with polymer materials. *ACS Macro Lett.* **2014**, *3*, 382–386. [CrossRef]

60. Kamyshny, A.; Steinke, J.; Magdassi, S. Metal-based inkjet inks for printed electronics. *Open Appl. Phys. J.* **2011**, *4*, 19–36. [CrossRef]

61. Zocca, A.; Gomes, C.M.; Staude, A.; Bernardo, E.; Günster, J.; Colombo, P. SiOC ceramics with ordered porosity by 3D-printing of a preceramic polymer. *J. Mater. Res.* **2013**, *28*, 2243–2252. [CrossRef]

62. Bucella, S.G.; Nava, G.; Vishunubhatla, K.C.; Caironi, M. High-resolution direct-writing of metallic electrodes on flexible substrates for high performance organic field effect transistors. *Org. Electron.* **2013**, *14*, 2249–2256. [CrossRef]

63. Munshi, A.S.; Martin, R.S. Microchip-based electrochemical detection using a 3-D printed wall-jet electrode device. *Analyst* **2016**, *141*, 862–869. [CrossRef] [PubMed]

64. Anderson, K.B.; Lockwood, S.Y.; Martin, R.S.; Spence, D.M. A 3D printed fluidic device that enables integrated features. *Anal. Chem.* **2013**, *85*, 5622–5626. [CrossRef] [PubMed]

65. Bhattacharjee, N.; Urrios, A.; Kang, S.; Folch, A. The upcoming 3D-printing revolution in microfluidics. *Lab Chip* **2016**, *16*, 1720–1742. [CrossRef] [PubMed]

66. Williams, J.M.; Adewunmi, A.; Schek, R.M.; Flanagan, C.L.; Krebsbach, P.H.; Feinberg, S.E.; Hollister, S.J.; Das, S. Bone tissue engineering using polycaprolactone scaffolds fabricated via selective laser sintering. *Biomaterials* **2005**, *26*, 4817–4827. [CrossRef] [PubMed]

67. Shirazi, S.F.S.; Gharehkhani, S.; Mehrali, M.; Yarmand, H.; Metselaar, H.S.C.; Kadri, N.A.; Osman, N.A.A. A review on powder-based additive manufacturing for tissue engineering: Selective laser sintering and inkjet 3D printing. *Sci. Technol. Adv. Mater.* **2015**, *16*, 033502. [CrossRef] [PubMed]

68. Kolan, K.C.R.; Leu, M.C.; Hilmas, G.E.; Velez, M. Effect of material, process parameters, and simulated body fluids on mechanical properties of 13-93 bioactive glass porous constructs made by selective laser sintering. *J. Mech. Behav. Biomed. Mater.* **2012**, *13*, 14–24. [CrossRef] [PubMed]

69. Yeong, W.Y.; Sudarmadji, N.; Yu, H.Y.; Chua, C.K.; Leong, K.F.; Venkatraman, S.S.; Boey, Y.C.F.; Tan, L.P. Porous polycaprolactone scaffold for cardiac tissue engineering fabricated by selective laser sintering. *Acta Biomater.* **2010**, *6*, 2028–2034. [CrossRef] [PubMed]

70. Wu, G.-H.; Hsu, S. Review: Polymeric-based 3D printing for tissue engineering. *J. Med. Biol. Eng.* **2015**, *35*, 285–292. [CrossRef] [PubMed]

71. Olakanmi, E.O.; Cochrane, R.F.; Dalgarno, K.W. A review on selective laser sintering/melting (SLS/SLM) of aluminium alloy powders: Processing, microstructure, and properties. *Prog. Mater. Sci.* **2015**, *74*, 401–477. [CrossRef]

72. Fuse 1: Benchtop Selective Laser Sintering (SLS) 3D Printer. Available online: https://formlabs.com/3d-printers/fuse-1/ (accessed on 10 July 2018).

73. Malinauskas, M.; Žukauskas, A.; Hasegawa, S.; Hayasaki, Y.; Mizeikis, V.; Buividas, R.; Juodkazis, S. Ultrafast laser processing of materials: From science to industry. *Light Sci. Appl.* **2016**, *5*, e16133. [CrossRef]

74. Marino, A.; Barsotti, J.; de Vito, G.; Filippeschi, C.; Mazzolai, B.; Piazza, V.; Labardi, M.; Mattoli, V.; Ciofani, G. Two-photon lithography of 3D nanocomposite piezoelectric scaffolds for cell stimulation. *ACS Appl. Mater. Interfaces* **2015**, *7*, 25574–25579. [CrossRef] [PubMed]

75. Mačiulaitis, J.; Deveikytė, M.; Rekštytė, S.; Bratchikov, M.; Darinskas, A.; Šimbelytė, A.; Daunoras, G.; Laurinavičienė, A.; Laurinavičius, A.; Gudas, R.; et al. Preclinical study of SZ2080 material 3D microstructured scaffolds for cartilage tissue engineering made by femtosecond direct laser writing lithography. *Biofabrication* **2015**, *7*, 015015. [CrossRef] [PubMed]

76. Ovsianikov, A.; Mühleder, S.; Torgersen, J.; Li, Z.; Qin, X.-H.; Van Vlierberghe, S.; Dubruel, P.; Holnthoner, W.; Redl, H.; Liska, R.; et al. Laser photofabrication of cell-containing hydrogel constructs. *Langmuir* **2014**, *30*, 3787–3794. [CrossRef] [PubMed]

77. Torgersen, J.; Ovsianikov, A.; Mironov, V.; Pucher, N.; Qin, X.; Li, Z.; Cicha, K.; Machacek, T.; Liska, R.; Jantsch, V.; et al. Photo-sensitive hydrogels for three-dimensional laser microfabrication in the presence of whole organisms. *J. Biomed. Opt.* **2012**, *17*, 105008. [CrossRef] [PubMed]

78. Gruene, M.; Deiwick, A.; Koch, L.; Schlie, S.; Unger, C.; Hofmann, N.; Bernemann, I.; Glasmacher, B.; Chichkov, B. Laser printing of stem cells for biofabrication of scaffold-free autologous grafts. *Tissue Eng. Part C Methods* **2011**, *17*, 79–87. [CrossRef] [PubMed]

79. Schizas, C.; Melissinaki, V.; Gaidukeviciute, A.; Reinhardt, C.; Ohrt, C.; Dedoussis, V.; Chichkov, B.N.; Fotakis, C.; Farsari, M.; Karalekas, D. On the design and fabrication by two-photon polymerization of a readily assembled micro-valve. *Int. J. Adv. Manuf. Technol.* **2010**, *48*, 435–441. [CrossRef]

80. Qu, J.; Kadic, M.; Naber, A.; Wegener, M. Micro-structured two-component 3D metamaterials with negative thermal-expansion coefficient from positive constituents. *Sci. Rep.* **2017**, *7*, 40643. [CrossRef] [PubMed]

81. Jonušauskas, L.; Skliutas, E.; Butkus, S.; Malinauskas, M. Custom on demand 3D printing of functional microstructures. *Lith. J. Phys.* **2015**, *55*, 227–236. [CrossRef]

82. Amato, L.; Gu, Y.; Bellini, N.; Eaton, S.M.; Cerullo, G.; Osellame, R. Integrated three-dimensional filter separates nanoscale from microscale elements in a microfluidic chip. *Lab Chip* **2012**, *12*, 1135–1142. [CrossRef] [PubMed]

83. Paiè, P.; Bragheri, F.; Carlo, D.D.; Osellame, R. Particle focusing by 3D inertial microfluidics. *Microsyst. Nanoeng.* **2017**, *3*, 17027. [CrossRef]

84. Rogers, C.I.; Qaderi, K.; Woolley, A.T.; Nordin, G.P. 3D printed microfluidic devices with integrated valves. *Biomicrofluidics* **2015**, *9*, 016501. [CrossRef] [PubMed]

85. Su, C.-K.; Hsia, S.-C.; Sun, Y.-C. Three-dimensional printed sample load/inject valves enabling online monitoring of extracellular calcium and zinc ions in living rat brains. *Anal. Chim. Acta* **2014**, *838*, 58–63. [CrossRef] [PubMed]

86. Su, C.-K.; Peng, P.-J.; Sun, Y.-C. Fully 3D-printed preconcentrator for selective extraction of trace elements in seawater. *Anal. Chem.* **2015**, *87*, 6945–6950. [CrossRef] [PubMed]

87. Su, C.-K.; Yen, S.-C.; Li, T.-W.; Sun, Y.-C. Enzyme-Immobilized 3D-Printed Reactors for Online Monitoring of Rat Brain Extracellular Glucose and Lactate. *Anal. Chem.* **2016**, *88*, 6265–6273. [CrossRef] [PubMed]

88. Rafeie, M.; Zhang, J.; Asadnia, M.; Li, W.; Warkiani, M.E. Multiplexing slanted spiral microchannels for ultra-fast blood plasma separation. *Lab Chip* **2016**, *16*, 2791–2802. [CrossRef] [PubMed]

89. Lee, W.; Kwon, D.; Choi, W.; Jung, G.Y.; Au, A.K.; Folch, A.; Jeon, S. 3D-printed microfluidic device for the detection of pathogenic bacteria using size-based separation in helical channel with trapezoid cross-section. *Sci. Rep.* **2015**, *5*, 7717. [CrossRef] [PubMed]

90. Yan, S.; Tan, S.H.; Li, Y.; Tang, S.; Teo, A.J.T.; Zhang, J.; Zhao, Q.; Yuan, D.; Sluyter, R.; Nguyen, N.T.; et al. A portable, hand-powered microfluidic device for sorting of biological particles. *Microfluid. Nanofluid.* **2018**, *22*, 8. [CrossRef]

91. Park, C.; Lee, J.; Kim, Y.; Kim, J.; Lee, J.; Park, S. 3D-printed microfluidic magnetic preconcentrator for the detection of bacterial pathogen using an ATP luminometer and antibody-conjugated magnetic nanoparticles. *J. Microbiol. Methods* **2017**, *132*, 128–133. [CrossRef] [PubMed]

92. Oh, S.; Kim, B.; Lee, J.K.; Choi, S. 3D-printed capillary circuits for rapid, low-cost, portable analysis of blood viscosity. *Sens. Actuators B Chem.* **2018**, *259*, 106–113. [CrossRef]

93. Santangelo, M.F.; Libertino, S.; Turner, A.P.F.; Filippini, D.; Mak, W.C. Integrating printed microfluidics with silicon photomultipliers for miniaturised and highly sensitive ATP bioluminescence detection. *Biosens. Bioelectron.* **2018**, *99*, 464–470. [CrossRef] [PubMed]

94. Tang, C.K.; Vaze, A.; Rusling, J.F. Automated 3D-printed unibody immunoarray for chemiluminescence detection of cancer biomarker proteins. *Lab Chip* **2017**, *17*, 484–489. [CrossRef] [PubMed]

95. Nie, J.; Gao, Q.; Qiu, J.; Sun, M.; Liu, A.; Shao, L.; Fu, J.; Zhao, P.; He, Y. 3D printed Lego®-like modular microfluidic devices based on capillary driving. *Biofabrication* **2018**, *10*, 035001. [CrossRef] [PubMed]

96. Symes, M.D.; Kitson, P.J.; Yan, J.; Richmond, C.J.; Cooper, G.J.T.; Bowman, R.W.; Vilbrandt, T.; Cronin, L. Integrated 3D-printed reactionware for chemical synthesis and analysis. *Nat. Chem.* **2012**, *4*, 349–354. [CrossRef] [PubMed]

97. Chen, C.; Wang, Y.; Lockwood, S.Y.; Spence, D.M. 3D-printed fluidic devices enable quantitative evaluation of blood components in modified storage solutions for use in transfusion medicine. *Analyst* **2014**, *139*, 3219–3226. [CrossRef] [PubMed]

98. Bishop, G.W.; Satterwhite-Warden, J.E.; Bist, I.; Chen, E.; Rusling, J.F. Electrochemiluminescence at bare and dna-coated graphite electrodes in 3D-printed fluidic devices. *ACS Sens.* **2016**, *1*, 197–202. [CrossRef] [PubMed]

99. Zangheri, M.; Cevenini, L.; Anfossi, L.; Baggiani, C.; Simoni, P.; Di Nardo, F.; Roda, A. A simple and compact smartphone accessory for quantitative chemiluminescence-based lateral flow immunoassay for salivary cortisol detection. *Biosens. Bioelectron.* **2015**, *64*, 63–68. [CrossRef] [PubMed]

100. Suh, Y.K.; Kang, S. A review on mixing in microfluidics. *Micromachines* **2010**, *1*, 82–111. [CrossRef]

101. Shallan, A.I.; Smejkal, P.; Corban, M.; Guijt, R.M.; Breadmore, M.C. Cost-effective three-dimensional printing of visibly transparent microchips within minutes. *Anal. Chem.* **2014**, *86*, 3124–3130. [CrossRef] [PubMed]

102. Bishop, G.W.; Satterwhite, J.E.; Bhakta, S.; Kadimisetty, K.; Gillette, K.M.; Chen, E.; Rusling, J.F. 3D-printed fluidic devices for nanoparticle preparation and flow-injection amperometry using integrated prussian blue nanoparticle-modified electrodes. *Anal. Chem.* **2015**, *87*, 5437–5443. [CrossRef] [PubMed]

103. Patrick, W.G.; Nielsen, A.A.K.; Keating, S.J.; Levy, T.J.; Wang, C.-W.; Rivera, J.J.; Mondragón-Palomino, O.; Carr, P.A.; Voigt, C.A.; Oxman, N.; et al. DNA assembly in 3D printed fluidics. *PLoS ONE* **2015**, *10*, e0143636. [CrossRef] [PubMed]

104. Kise, D.P.; Reddish, M.J.; Dyer, R.B. Sandwich-format 3D printed microfluidic mixers: A flexible platform for multi-probe analysis. *J. Micromech. Microeng.* **2015**, *25*, 124002. [CrossRef] [PubMed]

105. Plevniak, K.; Campbell, M.; Myers, T.; Hodges, A.; He, M. 3D printed auto-mixing chip enables rapid smartphone diagnosis of anemia. *Biomicrofluidics* **2016**, *10*, 054113. [CrossRef] [PubMed]

106. Mattio, E.; Robert-Peillard, F.; Vassalo, L.; Branger, C.; Margaillan, A.; Brach-Papa, C.; Knoery, J.; Boudenne, J.-L.; Coulomb, B. 3D-printed lab-on-valve for fluorescent determination of cadmium and lead in water. *Talanta* **2018**, *183*, 201–208. [CrossRef] [PubMed]

107. Guo, T.; Patnaik, R.; Kuhlmann, K.; Rai, A.J.; Sia, S.K. Smartphone dongle for simultaneous measurement of hemoglobin concentration and detection of HIV antibodies. *Lab Chip* **2015**, *15*, 3514–3520. [CrossRef] [PubMed]

108. Chan, K.; Weaver, S.C.; Wong, P.-Y.; Lie, S.; Wang, E.; Guerbois, M.; Vayugundla, S.P.; Wong, S. Rapid, affordable and portable medium-throughput molecular device for zika virus. *Sci. Rep.* **2016**, *6*, 38223. [CrossRef] [PubMed]

109. Fernandes, A.C.; Semenova, D.; Panjan, P.; Sesay, A.M.; Gernaey, K.V.; Krühne, U. Multi-function microfluidic platform for sensor integration. *New Biotechnol.* **2018**, in press. [CrossRef] [PubMed]

110. Li, K.; Wei, H.; Liu, W.; Meng, H.; Zhang, P.; Yan, C. 3D printed stretchable capacitive sensors for highly sensitive tactile and electrochemical sensing. *Nanotechnology* **2018**, *29*, 185501. [CrossRef] [PubMed]

111. Manzanares Palenzuela, C.L.; Novotný, F.; Krupička, P.; Sofer, Z.; Pumera, M. 3D-printed graphene/polylactic acid electrodes promise high sensitivity in electroanalysis. *Anal. Chem.* **2018**, *90*, 5753–5757. [CrossRef] [PubMed]

112. Lind, J.U.; Busbee, T.A.; Valentine, A.D.; Pasqualini, F.S.; Yuan, H.; Yadid, M.; Park, S.-J.; Kotikian, A.; Nesmith, A.P.; Campbell, P.H.; et al. Instrumented cardiac microphysiological devices via multimaterial three-dimensional printing. *Nat. Mater.* **2017**, *16*, 303–308. [CrossRef] [PubMed]

113. Muth, J.T.; Vogt, D.M.; Truby, R.L.; Mengüç, Y.; Kolesky, D.B.; Wood, R.J.; Lewis, J.A. Embedded 3D printing of strain sensors within highly stretchable elastomers. *Adv. Mater.* **2014**, *26*, 6307–6312. [CrossRef] [PubMed]

114. Lei, Z.; Wang, Q.; Wu, P. A multifunctional skin-like sensor based on a 3D printed thermo-responsive hydrogel. *Mater. Horiz.* **2017**, *4*, 694–700. [CrossRef]

115. Skin to E-Skin. Available online: https://www.nature.com/articles/nnano.2017.228 (accessed on 3 August 2018).

116. Shanmugam, A.; Usmani, M.; Mayberry, A.; Perkins, D.L.; Holcomb, D.E. Imaging systems and algorithms to analyze biological samples in real-time using mobile phone microscopy. *PLoS ONE* **2018**, *13*, e0193797. [CrossRef] [PubMed]

117. Scordo, G.; Moscone, D.; Palleschi, G.; Arduini, F. A reagent-free paper-based sensor embedded in a 3D printing device for cholinesterase activity measurement in serum. *Sens. Actuators B Chem.* **2018**, *258*, 1015–1021. [CrossRef]

118. Xiao, W.; Huang, C.; Xu, F.; Yan, J.; Bian, H.; Fu, Q.; Xie, K.; Wang, L.; Tang, Y. A simple and compact smartphone-based device for the quantitative readout of colloidal gold lateral flow immunoassay strips. *Sens. Actuators B Chem.* **2018**, *266*, 63–70. [CrossRef]

119. Knowlton, S.; Joshi, A.; Syrrist, P.; Coskun, A.F.; Tasoglu, S. 3D-printed smartphone-based point of care tool for fluorescence- and magnetophoresis-based cytometry. *Lab Chip* **2017**, *17*, 2839–2851. [CrossRef] [PubMed]

120. Kühnemund, M.; Wei, Q.; Darai, E.; Wang, Y.; Hernández-Neuta, I.; Yang, Z.; Tseng, D.; Ahlford, A.; Mathot, L.; Sjöblom, T.; et al. Targeted DNA sequencing and in situ mutation analysis using mobile phone microscopy. *Nat. Commun.* **2017**, *8*, 13913. [CrossRef] [PubMed]

121. Park, J.; Park, H. Thermal cycling characteristics of a 3D-printed serpentine microchannel for DNA amplification by polymerase chain reaction. *Sens. Actuators Phys.* **2017**, *268*, 183–187. [CrossRef]

122. Mulberry, G.; White, K.A.; Kim, B.N. 3D printed real-time PCR machine for infectious disease diagnostics. *Biophys. J.* **2017**, *112*, 462a. [CrossRef]

123. Bianchi, S.; Rajamanickam, V.P.; Ferrara, L.; Fabrizio, E.D.; Liberale, C.; Leonardo, R.D. Focusing and imaging with increased numerical apertures through multimode fibers with micro-fabricated optics. *Opt. Lett.* **2013**, *38*, 4935–4938. [CrossRef] [PubMed]

124. Malinauskas, M.; Žukauskas, A.; Belazaras, K.; Tikuišis, K.; Purlys, V.; Gadonas, R.; Piskarskas, A. Laser fabrication of various polymer microoptical components. *Eur. Phys. J. Appl. Phys.* **2012**, *58*, 20501. [CrossRef]

125. Liberale, C.; Cojoc, G.; Bragheri, F.; Minzioni, P.; Perozziello, G.; Rocca, R.L.; Ferrara, L.; Rajamanickam, V.; Fabrizio, E.D.; Cristiani, I. Integrated microfluidic device for single-cell trapping and spectroscopy. *Sci. Rep.* **2013**, *3*, 1258. [CrossRef] [PubMed]

126. Kirchner, R.; Chidambaram, N.; Schift, H. Benchmarking surface selective vacuum ultraviolet and thermal postprocessing of thermoplastics for ultrasmooth 3-D-printed micro-optics. *Opt. Eng.* **2018**, *57*, 041403. [CrossRef]

127. Thiele, S.; Arzenbacher, K.; Gissibl, T.; Giessen, H.; Herkommer, A.M. 3D-printed eagle eye: Compound microlens system for foveated imaging. *Sci. Adv.* **2017**, *3*, e1602655. [CrossRef] [PubMed]

128. Zhu, H.; Isikman, S.O.; Mudanyali, O.; Greenbaum, A.; Ozcan, A. Optical imaging techniques for point-of-care diagnostics. *Lab Chip* **2013**, *13*, 51–67. [CrossRef] [PubMed]

129. Singh, M.; Tong, Y.; Webster, K.; Cesewski, E.; Haring, A.P.; Laheri, S.; Carswell, B.; O'Brien, T.J.; Aardema, C.H.; Senger, R.S.; et al. 3D printed conformal microfluidics for isolation and profiling of biomarkers from whole organs. *Lab Chip* **2017**, *17*, 2561–2571. [CrossRef] [PubMed]

130. Jang, J.; Park, J.Y.; Gao, G.; Cho, D.-W. Biomaterials-based 3D cell printing for next-generation therapeutics and diagnostics. *Biomaterials* **2018**, *156*, 88–106. [CrossRef] [PubMed]

131. Chatzipetrou, M.; Massaouti, M.; Tsekenis, G.; Trilling, A.K.; van Andel, E.; Scheres, L.; Smulders, M.M.J.; Zuilhof, H.; Zergioti, I. Direct Creation of Biopatterns via a combination of laser-based techniques and click chemistry. *Langmuir* **2017**, *33*, 848–853. [CrossRef] [PubMed]

132. Chatzipetrou, M.; Milano, F.; Giotta, L.; Chirizzi, D.; Trotta, M.; Massaouti, M.; Guascito, M.R.; Zergioti, I. Functionalization of gold screen printed electrodes with bacterial photosynthetic reaction centers by laser printing technology for mediatorless herbicide biosensing. *Electrochem. Commun.* **2016**, *64*, 46–50. [CrossRef]

133. Schaffner, M.; Rühs, P.A.; Coulter, F.; Kilcher, S.; Studart, A.R. 3D printing of bacteria into functional complex materials. *Sci. Adv.* **2017**, *3*, eaao6804. [CrossRef] [PubMed]

134. Alam, M.N.H.Z.; Hossain, F.; Vale, A.; Kouzani, A. Design and fabrication of a 3D printed miniature pump for integrated microfluidic applications. *Int. J. Precis. Eng. Manuf.* **2017**, *18*, 1287–1296. [CrossRef]

135. 3D Printing: A New Era for Valve Manufacturing? Available online: http://www.valvemagazine.com/web-only/categories/trends-forecasts/7689-3d-printing-a-new-era-for-valve-manufacturing.html (accessed on 8 May 2018).

136. Building Atomic Force Microscope with 3D Printing, Electronics and LEGO. Available online: http://www.ro baid.com/tech/building-atomic-force-microscope-with-3d-printing-electronics-and-lego.htm (accessed on 8 May 2018).

137. Meloni, G.N.; Bertotti, M. 3D printing scanning electron microscopy sample holders: A quick and cost effective alternative for custom holder fabrication. *PLoS ONE* **2017**, *12*, e0182000. [CrossRef] [PubMed]

micromachines

MDPI

Article

A 3D-Printed Millifluidic Platform Enabling Bacterial Preconcentration and DNA Purification for Molecular Detection of Pathogens in Blood

Yonghee Kim †, Jinyeop Lee † and Sungsu Park *

School of Mechanical Engineering, Sungkyunkwan University, Suwon 16419, Korea;
hanskim723@gmail.com (Y.K.); softmemsljy@naver.com (J.L.)
* Correspondence: nanopark@skku.edu; Tel.: +82-31-290-7431
† The authors contributed equally to this work.

Received: 13 August 2018; Accepted: 14 September 2018; Published: 17 September 2018

Abstract: Molecular detection of pathogens in clinical samples often requires pretreatment techniques, including immunomagnetic separation and magnetic silica-bead-based DNA purification to obtain the purified DNA of pathogens. These two techniques usually rely on handling small tubes containing a few millilitres of the sample and manual operation, implying that an automated system encompassing both techniques is needed for larger quantities of the samples. Here, we report a three-dimensional (3D)-printed millifluidic platform that enables bacterial preconcentration and genomic DNA (gDNA) purification for improving the molecular detection of target pathogens in blood samples. The device consists of two millichannels and one chamber, which can be used to preconcentrate pathogens bound to antibody-conjugated magnetic nanoparticles (Ab-MNPs) and subsequently extract gDNA using magnetic silica beads (MSBs) in a sequential manner. The platform was able to preconcentrate very low concentrations (1–1000 colony forming units (CFU)) of *Escherichia coli* O157:H7 and extract their genomic DNA in 10 mL of buffer and 10% blood within 30 min. The performance of the platform was verified by detecting as low as 1 CFU of *E. coli* O157:H7 in 10% blood using either polymerase chain reaction (PCR) with post gel electrophoresis or quantitative PCR. The results suggest that the 3D-printed millifluidic platform is highly useful for lowering the limitations on molecular detection in blood by preconcentrating the target pathogen and isolating its DNA in a large volume of the sample.

Keywords: immunomagnetic separation (IMS); bacterial pathogen; 3D printing; preconcentration; DNA purification; molecular diagnostics

---

1. Introduction

It is important to accurately detect pathogens in clinical samples at very low concentrations [1,2]. Methods for detecting pathogens in clinical samples, such as blood and saliva, include bacterial culture and polymerase chain reaction (PCR) [3]. However, when detecting pathogens in clinical samples, there are limitations because of the presence of substances that inhibit PCR [3,4]. Thus, such methods of detection still require sample pretreatment to isolate the target microorganisms and purify their nucleic acids [3–5]. Among these pretreatment techniques, immunomagnetic separation (IMS) [5,6] and magnetic silica bead (MSB)-based DNA purification [7–9] are the most popular. However, these two technologies usually rely on handling small tubes containing several millilitres of the sample and manual operation; thus, there is an urgent need for automated systems that can process large volumes of the samples simultaneously because higher concentrations of purified DNA can be obtained by preconcentrating the pathogens and purifying DNA from larger volumes of samples.

In the past few decades, microfluidic devices (µFDs) have been developed as platforms that can detect pathogens [10–12]. In particular, µFDs offer several advantages for detection when integrating IMS. For example, μFDs have a large surface area, thus allowing antibody-conjugated magnetic nanoparticles (Ab-MNPs) and targeted bacterial cells to quickly bind to each other [13]. In addition, the magnetic interaction between the Ab-MNPs and permanent magnets is very strong in the thin microchannels of the μFDs, and the bacteria–Ab-MNP complexes can be trapped easily and quickly [14]. Recently, efforts have been made to extract and isolate target DNA using either MSBs or IMS in μFDs [7–9,15,16]. However, conventional μFDs [13–16] are typically not suitable for processing samples larger than 1 mL due to the small dimensions (~1 mm) of their microchannels. In addition, their fabrication requires multiple layers and several bonding steps, making them difficult to mass-produce. Most recently, we have demonstrated that a three-dimensional (3D)-printed μFD (3DpμFD) is an excellent platform for detecting pathogens because of its high speed, integration, and automation [17]. Compared to photolithography and soft lithography, 3D printing has many advantages when printing μFDs because this technique easily enables the printing of high aspect ratio structures and does not require complex binding steps to form a monolithic structure. In recent years, considerable efforts have been devoted to the development of 3D printing microfluidic platforms for separation and detection [17–19]. However, to the best of our knowledge, there have been no reports showing the integration of both IMS and DNA purification functions into a single device.

In the present study, we report a 3D-printed millifluidic device (3DpmFD) that can perform IMS and DNA purification of the target pathogen in 10 mL or higher volumes of samples. The performance of the 3DpmFD was tested with *Escherichia coli* O157:H7 and *Staphylococcus aureus* in a buffer and spiked blood samples. The performance of the device was verified by standard methods, such as colony counting, PCR, and quantitative PCR (qPCR).

2. Materials and Methods

2.1. 3D Printing of the Millifluidic Device

The millifluidic device was 3D-printed using a digital light processing (DLP) 3D printer (IM-96) (Carima Co., Seoul, Korea). This 3D-printing technique was based on photopolymerisation by emitting visible light at 405 nm onto a photocurable resin. The 3D sketch of the device was designed using the Student edition of Inventor® Professional (Autodesk Inc., Seoul, Korea). This 3D sketch was cut into 100-μm-thick layers in the z-axis direction using the Carima Slicer software and was separated into 115 layers. Each layer was irradiated for 1.5 s. After the 3D-printing, the printed structure was washed with 70% ethanol for 5 min to eliminate any uncured resin. Then, the structure was solidified for 10 min using visible light to improve its mechanical strength. The entire process, including the solidification step, takes 30 min and does not require any additional assembly steps.

Figure 1a,b show that the 3DpmFD (width × length × height dimensions of $20 \times 30 \times 10$ mm^3) consists of a cylindrical chamber (diameter: 5 mm, height: 2 mm) connected to two microchannels, a sample inlet (diameter: 2 mm), and DNA and waste outlets (diameter: 2 mm).

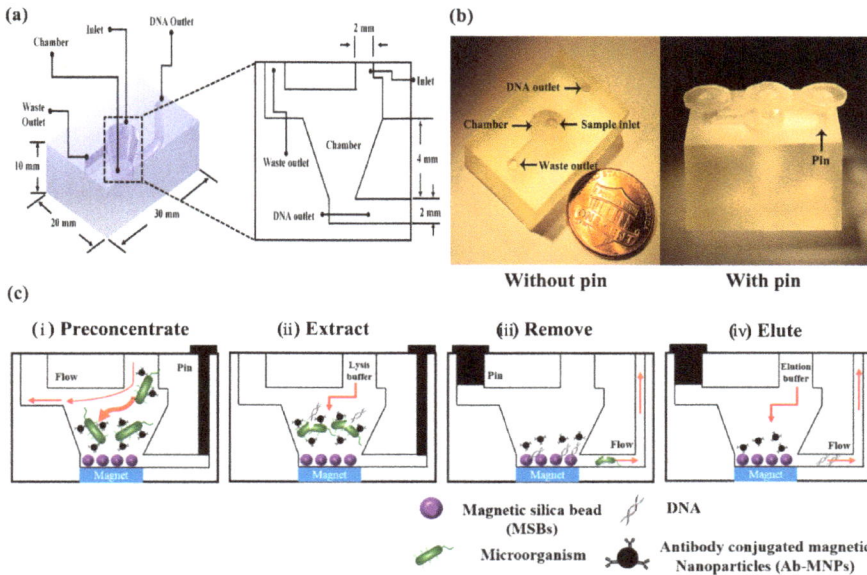

Figure 1. Bacterial preconcentration and genomic DNA (gDNA) purification on the three-dimensional (3D)-printed microfluidic device (μFD) (3DpmFD). (**a**) Design of the 3DpmFD. (**b**) A scale image of the 3DpmFD. (**c**) Schematic of the operational processes: (**i**) preconcentrating bacteria–Ab-MNP complexes; (**ii**) extracting gDNA from the complexes with lysis-binding buffer; (**iii**) removing buffer and bacterial debris by withdrawing flow using a pump; and (**iv**) eluting gDNA from MSBs with elution buffer.

2.2. Bacterial Culture

The bacterial strains used in this study were *E. coli* O157:H7 (ATCC 43894) (American Type Culture Collection, Bethesda, MD, USA) and *S. aureus* (ATCC 29213) (ATCC, The strains were grown overnight in Luria broth (LB) (Becton, Dickinson and Company, Franklin Lakes, NJ, USA) at 200 rpm and 37 °C. The culture was then diluted 100-fold with fresh LB and incubated again at 200 rpm and 37 °C until the optical density of the sample at 600 nm (OD_{600}) was 1. Before preconcentration, the samples were serially diluted 10 times with phosphate-buffered saline (PBS) (pH 7.4).

2.3. Synthesis of Ab-MNPs

Amine-modified superparamagnetic nanoparticles (MNPs) of 50 nm diameter were purchased from Chemicell Co. (Berlin, Germany). The MNPs were sonicated for about 40 s to prevent aggregation. A solution of the MNPs (1 mg/mL) in PBS was then made to react with glutaraldehyde (2.5% v/v) in PBS at room temperature (RT) for 1 h using a rotary incubator and then washed with borate buffer (10 mM, pH 7.0) [20]. Aldehyde-functionalised MNPs were then mixed with 50 μg/mL of affinity purified anti-*E. coli* O157:H7 antibody (SeraCare Life Sciences Inc., Milford, MA, USA) or affinity purified anti-*S. aureus* antibody (SeraCare Life Sciences Inc.) in borate buffer and incubated at RT overnight. Next, Ab-MNPs were washed with 500 μL of borate buffer and then mixed with 1% bovine serum albumin (BSA) (Thermo Fisher Scientific, Waltham, MA, USA) in PBS to block unreacted aldehyde groups on the Ab-MNPs at RT for 1 h. To remove the unbound BSA, the Ab-MNPs were washed with 500 μL of borate buffer again. Then, the Ab-MNPs were treated with 20 mg/mL of sodium cyanoborohydride (Sigma-Aldrich, St. Louis, MO, USA) in borate buffer. Finally, the Ab-MNPs were washed with Tris-HCl buffer (pH 8.0) and stored in PBS at 4 °C until their use.

2.4. Effect of Flow Rates on Bacteria Capturing Efficiency in the 3DpmFD

Ten millilitres (10 mL) of PBS containing *E. coli* O157:H7 at 10^4 colony-forming units (CFU)/mL were mixed with 200 μL of MNPs (10^{13} particles/mL, final concentration) conjugated with affinity-purified *E. coli* O157 antibodies and the mixture was incubated at 37 °C and 200 rpm for 20 min in a beaker. Then, the mixture was injected into the 3DpmFD through the sample inlet using a syringe pump (Harvard Apparatus, Boston, MA, USA) at various flow rates (1–10 mL/min) while placing a permanent magnet (diameter: 15 mm, height: 1.5 mm, magnetic flux density: 1720 G) underneath the chamber of the 3DpmFD.

The number of preconcentrated bacterial cells was estimated by counting the difference between the number of uncaptured bacterial cells and the total number of bacterial cells using the standard colony counting method [21]. Bacteria capturing efficiency was calculated using the following equation [17]:

$$\text{Capturing efficiency (\%)} = (Nt - Ne)/Nt \times 100\% \tag{1}$$

where Nt is the number of bacterial cells in the sample and Ne is the number of uncaptured bacterial cells in the sample.

2.5. Effect of MSB and Bacterial Concentration on DNA Purification

The 3DpmFD packed with MSBs (Chemicell Co., Berlin, Germany) was prepared by introducing 1 mL of PBS containing different concentrations (10^9 to 5×10^{10} particles/mL) of MSBs at 2 mL/min through the sample inlet while placing a permanent magnet under the chamber and closing the DNA outlet with a plastic pin. Different volumes (1–100 mL) of PBS containing *E. coli* O157:H7 or *S. aureus* at various concentrations (1–10^5 CFU/mL, final concentration) were first mixed with MNPs conjugated with antibodies specific to the pathogens and the mixture was incubated at 37 °C and 200 rpm for 20 min in a beaker. Then, the mixture was injected into the 3DpmFD packed with MSBs through the sample inlet using a syringe pump at 2 mL/min as shown in Figure 1c-i.

To lyse preconcentrated bacterial cells, 100 μL of Lysis and Binding buffer (Chemicell Co.) were loaded into the chamber as shown in Figure 1c-ii. The 3DpmFD was agitated by the automated mini-vibration system (DVM-N20 vibration motor, D&J WITH Co., Ltd., Seoul, Korea) at RT for 5 min and the blocking pins were removed from the 3DpmFD before placing it on the magnet. The buffer was removed from the chamber through the waste channel, and then 500 μL of washing buffer I (Chemicell Co.) and 500 μL of 70% ethanol were injected into the 3DpmFD sequentially. For the elution step, 50 μL of RNase free water was loaded into the 3DpmFD without placement on the magnet. After all the inlets and outlets were blocked with the pins again, the device was agitated on a thermo-shaker at 1200 rpm and 65 °C for 15 min. Finally, the DNA released from the MSBs was removed from the device through the DNA outlet.

2.6. Effect of Dilution Factor of Blood on Efficiency of Sample Preparation in 3DpmFD

The use of blood was approved by the Institutional Review Board (IRB) of the university (SKKU) and its approval number was SKKU 2017-11-006. Whole blood was purchased from Innovative Research, Inc. (Novi, MI, USA) and this product was treated with 0.1% K2 ethylenediaminetetraacetic acid (EDTA) to prevent blood coagulation. The hematocrit (hct) value was about 39%, which was measured using Fisherbrand™ Microhematocrit capillary tubes (Fisher Scientific Co LLC, Pittsburgh, PA, USA) and a MicroHematocrit Centrifuge (Thomas Scientific Inc., Swedesboro, NJ, USA). Then, 1 mL of whole blood containing 10^5 CFU/mL of *E. coli* O157:H7 was diluted to 10, 25, and 50% with PBS and 200 μL of Ab-MNPs.

2.7. Bacterial Preconcentration and DNA Purification in Spiked Blood Samples

Then, 1 mL of whole blood was mixed with 8 mL of PBS, 1 mL of *E. coli* O157:H7 (10–10^4 CFU/mL), and 200 μL of Ab-MNPs. The subsequent procedures were the same as those for the bacterial preconcentration and DNA purification steps.

2.8. Detection of Bacteria by PCR and qPCR Samples

Purified bacterial genomic DNA (gDNA) was amplified by PCR, and the amplification of target genes was verified by gel electrophoresis. The primers that were designed to amplify the 150-base pairs (bp) of *eae* gene coding intimin adherence protein in *E. coli* O157:H7 consist of a forward primer (GGCGGATTAGACTTCGGCTA) and a reverse primer (CGTTTTGGCACTATTTGCCC). The 207-bp of *nuc* gene coding the thermonuclease of *S. aureus* consists of a forward primer (ACACCTGAAACAAAGCATCC) and a reverse primer (TAGCCAAGCCTTGACGAACT). The conventional PCR was performed using a PCR reagent by MJ MINI™ thermocycler (Bio-RAD, Hercules, CA, USA). The PCR products were separated in a 1.5% TAE (Tris base, acetic acid, and EDTA) (50 ×) agarose gel at 100 V for 30 min.

The qPCR was performed using LightCycler® Nano (Roche, Basel, Switzerland), and its cycle threshold (Ct) value was determined. The same primer sets as those used for the PCR were used here.

3. Results and Discussion

3.1. Effect of Flow Rate on Bacteria Capturing Efficiency in the 3DpmFD

The capturing efficiencies at flow rates of 1, 2, 5, and 10 mL/min were 94%, 92.5%, 62.1%, and 47.9%, respectively (Figure 2). The results show that it is difficult to capture bacteria–Ab-MNPs complexes at high flow rates, such as 5 mL/min and 10 mL/min, with the magnetic force of the permanent magnet located at the bottom of the device. Since there was no statistical difference in 1 mL/min and 2 mL/min, we selected 2 mL/min as a flow rate for the following experiments.

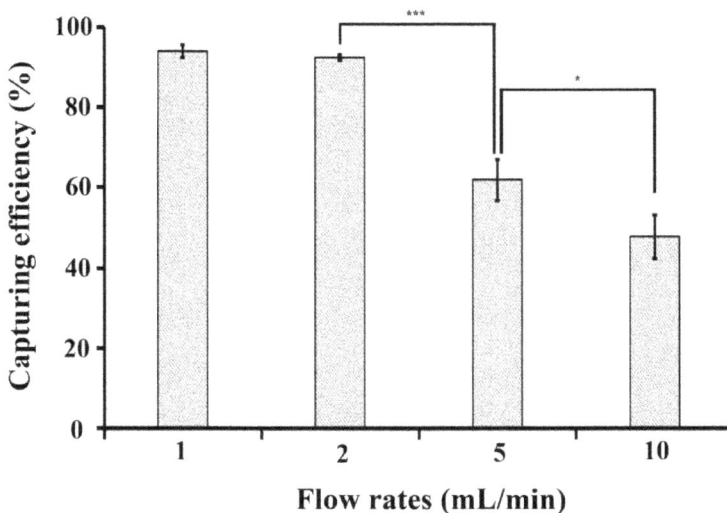

Figure 2. Effect of flow rate on bacteria capturing efficiency in 3DpmFD. Ten millilitres (10 mL) of PBS containing *E. coli* O157:H7 (10^4 CFU/mL) and Ab-MNPs (10^{13} particles/mL) were injected into 3DpmFD at different flow rates (1–10 mL/min). *: $p < 0.05$, ***: $p < 0.001$. Student *t*-test. Sample number = 3.

3.2. Optimisation of MSB Concentrations for DNA Purification using 3DpmFD

To find the optimal number for MSBs for DNA purification using the 3DpmFD, it was packed with different particle numbers (1×10^9 to 5×10^{10}) of MSBs before introducing 10 mL of PBS containing 10^3 CFU/mL and Ab-MNPs containing 10^{13} particles/mL. All the bacteria capturing efficiencies were about 90%, suggesting that the number of MSBs does not affect the bacterial capturing efficiency on the 3DpmFD.

Figure 3a shows that gDNA concentrations obtained using the 3DpmFDs increase as MSB concentrations in the range of 10^9 to 10^{10} particles/mL increase. The maximum gDNA concentration was obtained using the 3DpmFD packed with 10^{10} particles/mL of MSBs. However, the gDNA concentrations at 5×10^{10} MSBs/mL were lower than those at 10^{10} MSBs/mL. This may be attributed to the fact that 50 µL of the elution buffer was not sufficient to fully wet the MSBs in the 3DpmFD. Thus, some DNA might not have been released from some of the MSBs, resulting in lower gDNA concentrations.

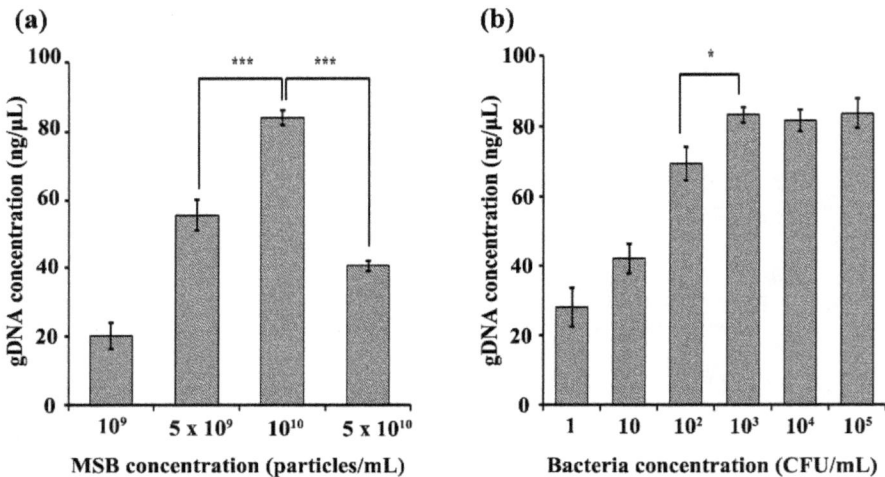

Figure 3. Effect of MSB and bacteria concentrations on DNA purification on the 3DpmFD: (a) gDNA concentration with different concentrations (10^9–5×10^{10} particles/mL) of MSBs for 10^3 CFU/mL of *E. coli* O157:H7; (b) gDNA concentration with different concentrations (1–10^5 CFU/mL) of *E. coli* O157:H7 using 10^{10} particles/mL of MSBs *: $p < 0.05$, ***: $p < 0.001$. Student *t*-test. Sample number = 3.

When the 3DpmFDs packed with 10^{10} MSBs/mL were injected along with 10 mL of PBS containing *E. coli* O157:H7 at different concentrations (1–10^5 CFU/mL) at a rate of 2 mL/min, the gDNA concentration obtained using the 3DpmFDs increased as the bacterial concentrations increased in the range of 1 to 10^3 CFL/mL, but did not increase any further at concentrations of 10^4 and 10^5 CFU/mL (Figure 3b), thus indicating that the surfaces of the MSBs were saturated with gDNA at such high concentrations. For such high concentrations, the sample solutions should be diluted to obtain accurate measurements.

3.3. Effect of Preconcentration and gDNA Purification Using 3DpmFD on Molecular Amplification of Genes in PBS

Once 10 mL of the Gram-negative pathogen *E. coli* O157:H7 at different concentrations (1–10^3 CFU/mL) was preconcentrated and its gDNA was purified in a sequential manner using the 3DpmFD, the gDNA was amplified using either PCR or qPCR to verify the yield. The results were compared to those obtained from either untreated samples or samples prepared with only the preconcentration step using 3DpmFD. In the samples with the preconcentration–purification steps, a concentration as low as 1 CFU/mL was detectable using PCR with gel electrophoresis (Figure 4c), whereas in the untreated samples and samples with only the preconcentration step, concentrations as low as 10^3 CFU/mL and 10 CFU/mL were detectable, respectively (Figure 4a,b). A similar trend was observed with qPCR. Considering that Ct values below 35 are reliable, a concentration as low as 1 CFU/mL was detectable in the samples with the preconcentration–purification steps using qPCR because its Ct value was 30.7 (Figure 4f). Concentrations as low as 10^3 CFU/mL and 10 CFU/mL were respectively detectable in the untreated samples and samples with only the preconcentration step because their respective Ct values were 30.8 and 33.4 (Figure 4d,e). This remarkable improvement in the samples with the preconcentration–purification steps in detection using PCR and qPCR can be explained as follows. In the sample with only the preconcentration step, the Ab-MNPs can improve the detection by providing higher numbers of bacterial cells to the PCR and qPCR. However, Ab-MNPs and cell debris can inhibit DNA polymerase, so that the benefit offered by the bacterial preconcentration is decreased due to the modest interference from Ab-MNPs and cell debris in DNA amplification [22]. By including both preconcentration and purification steps in the 3DpmFD, only purified gDNA can be provided to the PCR and qPCR. This results in the enhancement in detection using PCR and qPCR.

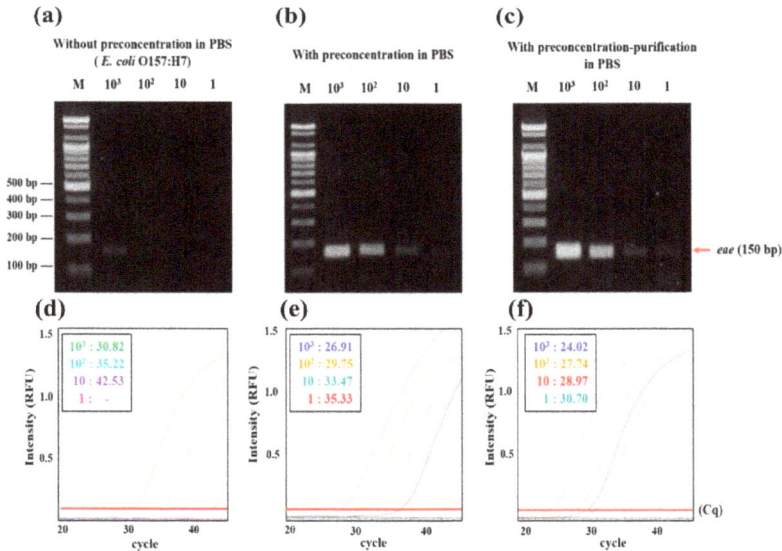

Figure 4. Confirmation of *E. coli* O157:H7 preconcentrating and DNA purifying efficiency using either PCR with gel electrophoresis and qPCR; (**a**) Gel electrophoresis of PCR products (*eae* gene) from 10 mL PBS containing *E. coli* O157:H7 at different concentrations (1–10^3 CFU/mL) before preconcentration; (**b**) after preconcentration; (**c**) after bacterial preconcentration and DNA purification; (**d**–**f**) qPCR results of (**a**–**c**), respectively.

The performance of the 3DpmFDs were further tested by preconcentrating the Gram-positive pathogen *S. aureus* at different concentrations (1–10^3 CFU/mL) and purifying its gDNA using the 3DpmFD. Similar to the results (Figure 4) for *E. coli* O157:H7, significant improvements in the detection of *S. aureus* by both molecular diagnostics techniques, PCR post gel electrophoresis and qPCR, were observed in the samples with the preconcentration–purification steps. The lowest concentration for the detection in the samples with both preconcentration–purification steps using

PCR post gel electrophoresis was 10^2 CFU/mL (Figure 5c), whereas those for the samples with only the preconcentration step were 10^3 CFU/mL (Figure 5b). However, *S. aureus* at all the concentrations (1–10^3 CFU/mL) in the untreated samples was not detectable (Figure 5a). A similar trend was observed using qPCR. The cell walls of Gram-positive bacteria are thicker than those of the Gram-negative ones [7] and are hard to lyse by thermocycling of PCR. As a result, even at 10^3 CFU/mL, the concentration of gDNA released from the lysed cells was not sufficient to be detected using PCR gel electrophoresis and qPCR. With the preconcentration step, bacterial numbers increased and their gDNA was detectable in the sample containing 10^3 CFU/mL. Further improvement in the detection was possible when the preconcentration and purification steps were included in the 3DpmFD because the Lysis and Binding buffer in the DNA purification step induces cell lysis. Together with the results in samples containing *E. coli* O157:H7, these results suggest that gDNA purification is required as an addition in the 3DpmFD to improve the detection of bacterial pathogens using molecular diagnostic methods.

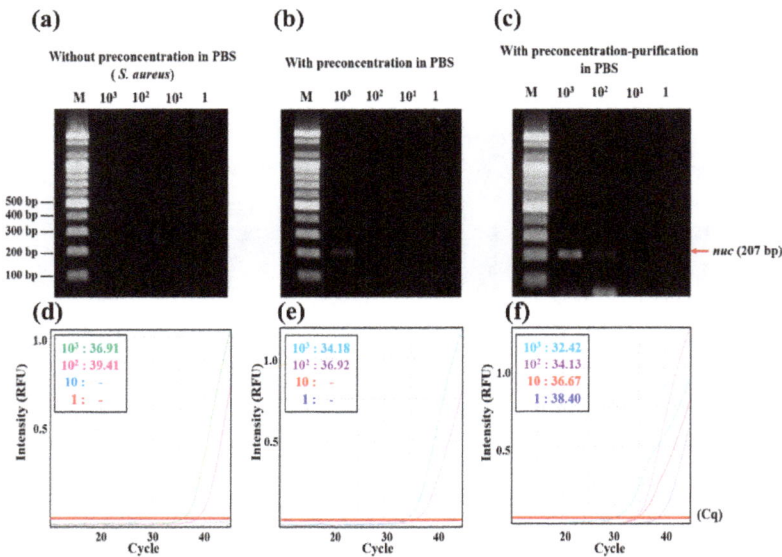

Figure 5. Confirmation of *Staphylococcus aureus* preconcentrating and DNA purifying efficiency using PCR with gel electrophoresis and qPCR; (**a**) Gel electrophoresis of PCR products (*nuc* gene) from 10 mL PBS containing *S. aureus* at different concentrations (1–10^3 CFU/mL) before preconcentration; (**b**) after preconcentration; (**c**) after bacterial preconcentration and DNA purification; (**d–f**) qPCR results of (**a–c**), respectively.

3.4. Dilution Effect on Preconcentration–Purification Efficiency in Blood Samples

We investigated the effect of dilution of blood on the preconcentration–purification efficiency by 3DpmFD. The bacteria capturing efficiency increased as the dilution factor increased. It was about 82% in the whole blood, while it was about 100% in 10% blood. These results suggest that the capturing efficiency was negatively affected by blood and blood dilution is required for the preconcentration step (Figure S1).

Post gel electrophoresis with PCR (Figure 6a) shows that blood should be diluted at least 50% to detect 10^5 CFU/ml of *E. coli* O157:H7 in the blood samples (Figure 6a). A similar trend was observed in the qPCR result (Figure 6b). There, 10^5 CFU/ml of *E. coli* O157:H7 in all the diluted blood samples was detectable because their Ct values were lower than 35. However, Ct values increased as the ratio of blood in the diluted blood samples increased, suggesting that the ratio of blood should be lowered to 10%.

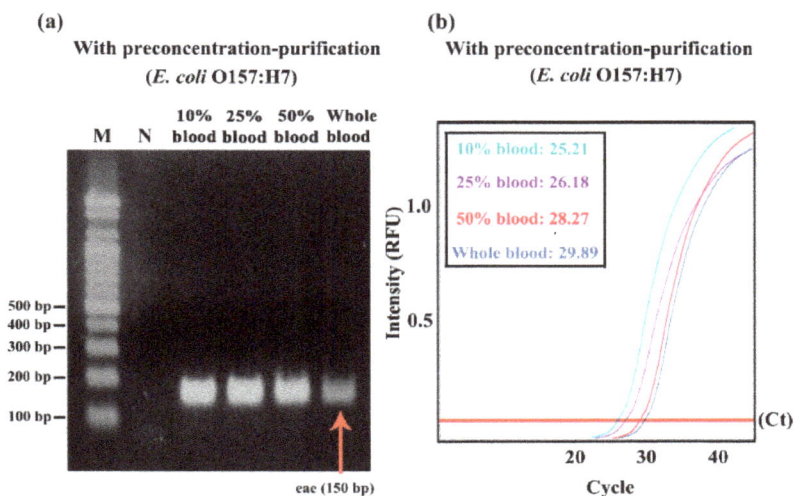

Figure 6. Confirmation of effect of dilution factor of blood on preconcentrating and gDNA purification efficiency with *E. coli* O157:H7 using PCR with gel electrophoresis and qPCR; (**a**) Gel electrophoresis of PCR products (*eae* gene) from different blood dilutions (10, 25, 50, and blood) containing 10^5 CFU/mL of *E. coli* O157:H7 after bacterial preconcentration and DNA purification; (**b**) qPCR result.

3.5. Spike Test in 10% Blood

Blood has a lot of molecular diagnostic inhibitors, such as hematin and haemoglobin, and they prevent the action of Taq polymerase [23]. IgG in the blood also interferes with the activation of Taq polymerase as it interacts with single-strand DNA at high temperatures [24]. To minimize this inhibitory effect on molecular amplification, users have to use commercial kits and systems for purifying bacterial DNA from blood [25–28]. The QIAamp blood mini kit can purify DNA in 200 μL of blood without any dilution but cannot differentiate bacterial DNA from the DNA of blood cells. Microfluidic devices are not suitable for processing a large volume of samples, such as 10 mL, due to its slow flow rates in the range of 1–10 μL/min [27,29]. The hollow spinning disk system does not require the dilution of blood samples but only about 40% of bacteria can be isolated from the plasma by the system [30]. Figure 7 shows that the performance of the 3DpmFD for the sample preparation from 10 mL of 10% diluted blood with *E. coli* O157:H7 is excellent because *E. coli* O157:H7 can be detected for a concentration of up to 1 CFU/mL with the preconcentration and purification steps, which is 100 and 10 times lower than without sample treatment (Figure 7a,b). These results are also confirmed by qPCR through the difference between the Cq values of the samples (Figure 7d–f). The Cq value of the 1 CFU/mL is 30.6 and that of the 10^3 CFU/mL without sample preparation is 33.8. These results suggest that the 3DpmFD can selectively preconcentrate the bacterial pathogen of interest and purify bacterial DNA from diluted blood samples.

For this reason, Figure 7a shows that the limit of detection is 10^3 CFU/mL without sample preparation. Thus, it is very hard to detect pathogens in blood samples using PCR without any sample preparation steps. However, Figure 7c shows that the 3DpmFD can eliminate the molecular diagnostic inhibitors effectively. This is equivalent to a 50-fold improvement of sensitivity compared with the microfluidic device for the detection of *E. coli* O157:H7 in 10% blood [31]. These results suggest that

3DpmFD is well-suited for improving the level of detection (LOD) in blood by including both bacterial preconcentration and DNA purification steps.

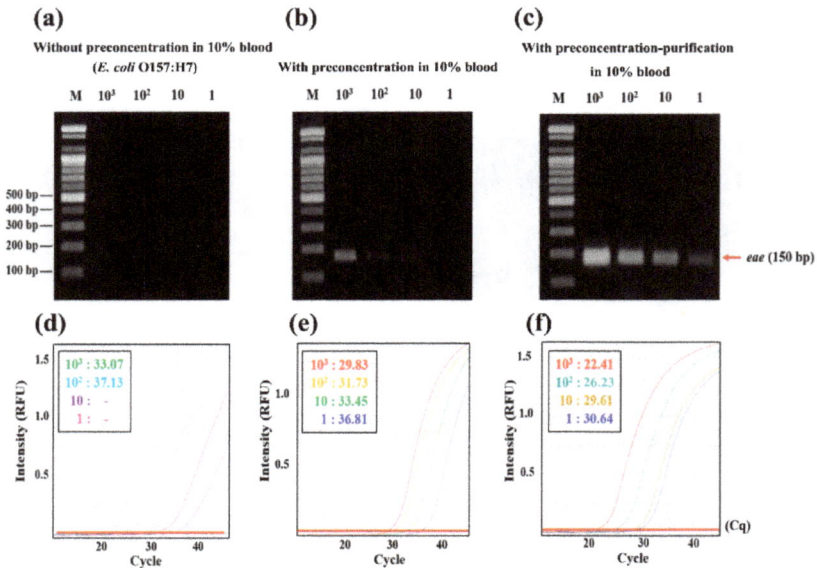

Figure 7. Confirmation of preconcentrating and DNA purification efficiency of *E. coli* O157:H7 in 10% blood using PCR with post gel electrophoresis after PCR (**a**–**c**) and qPCR (**d**–**f**). (**a**) Gel electrophoresis of PCR products (*eae* gene) from 10 mL of 10% blood containing *E. coli* O157:H7 at different concentrations $(1-10^3 \text{ CFU/mL})$ without preconcentration; (**b**) after preconcentrating; (**c**) after preconcentrating and DNA purification; (**d**–**f**) qPCR results of (**a**–**c**), respectively.

4. Conclusions

In this study, we report an improvement in detection for molecular diagnostics by successfully integrating both bacterial preconcentration and DNA purification in the 3DpmFD. The performance of the 3DpmFD shows that it is possible to preconcentrate very low concentrations of pathogens in 10 mL of 10% blood within 30 min, and 1 CFU/mL of *E. coli* O157:H7 can be detected by either PCR with post gel electrophoresis or qPCR. These results demonstrate that it is highly useful for improving molecular diagnostic methods in blood samples with the addition of the DNA purification step.

Supplementary Materials: The following are available online at t http://www.mdpi.com/2072-666X/9/9/472/s1, Figure S1: The effect of the dilution factor of blood on capturing efficiency with 10^3 CFU/mL of *E. coli* O157:H7 *: $p < 0.05$, Student's *t*-test. Sample number = 3.

Author Contributions: Conceptualisation, Y.K. and S.P.; Methodology, Y.K. and J.L.; Data Curation, Y.K. and J.L.; Writing (Original Draft Preparation), Y.K. and J.L.; Writing (Review & Editing), J.L. and S.P.; Supervision, S.P.

Acknowledgments: This research was supported by the BioNano Health-Guard Research Center as a Global Frontier Project (H-guard NRF-2018M3A6B2057299) through the National Research Foundation (NRF) of the Ministry of Education, Science and Technology (MEST) in Korea.

Conflicts of Interest: The authors declare no conflict of interest.

Micromachines **2018**, *9*, 472

References

1. Yang, L.; Banada, P.P.; Chatni, M.R.; Lim, K.S.; Bhunia, A.K.; Ladisch, M.; Bashir, R. A multifunctional micro-fluidic system for dielectrophoretic concentration coupled with immuno-capture of low numbers of listeria monocytogenes. *Lab Chip.* **2006**, *6*, 896–905. [CrossRef] [PubMed]

2. Hara-Kudo, Y.; Takatori, K. Contamination level and ingestion dose of foodborne pathogens associated with infections. *Epidemiol. Infect.* **2011**, *139*, 1505–1510. [CrossRef] [PubMed]

3. Varshney, M.; Li, Y.; Srinivasan, B.; Tung, S. Microfluidics and interdigitated array microelectrode-based impedance biosensor in combination with nanoparticles immunoseparation for detection of *Escherichia coli* O157:H7 in food samples. *Sens. Actuators B Chem.* **2007**, *128*, 99–107. [CrossRef]

4. Pengsuk, C.; Chaivisuthangkura, P.; Longyant, S.; Sithigorngul, P. Development and evaluation of a highly sensitive immunochromatographic strip test using gold nanoparticle for direct detection of Vibrio cholerae O139 in seafood samples. *Biosens. Bioelectron.* **2013**, *42*, 229–235. [CrossRef] [PubMed]

5. Weagent, S.D.; Bound, A.J. Evaluation of techniques for enrichment and isolation of *Escherichia coli* O157:H7 from artificially contaminated sprouts. *Int. J. Food. Microbiol.* **2001**, *71*, 87–92. [CrossRef]

6. Banada, P.P.; Chakravorty, S.; Shah, D.; Burday, M.; Mazzella, F.M.; Alland, D. Highly sensitive detection of Staphylococcus aureus directly from patient blood. *PLoS ONE* **2012**, *7*, 31126. [CrossRef] [PubMed]

7. Chung, Y.C.; Jan, M.S.; Lin, Y.C.; Lin, J.H.; Cheng, W.C.; Fan, C.Y. Microfluidic chip for high efficiency DNA extraction. *Lab Chip.* **2004**, *4*, 141–147. [CrossRef] [PubMed]

8. Karle, M.; Miwa, J.; Czilwik, G.; Auwärter, V.; Roth, G.; Zengerle, R.; von Stetten, F. Continuous microfluidic DNA extraction using phase-transfer magnetophoresis. *Lab Chip.* **2010**, *10*, 3284–3290. [CrossRef] [PubMed]

9. Geng, T.; Bao, N.; Sriranganathanw, N.; Li, L.; Lu, C. Genomic DNA extraction from cells by electroporation on an integrated microfluidic platform. *Anal. Chem.* **2012**, *84*, 9632–9639. [CrossRef] [PubMed]

10. Azimi, S.M.; Nixon, G.; Ahern, J.; Balachandran, W. A magnetic bead-based DNA extraction and purification microfluidic device. *Microfluid. Nanofluid.* **2011**, *11*, 157–165. [CrossRef]

11. Witek, M.A.; Llopis, S.D.; Wheatley, A.; McCarley, R.L.; Soper, S.A. Purification and preconcentration of genomic DNA from whole cell lysates using photoactivated polycarbonate (PPC) microfluidic chips. *Nucleic Acids Res.* **2006**, *34*, e74. [CrossRef] [PubMed]

12. Mahalanabis, M.; Al-Muayad, H.; Kulinski, M.D.; Altman, D.; Klapperich, C.M. Cell lysis and DNA extraction of gram-positive and gram-negative bacteria from whole blood in a disposable microfluidic chip. *Lab Chip.* **2009**, *9*, 2811–2817. [CrossRef] [PubMed]

13. Xia, N.; Hunt, T.P.; Mayers, B.T.; Alsberg, E.; Whitesides, G.M.; Westervelt, R.M.; Ingber, D.E. Combined microfluidic-micromagnetic separation of living cells in continuous flow. *Biomed. Microdevices* **2006**, *8*, 299. [CrossRef] [PubMed]

14. Ganesh, I.; Tran, B.M.; Kim, Y.; Kim, J.; Cheng, H.; Lee, N.Y.; Park, S. An integrated microfluidic PCR system with immunomagnetic nanoparticles for the detection of bacterial pathogens. *Biomed. Microdevices* **2016**, *18*, 116. [CrossRef] [PubMed]

15. Jayamohan, H.; Gale, B.K.; Minson, B.; Lambert, C.J.; Gordon, N.; Sant, H.J. Highly sensitive bacteria quantification using immunomagnetic separation and electrochemical detection of guanine-labeled secondary beads. *Sensors* **2015**, *15*, 12034–12052. [CrossRef] [PubMed]

16. Lein, K.Y.; Liu, C.J.; Lin, Y.C.; Kuo, P.L.; Lee, G.B. Extraction of genomic DNA and detection of single nucleotide polymorphism genotyping utilizing an integrated magnetic bead-based microfluidic platform. *Microfluid. Nanofluid.* **2009**, *6*, 539–555. [CrossRef]

17. Park, C.; Lee, J.; Kim, Y.; Lee, J.; Park, S. 3D-Printed microfluidic magnetic preconcentrator for the detection of bacterial pathogen using an atp luminometer and antibody-conjugated magnetic nanoparticles. *J. Microbiol. Methods* **2017**, *132*, 128–133. [CrossRef] [PubMed]

18. Lee, W.; Kwon, D.; Choi, W.; Jung, G.Y.; Au, A.K.; Folch, A.; Jeon, S. 3D-Printed microfluidic device for the detection of pathogenic bacteria using size-based separation in helical channel with trapezoid cross-section. *Sci. Rep.* **2015**, *5*, 7717. [CrossRef] [PubMed]

19. Lee, W.; Kwon, D.; Chung, B.; Jung, G.Y.; Au, A.; Folch, A.; Jeon, S. Ultrarapid detection of pathogenic bacteria using a 3D immunomagnetic flow assay. *Anal. Chem.* **2014**, *86*, 6683–6688. [CrossRef] [PubMed]

Micromachines **2018**, *9*, 472

20. Zhao, L.; Yang, B.; Dai, X.; Wang, X.; Gao, F.; Zhang, X.; Tang, J. Glutaraldehyde mediated conjugation of amino-coated magnetic nanoparticles with albumin protein for nanothermotherapy. *J. Nanosci. Nanotechnol.* **2010**, *10*, 7117–7120. [CrossRef] [PubMed]

21. Herigstad, B.; Hamilton, M.; Heersink, J. How to optimize the drop plate method for enumerating bacteria. *J. Microbiol. Methods* **2001**, *44*, 121–129. [CrossRef]

22. Higashi, T.; Minegishi, H.; Nagaoka, Y.; Fukuda, T.; Echigo, A.; Usami, R.; Maekawa, T.; Hanajiri, T. Effects of superparamagnetic nanoparticle clusters on the polymerase chain reaction. *Appl. Sci.* **2012**, *2*, 303–314. [CrossRef]

23. Akane, A.; Matsubara, K.; Nakamura, H.; Takahashi, S.; Kimura, K. Identification of the heme compound copurified with deoxyribonucleic acid (DNA) from bloodstrains, a major inhibitor of polymerase chain reaction (PCR) amplification. *J. Forensic Sci.* **1994**, *39*, 362–372. [CrossRef] [PubMed]

24. Al-Soud, W.A.; Jönsson, L.J.; Rådström, P. Identification and characterization of immunoglobulin G in blood as a major inhibitor of diagnostic PCR. *J. Clin. Microbiol.* **2000**, *38*, 345–350. [PubMed]

25. Hassan, M.M.; Ranzoni, A.; Cooper, M.A. A nanoparticle-based method for culture-free bacterial DNA enrichment from whole blood. *Biosens. Bioelectron.* **2018**, *99*, 150–155. [CrossRef] [PubMed]

26. Lopes, A.L.K.; Cardoso, J.; dos Santos, F.R.C.C.; Silva, A.C.G.; Stets, M.I.; Zanchin, N.I.T.; Soares, M.J.; Krieger, M.A. Development of a magnetic separation method to capture sepsis associated bacteria in blood. *J. Microbiol. Methods* **2016**, *128*, 96–101. [CrossRef] [PubMed]

27. Zelenin, S.; Hansson, J.; Ardabili, S.; Ramachandraiah, H.; Brismar, H.; Russom, A. Microfluidic-based isolation of bacteria from whole blood for sepsis diagnostics. *Biotechnol. Lett.* **2015**, *37*, 825–830. [CrossRef] [PubMed]

28. Jin, C.E.; Koo, B.; Lee, E.Y.; Kim, J.Y.; Kim, S.-H.; Shin, Y. Simple and label-free pathogen enrichment via homobifunctional imidoesters using a microfluidic (slim) system for ultrasensitive pathogen detection in various clinical specimens. *Biosens. Bioelectron.* **2018**, *111*, 66–73. [CrossRef] [PubMed]

29. Wu, Z.; Willing, B.; Bjerketorp, J.; Jansson, J.K.; Hjort, K. Soft inertial microfluidics for high throughput separation of bacteria from human blood cells. *Lab Chip.* **2009**, *9*, 1193–1199. [CrossRef] [PubMed]

30. Alizadeh, M.; Wood, R.L.; Buchanan, C.M.; Bledsoe, C.G.; Wood, M.E.; McClellan, D.S.; Blanco, R.; Ravsten, T.V.; Husseini, G.A.; Hickey, C.L.; et al. Rapid separation of bacteria from blood—Chemical aspects. *Colloids Surf. B* **2017**, *154*, 365–372. [CrossRef] [PubMed]

31. Wang, S.; Inci, F.; Chaunzwa, T.L.; Ramanujam, A.; Vasudevan, A.; Subramanian, S.; Chi Fai Ip, A.; Sridharan, B.; Gurkan, U.A.; Demirci, U. Portable microfluidic chip for detection of *Escherichia coli* in produce and blood. *Int. J. Nanomed.* **2012**, *7*, 2591.

micromachines

MDPI

Article

Fabrication of a Malaria-Ab ELISA Bioassay Platform with Utilization of Syringe-Based and 3D Printed Assay Automation

Christopher Lim [1], Yangchung Lee [1] and Lawrence Kulinsky [2,*]

[1]   Department of Chemical Engineering and Materials Science, University of California, Irvine,
      916 Engineering Tower, Irvine, CA 92627-2575, USA; cwplim@gmail.com (C.L.);
      Yangchung_Lee@amat.com (Y.L.)
[2]   Department of Mechanical and Aerospace Engineering, University of California, Irvine,
      5200 Engineering Hall, Irvine, CA 92627-2700, USA
*   Correspondence: lkulinsk@uci.edu; Tel.: +1-949-824-6769

Received: 30 August 2018; Accepted: 30 September 2018; Published: 2 October 2018

Abstract: We report on the fabrication of a syringe-based platform for automation of a colorimetric malaria-Ab assay. We assembled this platform from inexpensive disposable plastic syringes, plastic tubing, easily-obtainable servomotors, and an Arduino microcontroller chip, which allowed for system automation. The automated system can also be fabricated using stereolithography (SLA) to print elastomeric reservoirs (used instead of syringes), while platform framework, including rack and gears, can be printed with fused deposition modeling (FDM). We report on the optimization of FDM and SLA print parameters, as well as post-production processes. A malaria-Ab colorimetric test was successfully run on the automated platform, with most of the assay reagents dispensed from syringes. Wash solution was dispensed from an SLA-printed elastomeric reservoir to demonstrate the feasibility of both syringe and elastomeric reservoir-based approaches. We tested the platform using a commercially available malaria-Ab colorimetric assay originally designed for spectroscopic plate readers. Unaided visual inspection of the assay solution color change was sufficient for qualitative detection of positive and negative samples. A smart phone application can also be used for quantitative measurement of the assay color change.

Keywords: diagnostics; liquid handling; microfluidics; multiplex assays and technology; stereolithography; sample preparation; 3D printing

1. Introduction

1.1. Preface

Lab-on-a-chip systems are used to perform personalized health diagnostic tests (like bioassays) away from the lab [1]. Implementation of bioassay automation reduces the cost of medical personnel and diminishes the incidence of human errors [2,3]. Point-of-care (POC) diagnostic platforms are required to have low cost, low power use, be reliably automated, and free of sophisticated detection technologies [4]. POC platforms have already been realized in the form of lateral flow immunoassays (pregnancy tests) and paper fluidic colorimetric assays [5]. Unfortunately, the aforementioned POC devices are based on capillary flow, and therefore do not work well when more complex multi-step bioassays are performed [6]. In these cases, some other fluid propulsion mechanisms instead of capillary flow are required. There are many POC devices that rely on active propulsion techniques, like centrifugal propulsion, electrowetting, and magnetic beads [7,8]. These tests typically entail increased fabrication and material cost, complex automation schemes, and sophisticated hardware.

In this work, we developed two distinct approaches for the realization of an inexpensive automated colorimetric immunoassay with multiple wash steps: (1) A fused deposition modeling (FDM)-printed (non-disposable) frame was used and a disposable fluidic chip that includes an elastomeric dome and fluidic channels fabricated using SLA, was printed, where the fluidic movement was facilitated by servomotors pushing on the elastomeric dome, and propelling the reagents or wash from the domed reservoirs through fluidic channels to the test chamber, or (2) an FDM-printed non-disposable frame was used, with servomotors connected to standard inexpensive and readily-available disposable plastic syringes filled with wash and reagents to automate the steps of the assay.

Draining of the test chamber was performed with a syringe attached to a servomotor, where negative pressure was created by the pulling on the syringe plunger. All automation was controlled by a program uploaded to the Arduino-based electronic board of the platform via a computer. The test results of the colorimetric bioassay can be assessed by eye (for qualitative measurements) or with a smart phone for quantitative measurement, and e-mailed or texted to a hospital or doctor's office [9,10]. In this work, we created a proof-of-principle platform that utilized elements of both approaches (1) and (2), as outlined above.

We used plastic syringes for most of the platform reagent reservoirs, including drainage and waste. The wash reservoir consisted of an SLA-printed elastomeric dome. Thus, the feasibility of both approaches was tested at the same time. The automated platform demonstrated in the presented work proved the viability of both approaches: (1) Construction of an automated bioassay platform using syringes only, or (2) a printed dome-based bioassay platform. To our knowledge, this study constitutes the first demonstration of an SLA-based dome bioassay platform and a syringe-based automated bioassay platform. The concepts developed in the present work build on prior research that utilized FDM to fabricate a fluidic chip with embedded microchannels [11]. Fluid leakage was a recurring issue in the design, due to the layer-by-layer deposition nature of FDM. Our present fabrication approach utilizes syringes and photocurable resin crosslinked using SLA, thus mitigating the problem of fluid leakage. The present platform employs FDM printing for the fabrication of parts that are not in contact with fluids, such as the non-disposal platform frame and the gear/rack mechanisms. Additionally, the prior fluidic platform design used molded silicone domes that needed to be sealed to the body of 3D printed fluidic chip. Our present fabrication route avoids the need to seal the domes to the plastic chip by 3D-printing the integrated dome.

The proposed approach is useful in making POC bioassays more widespread. All of the parts and materials required for the demonstrated platforms are inexpensive and widely available. For example, a hospital in rural India with an FDM printer could be sent a stereolithography file (STL file) to print a frame. The Arduino electronic board and other parts, such as servomotors, can be easily acquired. All the parts and materials to produce our automated platform cost less than $100, while all disposable materials, including syringes, tubing, and reagents, cost around $5 per test.

If an exclusively syringe-based platform is utilized, the only disposable parts needed are plastic syringes. The syringes would be filled with a prescribed volume of reagents, and the required program uploaded to the Arduino board. This approach provides the most flexibility to perform a wide range of bioassays on the spot, as it does not require all the reagents in typical pre-packaged volumes. This simplifies assay logistics, including that of cold storage of reagents [12]. We expect that our proposed recipes and programs will be readily available for download by non-profit organizations and reagent makers.

There is a growing interest in manufacturing functional parts inexpensively with high-aspect-ratio geometry and complex topologies [13]. Additive manufacturing (AM), including various types of 3D printing, is an increasingly popular form of fabrication [14]. The AM process requires a part to be designed using a CAD program, such as SolidWorks, followed by the extrapolation of the structural information into an STL file. The STL file partitions the CAD drawing into a series of volumetric pixels, which are digitally sliced into layers along the Z direction [15]. These layers are physically deposited onto a substrate in layer-by-layer manner, using various AM methods, including FDM and SLA. FDM and SLA accommodate a variety of material choices, including elastomeric materials, allowing for the fabrication of flexible parts such as the dome-based elastic pumps presented in this work.

In this work, we conducted a brief review of microfluidic device fabrication methods, and emphasized recent developments in additive manufacturing. We followed this review with a description of the fabrication processes selected for our bioassay platform. Lastly, we detailed an outline of our experimental procedures, and described the experimental results of a malaria bioassay performed on our automated platform.

1.2. Fabrication Techniques for Microfluidic Devices

The first traditional microfluidic devices were fabricated using glass and silicon, utilizing mostly a toolbox of lithographic techniques employed in the semiconductor industry [16]. Equipment selection was subsequently further developed to include femtosecond lasers to fabricate glass microfluidic chips [17,18], and the selection of materials employed for fluidic chips was also expanded to include such materials as "liquid Teflon" [19] and environmentally-sensitive materials such as hydrogels [20]. With the advent of soft lithography and 3D printing, there was a push to develop inexpensive and reliable microfluidic platforms with biocompatible plastics and resins, demonstrated by the recent fabrication trends outlined below [21–24].

1.2.1. Soft Lithography

Soft lithography is based on a three-step process where microfabrication techniques are first used to produce a reusable mold, which is then filled with a curable epoxy Polydimethylsiloxane (PDMS) and subsequently separated from the mold after an appropriate curing time. The molded part is typically attached to a glass substrate using plasma treatment [25]. The soft lithography approach allows for the usage of more expensive high-precision cleanroom lithographic techniques to produce a mold. Once the mold is produced, it can be used multiple times to fabricate identical fluidic chips, dramatically reducing the cost of individual fluidic microdevices.

The soft lithography approach has allowed for much more widespread adoption of lab-on-a-chip devices, and for the production of intricate parts such as elastomeric valves. For example, the PDMS valves used by the Quake group have facilitated the implementation of large-scale microfluidic bioreactors for drug discovery and other applications [26]. However, when the device is assembled from separately produced parts, the manufacturability is restricted, due to cumbersome alignment, bonding, and assembly [24,27].

3D printing presents an approach where a complete microfluidic device is produced without post-processing alignment and assembly steps. Commercial 3D printers are beginning to challenge the resolutions achievable by soft lithography [28].

1.2.2. Inkjet 3D Printing (i3DP)

Inkjet 3D printing is based on inkjet technology and can operate in continuous or drop-on-demand mode. To achieve high accuracy performance in i3DP, there are four critical elements that must be considered: ink material, substrate properties, printing platform, and droplet generation [29]. Inconsistency or change in any of the parameters affects the process reliability. To maintain steady performance, i3DP needs to be used regularly; the clogging of apertures in a print head due to drying of the ink is a persistent problem. There is considerable cost involved in changing from one material to another, since the previous material must be flushed out by the new resin.

One of the greatest advantages of i3DP is its ability to deliver multiple materials at the same time during the fabrication process, allowing for a wide range of material properties (hard and soft plastics, elastomers), and the utilization of inks of different color [30,31]. Creating prototypes with smooth finishes and complex shapes is possible with i3DP; flow channels have been integrated with porous semi-permeable membranes supporting cell culture to study the transport and profiling of drugs [32,33].

Hwang et al. utilized i3DP to print pillars with a diameter of 250 µm and found that the resolution of the process depended on the droplet size, the printer nozzle spacing, and the reflow of the material prior to UV curing. These factors affected the droplet spreading, changing the final dimensions of the printed devices [34].

Paydar et al. examined multi-material 3D printing for microfluidic interconnects. They fabricated a specialized interconnector part, composed of two rigid clamps for the mechanical attachment of a flexible elastomeric O-ring gasket to a microfluidic device. The parts were printed in a single step, eliminating the need for adhesives or additional assembly. While the cost of manufacturing was low, the interconnector was characterized by low maximum sealing pressure due to material fatigue [35].

1.2.3. Two-Photon Polymerization (2PP)

Two-photon polymerization (2PP) is a laser-based technique that utilizes a femtosecond laser to create 3D structures in the bulk of photocurable epoxy resin, employing a highly localized process where two photons are absorbed simultaneously by the molecule being cured [36]. The synergistic effects of optical, chemical, and material non-linearity make it possible to achieve reproducible resolution of tens of nanometers [37].

2PP has shown great potential for fabrication in microfluidics. Kumi et al. described the fabrication of a master for casting PDMS with rectangular microchannels of high aspect ratios by modifying SU-8 resist (material for the master) with a photoacid generator that then allows for the use of 2PP on SU-8 resist. By using the modified resin, the fabrication speed was also increased from 200 $\mu m \cdot s^{-1}$ to 10,000 $\mu m \cdot s^{-1}$ with a print time of 1 h. This technique required extensive resin preparation and had a slow build speed [38].

Kawata et al. achieved resolutions of 120 nm, and other attempts included the use of new photo-initiators, a continuous scanning mode, a shorter wavelength, a longer exposure time, and confining the polymerization phenomenon using a quencher molecule [39]. Sugioka et al. conducted a comprehensive review of the fundamentals and fabrication of 3D micro- and nano-components based on 2PP [40].

While 2PP currently produces the highest resolution for microfabricated three-dimensional structures, it is a very time-consuming process: the time required to fabricate a 1 mm^3 volume microfluidic structure exceeds 104 days [41]. The high cost of femtosecond lasers, positioning systems, optics, and the difficulty of working with multi-material systems are some factors hindering the utilization of 2PP for the mass-production of microfluidic devices.

1.2.4. Fused Deposition Modeling (FDM)

FDM is one of the most widely used additive manufacturing techniques. Many polymers used with FDM techniques are inherently biocompatible [42]. In FDM, filament material is extruded through nozzles and deposited onto a heated substrate [43]. Due to the inherent propensity of melted fiber to solidify as a line, there are limitations in the dimensional accuracy and the surface texture of the produced parts, resulting in a staircase pattern that is ubiquitous on many parts that are produced with FDM.

Lee et al. printed microfluidic features via FDM and evaluated the printing resolution, accuracy, biocompatibility, and surface roughness of acrylonitrile butadiene styrene (ABS) with P430 filament. They found that the accuracy of the printed features had an average deviation of 60.8 µm and 71.5 µm along the Y and X axis, respectively. The surface of the printed channels was rough with protruding filament strands [44].

Kitson et al. fabricated polypropylene reactionware with cylindrical channels 0.8 mm in diameter. The devices could be fabricated in a few hours and could avoid blockages due to the formation of precipitates. The potential of 3D manufacturing was demonstrated by stopping mid-point in the fabrication process to deposit solid reagents into a chamber, which was then sealed with the printer. This is a valuable feature not so easily realized with i3DP or stereolithography [45].

Bishop et al. created a semi-transparent fluidic device using poly(ethyleneterephthalate) with threaded ports, enabling the integration of commercial tubing as well as specially designed 3D fittings [46]. The transparent device included 800 μm × 800 μm square channels. A low-cost desktop Makerbot (New York, NY, USA) 3D printer was used for fabrication of the device. Prussian blue nanoparticles were synthesized in their lab and mixed in a 3D printed channel and applied to electrode surfaces for sensing of H_2O_2.

Recently, Dolomite (Royston, UK) launched a production line that offered an FDM printer specifically designed for fabrication of microfluidic platforms [47]. The printer head design allowed for reliable printing of cyclic olefin copolymer (COC). Their new software guided the production to guarantee smooth surface finish inside the channels. This is contrasted with conventional 3D printing, which emphasizes outside surface texture.

1.2.5. Stereolithography (SLA)

SLA is an attractive option for microfluidics due to increasing availability of SLA printers utilizing inexpensive micromirror-based projectors, including some printer kits costing as little as $100 [48]. The performance of SLA printers is quantified by the dimensional accuracy and the surface roughness of the printed object [49,50]. These factors are influenced by the fabrication settings: object orientation, layer thickness, resin properties, and build style. The minimum cross-sectional area of a microchannel made by SLA depends on the laser spot size and the resin viscosity. This resin must be drained post-print [51].

Recent developments have expanded the SLA material selection for a single print to include elastomers and ceramics, while some printers are capable of using multiple resins. It is predicted that in the future it will be possible to use SLA to fabricate metallic sensors and actuators on flexible membranes [51–53].

Comina et al. used a Miicraft (Hsinchu, Taiwan) 3D printer to fabricate a reusable mold for PDMS casting [54]. The molds had structures of multiple feature sizes ranging from 50 μm to several mm. The resin had to be manually coated with protective ink to be properly used with PDMS. The Miicraft was also used to print a device with open fluidic channels that are subsequently sealed on top with the adhesive tape. In another study, Comina et al. printed a unibody lab-on-a-chip (LOC) consisting of a separate microfluidic level, and a layer with optical components. The integrated finger pump was used to initiate the preparatory sequence of mixing two reagents and three analytes. The colorimetric glucose sensing assay was read by a smart phone [55].

Wang et al. demonstrated an effective approach to fabricating structural devices using 3D printing. A monomer initiator was added to the Miicraft resin to allow for modification of the surface properties such as hydrophobicity and hydrophilicity [56].

Shallan et al. used a Miicraft printer for the fabrication of a transparent microfluidic device with enclosed 250 μm diameter channels [57]. The dimension of the printed channels had a deviation of 50 to 100 μm from the designed dimensions, and the roof of the sealing channel was rough. This could be improved through changing the curing depth, intensity, exposure wavelength, and time. They argued that inexpensive Miicraft provides a sufficient resolution for most microfluidic devices.

Patrick et al. printed fluidic open channels using a laser-rastering SLA printer (Form1+, Formlabs, Somerville, MA, USA) [58]. They found that the smallest achievable diameter of a circular channel was 900 μm, and the smallest channel side for a square channel was 650 μm. The surface topology was inspected using SEM and visible striations were discovered. Overall, the cost of the materials for each fluidic chip was around $6.

A sample library of standardized microfluidic components was manufactured using SLA by Lee et al. and Bhargava et al. [59,60]. These parts were then used to create a number of modular and reconfigurable microfluidic units. This approach allowed for the creation of complex microfluidic designs based on simple interlocking fluidic modules. The library allowed for the further miniaturization of microfluidic elements and materials.

The Folch lab has printed diaphragm valves and a peristaltic pump integrated within a LOC device printed using SLA with biocompatible Somos WaterShed XC 11122 resin (Elgin, IL, USA) [61]. The valves were leakage free at a closing pressure of 6 psi (provided by the compressed air), and they were operated over many cycles. These valves are regarded as functional modules: two valves can be paired to build a switch, or three valves can be put together in a series to build a pump [62]. However, the portability of these microfluidic devices was limited by the need to use peripherals such as gas canisters.

1.3. Combining Several Fabrication Techniques to Manufacture an Automated Fluidic Malaria Enzyme-Linked Immunosorbent Assay (ELISA) Bioassay Platform

Each of the techniques discussed thus far has its own set of advantages and disadvantages. For example, stereolithography, while allowing for higher resolution, is more expensive and less accessible than fused deposition modelling [63]. Therefore, in this work, we applied a combination of fabrication techniques.

In one approach when only an FDM printer is available, we developed a system based on disposable plastic syringes coupled to programmable servomotors. FDM was used to print the non-disposable frame for an automated colorimetric malaria detection test based on enzyme-linked immunosorbent assay (ELISA). In another approach, when the end user has access to an SLA system, we developed a fabrication approach for SLA to produce elastomeric domes with integrated microfluidic channels. Flow of reagents was facilitated by servomotors compressing the elastomeric dome, and propelling the reagents through a microfluidic channel into a connected test chamber. In this approach, the frame was still printed using an FDM printer. We avoided fabricating microfluidic channels using FDM, because filament-based deposition often resulted in a "staircase effect" as discussed previously, causing fabricated structures to be prone to leakages.

2. Materials and Methods

2.1. Fused Deposition Modeling and Stereolithography

The FDM parts were manufactured using a acrylonitrile butadiene styrene (ABS) filament on a dual extruder AirWolf HD2x printer (Costa Mesa, CA, USA). The printing parameters, including extrusion, travel speeds, and printer head temperature, were optimized as discussed in the "Results and Discussion" section below. A Formlabs' (Somerville, MA, USA) Form 1+ SLA printer was used to produce the elastomeric domes and integrated fluidic microchannel from proprietary flexible clear resin GPCL02 sold by Formlabs.

2.2. Automation of Malaria-Ab Colorimetric ELISA

A malaria-Ab ELISA colorimetric detection kit (IBL International GmbH, Hamburg, Germany) was used to test functionality of the fabricated automated platform. The colorimetric assay was designed to detect antibodies in subjects infected with four *Plasmodium* species that cause malaria. The most significant parasitic diseases in humans are: *P. falciparum*, *P. vivax*, *P. ovale*, and *P. malariae* [64].

The ELISA kit contained clear polystyrene wells coated with recombinant antigens. When the test sample (either serum or plasma) was added to the well, the specific antibodies in the sample combined with antigens in the well. Subsequently, a conjugate solution of recombinant antigens conjugated to horseradish peroxidase was added to the well, and these antigens reacted with the specific antibodies, if they were present.

When the substrate solution of urea peroxide and tetramethyl benzidine was added to the well, it led to a change in solution color from colorless (in case of no specific antibodies present) to blue (when antibodies were present), and finally to yellow when the stop solution was added. It is possible to determine the concentration of the antibodies in the sample by measuring the color intensity with spectral analysis, but in this proof-of-concept study, we limited our experiments to the positive and negative controls provided with the kit. The full details on the recommended volumes of reagents and incubation temperatures were provided by the kit manufacturers and they are summarized in Table 1 [65].

Table 1. Malaria-Ab Enzyme-linked Immunosorbant Assay (ELISA) instructions as provided by the kit manufacturer and modified steps executed by the automated bioassay platform.

ELISA Step	Malaria-Ab ELISA Kit Instruction	Automated Assay Modifications
Sample	Add 50 µL of the undiluted sample to a coated well. Mix on a plate shaker for 30 s. Incubate (covered) at 37 °C for 30 min.	50 µL of sample is pipetted into the wells (O). Incubation occurs for 1 hr at room temperature.
Wash	Wash five times with 300 µL of working strength wash buffer. A short soak time of about 30 s is recommended between each wash cycle. Tap out excess liquid.	The well is flushed with 300 µL wash solution contained within the elastomeric dome (N) as the movable arm (M) compresses the dome. The arm is controlled by the servomotor (L) and Arduino board (J). The dirty wash solution is aspirated from the well using an aspiration syringe located on the bottom platform. The syringe is driven by 3D printed rack-and-gear set (B) and the servomotor (C). The wash/aspiration cycle is repeated three times.
Conjugate Incubation	Add 50 µL of diluted (1:10) conjugate to each well. Incubate (covered) at 37 °C for 30 min.	50 µL of diluted malaria conjugate is dispensed into the reagent well (O) from a syringe (F). The barrel of this syringe is pushed using a servomotor (H) controlled by the Arduino board (D). Incubation occurs for 30 min at room temperature.
Wash	See wash step above.	See wash step above.
Substrate Incubation	Add 50 µL substrate/chromogen Tetramethylbenzidine (TMB) mixture to each well. Incubate at room temperature for 30 min. As the substrate is photosensitive, it is recommended that the plate be protected from light during this incubation.	50 µL of TMB substrate solution is dispensed into the reagent well (O) from a syringe on the syringe platform (F). The barrel of this syringe is pushed using a servomotor (G) controlled by the Arduino board (D). Incubation occurs for 15 min.
Stop	Add 50 µL stop solution to each well. (Blue color changes to yellow).	50 µL of stop solution is dispensed into the reagent well (O) from a syringe (F). The barrel of this syringe is pushed using a servomotor (E) controlled by the Arduino board (D).
Detection	Read with fluorescent microplate reader at 450 nm (A450). Use of a reference filter at 620–690 nm will eliminate the effects of scratches, bubbles, etc.	A color change is accessed visually or via a cell phone positioned on the frame (P) with phone camera aligned to the reagent wells (O) to read the results using a colorimetric detection application.

We performed incubation not at the recommended 37 °C, but at a room temperature, usually 20 °C to 23 °C. Where the kit instructions recommended draining the test tube and tapping it to make sure that all fluid exited the tube, we implemented automation where a plastic tube is placed within the test well to reach the bottom. This tube was connected to the aspiration syringe (labeled "I" on Figure 1) whose plunger was pulled out by the gear and rack hardware controlled by the servomotor. Pulling of the plunger created the suction necessary to aspirate solutions from the test well.

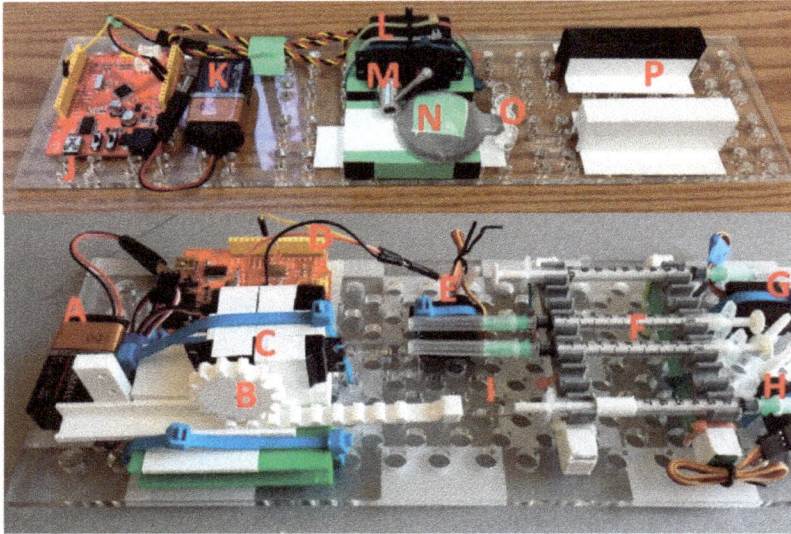

Figure 1. The automated bioassay platform is composed of two stackable boards: The bottom platform contains a battery (A), an aspiration syringe (I) connected via a printed rack-and-gear set (B) to the servomotor (C), which is controlled with an Arduino board (D) that also controls other servomotors (E, G, H) that push on syringes (F) with conjugate, substrate, and stop solutions. The syringes on the bottom board are connected via plastic tubes (4 mm outer diameter, 2 mm inner diameter) to the well (O) on the top platform, where another battery (K) and Arduino board (J) power and control the servomotor (L) that through arm (M), presses on an elastomeric dome (N) containing wash. The top platform also contains a printed frame (P) onto which a smart phone can be placed to detect color changes in the well.

Each fluidic step of the assay was performed by an action of a servomotor controlled through the Arduino board. The servomotors' plastic arms pushed onto the plungers of plastic syringes filled with reagents (conjugate solution, substrate, stop solution). We elected to dispense the wash solution from an SLA-printed elastomeric dome containing integrated microfluidic channels, in order to demonstrate an alternative part to syringes for all steps of the assay. The elastomeric dome is easily substituted with another syringe to transition to a completely syringe-based automated fluidic ELISA platform. The complete automated bioassay platform is presented in Figure 1.

The sequence of the bioassay steps is summarized below. Steps of the malaria-Ab ELISA assay as recommended by the kit manufacturer, and the modified steps are executed by the automated platform are listed in Table 1.

(1) Pipette 50 µL of sample into the wells (O). Incubate for 1 hr at room temperature.
(2) Aspirate the sample from the well. Flush the well with wash solution contained within the elastomeric dome (N). The wash solution is dispensed using the movable arm (M) to compress the dome.
(3) Aspirate the dirty wash solution from the well using an aspiration syringe located on the bottom platform. Repeat three times.

(4) Dispense 50 μL of malaria conjugate into the reagent well (O) from a syringe (F). The barrel of this syringe is pushed using a servomotor (H). Incubate for 30 min at room temperature.

(5) Aspirate the conjugate from the well. Repeat step 3.

(6) Dispense 50 μL of the TMB substrate solution into the reagent well (O) from a syringe on the syringe platform (F). The barrel of this syringe is pushed using a servomotor (G). Incubate for 15 min at room temperature.

(7) Dispense 50 μL of stop solution into the reagent well (O) from a syringe on the syringe platform (F). The barrel of this syringe is pushed using a servomotor (E).

(8) Visually inspect the color change if the platform is used for qualitative, rather than quantitative analysis. Alternatively, position a cell phone on the frame (P) and align the phone camera to the reagent wells (O) to read the results using a special colorimetric detection app, Color Catcher (Cloud Innovation Team, Austin, TX, USA) [66]. This app reads RBG values from scanned or photographed images.

2.3. Hardware

There were five servomotors used with this automated assay platform: three TowerPro-SG5010 (Shenzhen Hao Qi Core Technology Co., Ltd., Shenzhen, China) servomotors (E, G, and H on Figure 1). These servomotors were used to push onto the plungers of syringes with the stop solution, with TMB substrate, and with the malaria conjugate. A continuous Airtronics Servo 94102 (Sanwa Denshi Co., Tokyo, Japan) (part C in Figure 1) was used in conjunction with a 3D printed rack-and-gear set (B) to pull out the plunger of the aspiration syringe. An Ultra Torque HS-645MG (Hitec RCD USA Inc., Poway, CA, USA) (M) servomotor was used to rotate the arm that pushed on the elastomeric dome (N) containing the wash. The servomotors were programmed to follow the ELISA steps, dispensing the appropriate amount of reagent between each step. These directions were uploaded to the Arduino boards (D, J) (Seeedstudio, Shenzen, China) controlling the servomotors. The arm position of a servomotor was determined by the pulse width-modulated (PWM) signal sent via the control board.

3. Results

3.1. Fused Deposition Modeling Optimization

The extrusion temperature for FDM was tested in the range from 225 °C to 240 °C in 5 °C increments. At 225 °C, the filament did not have sufficient adhesion to the base plane of the printer, while at 240 °C, the filament extruded out of the nozzle was difficult to control, as it was not viscous enough. The optimal extrusion temperature was found to be in the range between 230 °C and 235 °C.

The travel speed of the printer head significantly influenced the quality of the final build. Figure 2 presents the test samples produced with the optimized extrusion temperature, and with travel speeds of 15, 20, and 30 mm/s (left to right). The test samples had a geometry of 2.5 cm × 2.5 cm square, with a thickness of 1.5 mm. On top of the sample, there were two structures: a T-shaped wall placed at the edge of the test sample, and a microfluidic channel in the middle of the test sample. The microfluidic channel had an inner diameter of 1.5 mm and an outer diameter of 3 mm. As Figure 2 indicates, the best quality of the printed structures resulted with the extruder's travel speed of 15 mm/s.

Figure 2. Test samples produced with fused deposition modeling (FDM) using travel speeds of the extruder (left to right) of 15, 20, and 30 mm/s.

3.2. Stereolithographic Fabrication Optimization

For microfluidic parts produced with SLA, one of the major challenges was the process of clearing out the uncrosslinked resin from the enclosed microfluidic channel [67]. We evaluated the smallest channel diameter achievable with SLA by printing chips with 1 cm, 3 cm, and 5 cm long microchannels. Each sample contained five channels with different hydraulic diameters of 0.7 mm, 1.0 mm, 1.4 mm, 1.7 mm, and 2.1 mm (Figure 3a). The longer the channel and the smaller its diameter, the more difficult it was to clear the channel from uncross-linked resin. For example, all of the channels in the 5 cm long test chip were clogged, even with 30 min of soaking time in isopropyl alcohol (IPA).

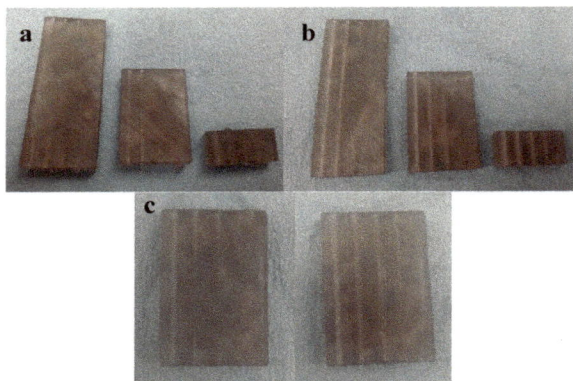

Figure 3. Test Samples containing microchannels with diameters of 2.1 mm, 1.7 mm, 1.4 m, 1.0 mm, and 0.7 mm (right to left). The test chip lengths are 5 cm, 3 cm, and 1 cm. Prolonging the isopropyl alcohol (IPA) soak from 30 min (**a**) to 6 h (**b**) helps with clearing the microchannels from uncrosslinked resin. (**c**) A comparison between IPA soaking cleaning (left) and IPA injection cleaning (right) for the 3 cm long microchannels. IPA injection cleaning achieved superior results in cleaning the channels from uncrosslinked resin.

Extending the IPA soak time of the SLA fabricated sample from 30 min (Figure 3a) to 6 h (Figure 3b) helped to clear out channels better. Further extending the soak time in IPA past 6 h did not result in any noticeable change, while the cross-linked resin started to deteriorate. The channels with diameters of 1.75 mm and larger were cleared out even in the 5 cm long test pieces.

For straight channels, it was possible to remove the uncross-linked resin by feeding a thin wire through the channel, but in case of tortuous channels (of sinusoidal or zig-zag geometry), this technique was difficult to execute successfully, as the wire could not navigate the bends of the tortuous channels.

Instead, we implemented the manual injection of IPA with a syringe directly into the channels. This technique was more efficient and superior to the IPA soak method.

As can be seen on Figure 3c, the 3 cm long microchannels were still clogged after a 6 h IPA soak, while the IPA injection successfully cleared the microchannels in another sample of the same geometry. IPA injection is capable of clearing channels all the way down to microchannels with 1 mm diameter.

3.3. Channel Deformation

Channel cross-sections were distorted during FDM and SLA fabrication. Measurements in the X and Y directions (width and height of the channels' cross-sections, respectively) were obtained for all channels and compared to the designed values. The cross-sections of all the channels were oval, meaning that the diameter in the Y direction was larger than in the X direction. Figure 4 compares the designed and measured diameters in the X and Y direction for channels made with both FDM and SLA.

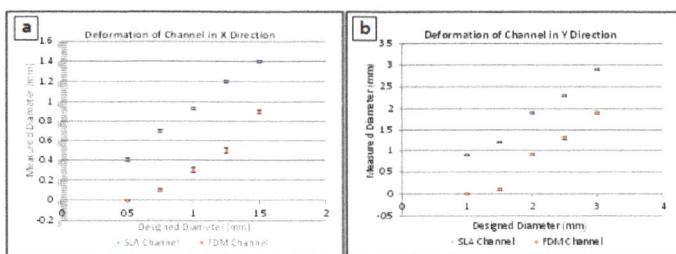

Figure 4. Deformations of hollow channels printed with Stereolithography (SLA) and Fused Deposition Modeling (FDM) technologies. A total of five samples were measured for each experimental point, and error bar heights represent one standard deviation. Designed vs measured channel diameter measurements are presented for (**a**) the deformation of channels in the X direction (width of the channels) and for (**b**) the deformation of channels in the Y direction (height of the channels).

It was seen that in all cases, SLA produced channel diameters that were more representative of the initial design, while FDM was far less accurate due to spreading of the liquid filament during printing. Both methods produced distorted channel heights due to the sagging of unsupported filaments. However, SLA-fabricated samples channels tended to sag less than the FDM-fabricated channels. The highest deviation observed corresponded to FDM-printed channels that were larger than 1 mm diameter in the X direction.

3.4. Performance of the Automated System

The finalized printed LOC platform was assembled using tubes to connect the syringes to the reservoir wells. The reagents contained within the syringes were injected into the wells (boxed in red in Figure 5). The dead volume of the system, comprising the inner space of the tubes connecting the syringes and the test well, equaled 0.14 ± 0.02 cm^3 for each syringe. In order to compensate for that dead volume, correspondingly larger volumes were dispensed from the syringes.

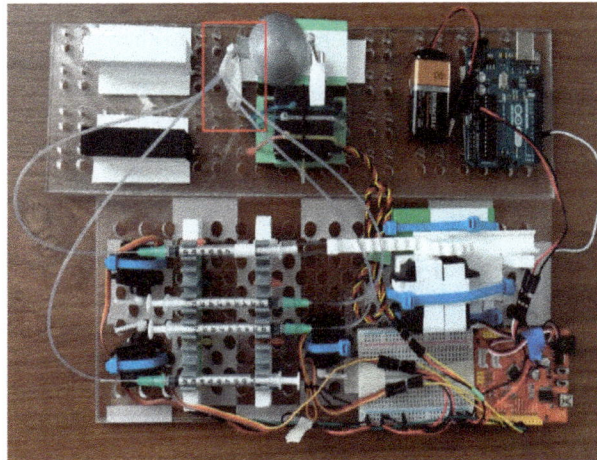

Figure 5. Top view of the assembled automated ELISA bioassay platform. Test wells are indicated by the red frame on the picture. See the caption of Figure 1 for a description of the individual components of the automated platform.

An assay was performed using the automation process described in the Section 2.2 above. A close-up photo of the positive and negative results is provided in Figure 6. The wells with the blue solution were the positive controls, and the colorless wells represented the negative control. The positive control solution changed from blue to yellow when the stop solution was added, while the colorless negative controls did not change color. The positive and negative control color change was correct, signifying the successful demonstration of the automated malaria-Ab assay.

Figure 6. Colorimetric bioassay results after the stop solution is added. The wells with the blue solution are the positive controls and the colorless wells represent the negative controls.

Five assays were performed on the automated platform. Each assay consisted of four test wells: two wells with positive controls and two wells with negative controls. The tests were performed using one well at a time. Subsequent test runs were performed by replacing the tubes and manually positioning them into another test well. All of the assays were performed successfully without any false readings. Minor test-to-test variations (within 0.05 cm^3) of the dispensed wash did not affect the test results.

A Windows Nokia Lumia 521 (Nokia, Espoo, Finland) smart phone with the installed Color Catcher (Cloud Innovation Team, Austin, TX, USA) app was used to capture the RBG color code of the solution in the well at the end of each test. There was no need to withdraw the tubes from the test well during this image acquisition. While the color change was easily distinguishable with the naked eye, such captured color codes are useful for performing quantitative bioassays. The video of the operation of the automated bioassay platform during the test can be viewed online as part of the Supplemental Materials for this article.

4. Conclusions

We have reported on the fabrication of an automated colorimetric Malaria-Ab immunoassay platform based on readily-available parts, such as servomotors and disposable syringes. We have also outlined an alternative syringe-based automated assay that utilizes elastomeric reservoirs (printed using SLA) with programmable servomotors compressing elastomeric reservoirs to inject reagents or to wash a test well. The aspiration syringe actuated by a continuous servomotor facilitated the draining of test wells after specific assay steps.

To demonstrate the validity of our approach, we fabricated a hybrid system that included syringes for all of the reagents except for the wash, which was contained in an elastomeric dome reservoir. Successful validation of the automated platform was performed with positive and negative controls, and with the reagents included in the commercial malaria-Ab immunoassay kit.

The demonstrated syringe-based automation is easily expanded to a wide variety of assays to create flexible and affordable point-of-care platforms. Implementation of the platform is simplified by providing rural clinics with syringe kits with pre-measured reagent syringes sealed in metallized pouches. These pouches can be punctured by the syringe plunger during the corresponding assay step when conducting the bioassay. Additional validation is required before the application of the presented automated platform for qualitative (rather than quantitative) tests.

Supplementary Materials: The following are available online at http://www.mdpi.com/2072-666X/9/10/502/s1,Video S1: Validation of a syringe-based automated point-of-care malaria-Ab immunoassay.

Author Contributions: Conceptualization, C.L., Y.L., and L.K.; methodology, C.L. and L.K.; software, Y.L.; validation, C.L. and Y.L.; formal analysis, C.L., Y.L., and L.K.; investigation, C.L. and Y.L.; resources, L.K.; data curation, C.L. and Y.L.; writing—original draft preparation, C.L.; writing—review and editing, C.L., Y.L., and L.K.; visualization, C.L.; supervision, L.K.; project administration, L.K.; funding acquisition, L.K.

Funding: This research received no external funding.

Conflicts of Interest: The authors declare no conflict of interest.

References

1. Urrios, A.; Parra-Cabrera, C.; Bhattacharjee, N.; Gonzalez-Suarez, A.M.; Rigat-Brugarolas, L.G.; Nallapatti, U.; Samitier, J.; DeForest, C.A.; Posas, F.; Garcia-Cordero, J.L.; et al. 3D-printing of transparent bio-microfluidic devices in PEG-DA. *Lab Chip* **2016**, *16*, 2287–2294. [CrossRef] [PubMed]
2. Crowther, J. *The Elisa Guidebook*; Springer Science & Business Media: Berlin, Germany, 2009; p. 54.
3. Wiltbank, T.; McCaroll, D.; Wartick, M. An undetectable source of technical error that could lead to false negative results in enzyme linked immunosorbent assay of antibodies to HIV-1. *Rapid Commun.* **1988**, *29*, 75–77. [CrossRef]
4. Knowlton, S.; Yu, C.H.; Ersoy, F.; Emadi, S.; Khademhosseini, A.; Tasoglu, S. 3D-printed microfluidic Chips with patterned, cell-laden hydrogel constructs. *Biofabrication* **2016**, *8*, 025019. [CrossRef] [PubMed]
5. Ellerbee, A.K.; Phillips, S.T.; Siegel, A.C.; Mirica, K.A.; Martinez, A.W.; Striehl, P.; Jain, N.; Prentiss, M.; Whitesides, G.M. Quantifying colorimetric assays in paper-based microfluidic devices by measuring the transmission of light through paper. *Anal. Chem.* **2009**, *81*, 8447–8452. [CrossRef] [PubMed]
6. Lutz, B.; Liang, T.; Fu, E.; Ramachandran, S.; Kauffman, P.; Yager, P. Dissolvable fluidic time delays for programming multi-step assays in instrument-free paper diagnostics. *Lab Chip* **2013**, *13*, 2840–2847. [CrossRef] [PubMed]

7. Schaller, V.; Sanz-Velasco, A.; Kalabukhov, A.; Schneiderman, J.F.; Öisjöen, F.; Jesorka, A.; Astalan, A.P.; Krozer, A.; Rusu, C.; Enoksson, P.; et al. Towards an electrowetting-based digital microfluidic platform for magnetic immunoassays. *Lab Chip* **2009**, *9*, 3433–3436. [CrossRef] [PubMed]

8. Nguyen, N.T.; Hejazian, M.; Ooi, C.H.; Kashaninejad, N. Recent advances and future perspectives on microfluidic liquid handling. *Micromachines* **2017**, *8*, 186. [CrossRef]

9. Cevenini, L.; Calabretta, M.M.; Tarantino, G.; Michelini, E.; Roda, A. Smartphone-interfaced 3D printed toxicity biosensor integrating bioluminescent "sentinel cells". *Sens. Actuators* **2016**, *225*, 249–257. [CrossRef]

10. Comina, A.; Suska, A.; Filippini, G. A 3D printed device for quantitative enzymatic detection using cell phones. *Anal. Methods* **2016**, *8*, 6135–6142. [CrossRef]

11. Bauer, M.; Kulinksy, L. Fabrication of a lab-on-chip device using material extrusion (3D printing) and demonstration via Malaria-Ab ELISA. *Micromachines* **2018**, *9*, 27. [CrossRef]

12. Crowther, J. *The Elisa Guidebook*; Springer Science & Business Media: Berlin, Germany, 2009; p. 332.

13. Becker, H.; Locascio, L. Polymer microfluidic devices. *Talanta* **2002**, *56*, 267–287. [CrossRef]

14. Zhang, Y.; Wu, L.; Kane, S.; Deng, Y.; Jung, Y.G.; Lee, G.H.; Zhang, J. Additive manufacturing of metallic materials: A review. *J. Mater. Eng. Perform.* **2018**, *27*, 1–13. [CrossRef]

15. Parra-Cabrera, C.; Achille, C.; Kuhn, S.; Ameloot, R. 3D printing in chemical engineering and catalytic technology: Structured catalysts, mixers and reactors. *Chem. Soc. Rev.* **2018**, *47*, 209–230. [CrossRef] [PubMed]

16. Betancourt, T.; Brannon-Peppas, L. Micro- and nanofabrication methods in nanotechnological medical and pharmaceutical devices. *Int. J. Nanomed.* **2006**, *1*, 483–495. [CrossRef]

17. Sugioka, K.; Cheng, Y. Femtosecond laser processing for optofluidic fabrication. *Lab Chip* **2012**, *12*, 3576. [CrossRef] [PubMed]

18. Yalikun, Y.; Hosokawa, Y.; Lino, T.; Tanaka, Y. An all-glass 12 µm ultra-thin and flexible micro-fluidic chip fabricated by femtosecond laser processing. *Lab Chip* **2016**, *16*, 2427–2433. [CrossRef] [PubMed]

19. Rolland, J.P.; Van Dam, R.M.; Schorzman, D.A.; Quake, S.R.; DeSimone, J.M. Solvent-resistant photocurable "liquid teflon" for microfluidic device fabrication. *J. Am. Chem. Soc.* **2004**, *126*, 232. [CrossRef] [PubMed]

20. Beebe, D.J.; Moore, J.S.; Yu, Q.; Liu, R.H.; Kraft, M.L.; Jo, B.H.; Devadoss, C. Microfluidic tectonics: A comprehensive construction platform for microfluidic systems. *Proc. Natl. Acad. Sci. USA* **2000**, *97*, 13488–13493. [CrossRef] [PubMed]

21. Lee, S.; Kim, H.; Lee, W.; Kim, J. Finger-triggered portable PDMS suction cup for equipment-free microfluidic pumping. *Micro Nano Syst. Lett.* **2018**, *6*, 1–5. [CrossRef]

22. Yazdi, A.; Popma, A.; Wong, W.; Nguyen, T.; Pan, Y.; Xu, J. 3D printing: an emerging tool for novel microfluidics and lab-on-a-chip applications. *Microfluid. Nanofluid.* **2016**, *20*, 50. [CrossRef]

23. He, Y.; Wu, Y.; Fu, J.Z.; Gao, Q.; Qiu, J.J. Developments of 3D printing microfluidics and applications in chemistry and biology: A review. *Electroanalysis* **2016**, *28*, 1658–1678. [CrossRef]

24. Zhang, C.; Bills, B.; Manicke, N. Rapid prototyping using 3D printing in bioanalytical research. *Bioanalysis* **2017**, *9*, 329–331. [CrossRef] [PubMed]

25. Tang, S.; Whitesides, G. Basic Microfluidic and Soft Lithographic Techniques. Available online: https://gmwgroup.harvard.edu/pubs/pdf/1073.pdf (accessed on 25 May 2018).

26. Unger, M.A.; Chou, H.P.; Thorsen, T.; Scherer, A.; Quake, S.R. Monolithic microfabricated valves and pumps by multilayer soft lithography. *Science* **2000**, *288*, 113–116. [CrossRef] [PubMed]

27. Thorsen, S.; Maerkl, S.J.; Quake, S.R. Microfluidic large-scale integration. *Science* **2002**, *298*, 580–584. [CrossRef] [PubMed]

28. Waheed, S.; Cabot, J.M.; Macdonald, N.P.; Lewis, T.; Guijt, R.M.; Paull, B.; Breadmore, M.C. 3D printed microfluidic devices: Enablers and barriers. *Lab Chip* **2016**, *16*, 1993–2013. [CrossRef] [PubMed]

29. Hopkinson, N.; Smith, P. *Handbook of Industrial Inkjet Printing: A full System*; John Wiley & Sons: Hoboken, NJ, USA, 2018.

30. Pilipović, A.; Raos, P.; Šercer, M. Experimental analysis of properties of materials for rapid prototyping. *Int. J. Adv. Des. Manuf. Technol.* **2009**, *40*, 105–115. [CrossRef]

31. Mueller, J.; Courty, D.; Spielhofer, M.; Spolenak, R.; Shea, K. Mechanical properties of interfaces in inkjet 3D printed single- and multi-material parts. *3D Print. Addit. Manuf.* **2017**, *4*, 193–199. [CrossRef]

32. Anderson, K.B.; Lockwood, S.Y.; Martin, R.S.; Spence, D.M. A 3D printed fluidic device that enables integrated features. *Anal. Chem.* **2013**, *85*, 5622–5626. [CrossRef] [PubMed]

33. Lockwood, S.Y.; Meisel, J.E.; Monsma, F.J., Jr.; Spence, D.M. A diffusion-based and dynamic 3D-printed device that enables parallel in vitro pharmacokinetic profiling of molecules. *Anal. Chem.* **2016**, *88*, 1864–1870. [CrossRef] [PubMed]

34. Hwang, Y.; Paydar, O.; Candler, R. 3D printed molds for non-planar PDMS microfluidic channels. *Sens. Actuators A Phys.* **2015**, *226*, 137–142. [CrossRef]

35. Paydar, O.H.; Paredes, C.N.; Hwang, Y.; Paz, J.; Shah, N.B.; Candler, R.N. Characterization of 3D-printed microfluidic chip interconnects with integrated O-rings. *Sens. Actuators A Phys.* **2014**, *205*, 199–203. [CrossRef]

36. Staudinger, U.; Zyla, G.; Krause, B.; Janke, A.; Fischer, D.; Esen, C. Development of electrically conductive microstructures based on polymer/CNT nanocomposites via two-photon polymerization. *Microelectron. Eng.* **2017**, *179*, 48–55. [CrossRef]

37. Xing, J.F.; Zheng, M.L.; Duan, X.M. Two-photon polymerization microfabrication of hydrogels: An advanced 3D printing technology for tissue engineering and drug delivery. *Chem. Soc. Rev.* **2015**, *44*, 5031–5039. [CrossRef] [PubMed]

38. Kumi, G.; Yanez, C.; Belfield, K.; Fourkas, J.T. High-speed multiphoton absorption polymerization: Fabrication of microfluidic channels with arbitrary cross-sections and high aspect ratios. *Lab Chip* **2010**, *10*, 1057–1060. [CrossRef] [PubMed]

39. Kawata, S.; Sun, H.B.; Tanaka, T.; Takada, K. Finer features for functional microdevices. *Nature* **2001**, *412*, 697–698. [CrossRef] [PubMed]

40. Sugioka, K.; Cheng, Y. Femtosecond laser three-dimensional micro- and nanofabrication. *Appl. Phys. Rev.* **2014**, *1*, 041303. [CrossRef]

41. Ziemian, C.M.; Crawn, P.M. Computer aided decision support for fused deposition modelling. *Rapid Prototyp. J.* **2001**, *7*, 138–147. [CrossRef]

42. Modjarrad, K.; Ebnesajjad, S. *Handbook of Polymer Applications in Medicine and Medical Devices*; William Andrew Publishing: Norwich, NY, USA, 2014.

43. Chia, H.; Wu, B. Recent advances in 3D printing of biomaterials. *J. Biol. Eng.* **2015**, *9*, 4. [CrossRef] [PubMed]

44. Lee, J.; Zhang, M.; Yeong, W. Characterization and evaluation of 3D printed microfluidic chip for cell processing. *Microfluid. Nanofluid.* **2016**, *20*, 1–15. [CrossRef]

45. Kitson, P.J.; Symes, M.D.; Dragone, V.; Cronin, L. Combining 3D printing and liquid handling to produce user-friendly reactionware for chemical synthesis and purification. *Chem. Sci.* **2013**, *4*, 3099–3103. [CrossRef]

46. Bishop, G.W.; Satterwhite, J.E.; Bhakta, S.; Kadimisetty, K.; Gillette, K.M.; Chen, E.; Rusling, J.F. 3D-printed fluidic devices for nanoparticle preparation and flow-injection amperometry using integrated prussian blue nanoparticle-modified electrodes. *Anal. Chem.* **2015**, *87*, 5437–5443. [CrossRef] [PubMed]

47. Sher, D. "Dolomite's Fluidic Factory 3D Prints $1 Microfluidic Chips." 3D Printing Industry, 2015. Available online: https://3dprintingindustry.com/news/dolomites-fluidic-factory-3d-prints-1-microfluidic-chips-60883/ (accessed on 25 May 2018).

48. 3D Systems, Inc. Available online: https://www.3dsystems.com/ (accessed on 25 May 2018).

49. Au, A.; Lee, W.; Folch, A. Mail-order microfluidics: Evaluation of stereolithography for the production of microfluidic devices. *Lab Chip* **2014**, *14*, 1294–1301. [CrossRef] [PubMed]

50. Zhou, C.; Chen, Y.; Yang, Z.G.; Khoshnevis, B. Development of a multi-material mask-image-projection-based stereolithography for the fabrication of digital materials. *Rapid Prototyp. J.* **2013**, *19*, 153–165. [CrossRef]

51. Zhang, X.; Jiang, X.N.; Sun, C. Micro-stereolithography of polymeric and ceramic microstructures. *Sens. Actuators A Phys.* **1999**, *77*, 149–156. [CrossRef]

52. Amend, P.; Hentschel, O.; Scheitler, C.; Baum, M.; Heberle, J.; Roth, S.; Schmidt, M. Fast and flexible generation of conductive circuits. *J. Laser Micro/Nanoeng.* **2013**, *8*, 276–286. [CrossRef]

53. Comina, G.; Suska, A.; Filippini, D. PDMS lab-on-a-chip fabrication using 3D printed templates. *Lab Chip* **2014**, *14*, 424–430. [CrossRef] [PubMed]

54. Comina, G.; Suska, A.; Filippini, D. Low cost lab-on-a-chip prototyping with a consumer grade 3D printer. *Lab Chip* **2014**, *14*, 2978–2982. [CrossRef] [PubMed]

55. Comina, G.; Suska, A.; Filippini, D. Autonomous chemical sensing interface for universal cell phone readout. *Chem. Int. Ed.* **2015**, *54*, 8708–8712. [CrossRef] [PubMed]

56. Wang, X.; Cai, X.; Guo, Q.; Zhang, T.; Kobe, B.; Yang, J. i3DP, a robust 3D printing approach enabling genetic post-printing surface modification. *Chem. Commun.* **2013**, *49*, 10064–10066. [CrossRef] [PubMed]

57. Shallan, A.I.; Smejkal, P.; Corban, M.; Guijt, R.M.; Breadmore, M.C. Cost-Effective Three-Dimensional Printing of Visibly Transparent Microchips within Minutes. *Anal. Chem.* **2014**, *86*, 3124–3130. [CrossRef] [PubMed]

58. Patrick, W.G.; Nielsen, A.A.; Keating, S.J.; Levy, T.J.; Wang, C.W.; Rivera, J.J.; Mondragón-Palomino, O.; Carr, P.A.; Voigt, C.A.; Oxman, N.; et al. DNA assembly in 3D printed fluidics. *PLoS ONE* **2015**, *10*, 1–18. [CrossRef] [PubMed]

59. Lee, K.G.; Park, K.J.; Seok, S.; Shin, S.; Park, J.Y.; Heo, Y.S.; Lee, S.J.; Lee, J.L. 3D printed modules for integrated microfluidic devices. *RSC Adv.* **2014**, *4*, 32876–32880. [CrossRef]

60. Bhargava, K.; Thompson, B.; Malmstadt, N. Discrete elements for 3D microfluidics. *PNAS* **2014**, *111*, 15013–150138. [CrossRef] [PubMed]

61. Lai, H.; Folch, A. Design and dynamic characterization of "single-stroke" peristaltic PDMS micropumps. *Lab Chip* **2011**, *11*, 336–342. [CrossRef] [PubMed]

62. Bhushan, B.; Caspers, M. An overview of additive manufacturing (3D printing) for microfabrication. *Microsyst. Technol.* **2017**, *23*, 1117. [CrossRef]

63. Garcia, J.; Yang, Z.; Mongrain, R.; Leask, R.L.; Lachapelle, K. 3D printing materials and their use in medical education: A review of current technology and trends for the future. *BMJ. Simul. Technol. Enhanc. Learn.* **2018**, *4*, 27–40. [CrossRef] [PubMed]

64. World Health Organization. Malaria. Available online: http://www.who.int/mediacentre/factsheets/fs094/en (accessed on 26 May 2018).

65. Malaria-Ab-ELISA. Available online: http://www.ibl-international.com/media/catalog/product/R/E/RE58601_IFU_en_Malaria-Ab_ELISA_V4_2011-08-03_sym3.pdf (accessed on 26 May 2018).

66. Color Catcher. Available online: http://appcrawlr.com/android/color-catcher-free (accessed on 26 May 2018).

67. Zhou, Y. The recent development and applications of fluidic channels by 3D printing. *J. Biomed. Sci.* **2017**, *24*, 80. [CrossRef] [PubMed]

Article

Assessing the Reusability of 3D-Printed Photopolymer Microfluidic Chips for Urine Processing

Eric Lepowsky [1], Reza Amin [1] and Savas Tasoglu [1,2,3,4,5,*]

[1]   Department of Mechanical Engineering, University of Connecticut, Storrs, CT 06269, USA;
      eric.lepowsky@uconn.edu (E.L.); reza.amin@uconn.edu (R.A.)
[2]   Department of Biomedical Engineering, University of Connecticut, Storrs, CT 06269, USA
[3]   Institute of Materials Science, University of Connecticut, Storrs, CT 06269, USA
[4]   Institute for Collaboration on Health, Intervention, and Policy, University of Connecticut,
      Storrs, CT 06269, USA
[5]   The Connecticut Institute for the Brain and Cognitive Sciences, University of Connecticut,
      Storrs, CT 06269, USA
*   Correspondence: savas.tasoglu@uconn.edu; Tel.: +1-860-486-5919

Received: 26 September 2018; Accepted: 14 October 2018; Published: 15 October 2018

Abstract: Three-dimensional (3D) printing is emerging as a method for microfluidic device fabrication boasting facile and low-cost fabrication, as compared to conventional fabrication approaches, such as photolithography, for poly(dimethylsiloxane) (PDMS) counterparts. Additionally, there is an increasing trend in the development and implementation of miniaturized and automatized devices for health monitoring. While nonspecific protein adsorption by PDMS has been studied as a limitation for reusability, the protein adsorption characteristics of 3D-printed materials have not been well-studied or characterized. With these rationales in mind, we study the reusability of 3D-printed microfluidics chips. Herein, a 3D-printed cleaning chip, consisting of inlets for the sample, cleaning solution, and air, and a universal outlet, is presented to assess the reusability of a 3D-printed microfluidic device. Bovine serum albumin (BSA) was used a representative urinary protein and phosphate-buffered solution (PBS) was chosen as the cleaning agent. Using the 3-(4-carboxybenzoyl)quinoline-2-carboxaldehyde (CBQCA) fluorescence detection method, the protein cross-contamination between samples and the protein uptake of the cleaning chip were assessed, demonstrating a feasible 3D-printed chip design and cleaning procedure to enable reusable microfluidic devices. The performance of the 3D-printed cleaning chip for real urine sample handling was then validated using a commercial dipstick assay.

Keywords: microfluidics; 3D printing; reusability; biofouling

1. Introduction

There is significant interest in microfluidics due to its large variety of applications, including cancer screening [1–3], micro-physiological system engineering [4,5], high-throughput drug testing [6,7], and point-of-care diagnostics [8–13]. In particular, point-of-care devices can help to enable routine health monitoring and preventative care, which can improve public health while increasing healthcare savings. In a study reported by Health Affairs, it has been shown that a 90% increase in specific preventative screenings back in 2006 would have saved more than 2 million lives without a significant increase in healthcare costs—in fact, a 0.2% decrease in costs was estimated [14]. Microfluidic devices have a proven record of being effective analytical devices, capable of controlling the flow of fluid samples, containing reaction and detection zones, and displaying results, all within a compact footprint, and are therefore appropriate for addressing this need for routine and preventative care [15–20].

While microfluidic devices are low-cost in terms of material cost, conventional fabrication is challenging and costly. Furthermore, there is an unmet need to address the convenience of implementing microfluidic devices in sophisticated medical devices: users should not need to replace the chips routinely. To address the burdensome and expensive fabrication of microfluidic devices, as well as the need for convenience with respect to medical devices, there is an increasing need to develop microfluidic devices for long-term use [21–24]. While single-use microfluidic chips may be practical for some applications, such as for prototyping and experimental testing, devices that allow for multiple uses are better suited for high-throughput testing and point-of-care diagnostics, as the limited lifetime of microfluidic devices has been cited as a barrier to commercialization [25]. Furthermore, market trends indicate that by 2018, the field of microfluidics was estimated to reach \$3.6–5.7 billion [26], demonstrating the continued growth of the field, particularly within clinical diagnostics, pharmaceutical research, point-of-care diagnostics, and analytical applications. Thus, multiuse devices will prove very useful while combatting the waste that accumulates with disposable chips in these applications. For these reasons, the reusability of conventional poly(dimethylsiloxane (PDMS) microfluidic devices has previously been assessed [24].

Microfluidic devices are traditionally fabricated by combining photolithography techniques for making a master mold and soft lithography using PDMS and bonding. The process as a whole is complex, expensive, laborious, as well as time-consuming [27]. There exist several alternative approaches, one of which being 3D printing. Using hydrogels, Beebe et al. developed an ultrafast fabrication platform combining liquid-phase polymerization and lithography compatible with a variety of geometries and valve designs [28]. Another alternative is Teflon, which was demonstrated by Rolland et al. [29]. Unlike PDMS, the liquid-Teflon, a type of photocurable perfluoropolyether (PFPE), is resistant to swelling in common organic solvents, in addition to being liquid at room temperature and exhibiting low surface energy, low modulus, high gas permeability, and low toxicity. Glass can also be used for microfluidic device fabrication. For instance, femtosecond laser direct writing can modify the interior of glass through multiphoton absorption to form microfluidic channels and optofluidic components [30]. Femtosecond laser processing can even be used to fabricated ultrathin, flexible microfluidic chips [31]. Microfluidic devices can also be fabricated using plastics. A study by Wan et al. indicated that poly(methyl methacrylate) (PMMA) can be recycled for biological microfluidic device applications. In this experiment, researchers pumped bleach and ethanol through the microfluidic channels after the chip was used, and then they melted the PMMA so that the plastic could be reused for multiple chip iterations [32]. While studies have indicated that the recycling of plastic-based microfluidic chips is possible, there is a lack of research being conducted on the reusability of microfluidic chips. Reusability is advantageous over recyclability since recycling involves the remanufacturing of the device. Investigation is needed to determine if merely rinsing the channels with a cleaning agent, such as a phosphate-buffered solution (PBS) or deionized (DI) water, will allow for the microfluidic chip to be reused with negligible protein absorption and cross-contamination between samples.

3D printing is a new technique that can bypass many limitations seen with PDMS-based approaches, such as expensive equipment and clean room facilities, thus making the fabrication process easier, cheaper, and more flexible [26,27,33,34]. There are various 3D printing techniques that can be used to fabricate microfluidics. The most applicable ones to microfluidics are stereolithography (SLA), multijet modeling (MJM), and fused deposition modeling (FDM) [27,33–35]. All techniques use a computer-aided design (CAD) sketch as a blueprint and build the structure layer-by-layer from printing material, which can be plastic filaments, liquid resin, or powder [27,36]. The most commonly used commercialized 3D printing technology for micro-fabrication is the SLA technique, which is defined as a method for making solid objects by successively printing thin layers of a curable material (i.e., a UV-curable liquid resin) on top of the other [35]. In SLA printing, a laser beam or projector is used to crosslink the resin in the predetermined pattern [33]. For microfluidic devices, a microchannel is built by photo-polymerizing the channel walls and then draining the uncured resin from the channel cavity after the printing is complete [37]. For the purpose of microfluidic devices which involve imaging, clear resin is available.

When a microfluidic device is first manufactured and put into use, it is assumed that the channels will provide a clean environment for the samples to flow through. However, the size and composition of these devices can make it unusable after just a few experiments if not designed with reusability in mind [38]. Within the channels of a microfluidic device, the surface area-to-volume ratio is high, meaning more of the fluid is exposed to the surfaces of the channel per unit volume. If the device is made of a porous or hydrophobic material, it can allow molecules to stick to the channel walls, rendering it unusable. In addition, the charge on a molecule can cause them to stick to the channel walls if the chip is made of a polar material [39]. Further, if there are areas in the channel where the fluid flow becomes static, especially around channel intersections, there is a high likelihood that a molecule buildup will occur, causing elevated levels of absorption or contamination by the molecule, or even reducing or completely inhibiting the flow through that channel [38]. Nonetheless, the change in flow through the channel should not overshadow the larger issue: the buildup of molecules, and possibly cells, can cause largely inaccurate results in sample testing. A primary challenge in producing microfluidic devices for long-term use is the biofouling that often occurs on the surface of integrated channels and features [39,40]. This phenomenon occurs, due to surface interactions between the walls of the channels and the biological sample flowing through the channels. To address the long-term use of microfluidic devices, there has been a growing research interest in developing new anti-fouling methods and materials [21,22,41–43]. Additionally, other alternatives, such as recyclable chips or reusable chips, are actively being explored [23,24,32,44].

Leveraging the advantages of 3D printing with the aim of addressing the need for reusable microfluidic devices due to their potential societal impact, herein, a 3D-printed microfluidic chip was designed to assess the reusability of 3D-printed chips by means of quantifying its absorption characteristics. The proposed solution serves to validate the ability to deliver a sample and clean the device for reusability. The core function of the demonstrated 3D-printed microfluidic cleaning chip is to integrate a cleaning procedure into a pre-existing single-use device such that the chip/device may be reused (Figure 1b). To facilitate the cleaning process, the cleaning chip is connected to a syringe pump to deliver either the sample, cleaning solution, or air, at a controlled rate, depending upon which inlet the pump is attached to; the inlet channels converge to a single outlet. In a real-world application, the sample would flow from the outlet of the cleaning chip into the sample handling portion of the pre-existing device (Figure 1a,b). For the purpose of assessing the protein uptake by the chip, a calibration curve (Figure 1c) was created by injecting protein samples of known concentration into channels of equal dimensions to the cleaning chip, performing fluorescent assays, and using a fluorescence microscope to quantify the fluorescent light intensity, in which a greater light intensity corresponds to a greater concentration of protein present. Specifically, the 3-(4-carboxybenzoyl)quinoline-2-carboxaldehyde (CBQCA) fluorescence detection method was used, in which the CBQCA fluoresces green in the presence of protein when excited by a blue light source. For assessing the cross-contamination between samples, the sample from the outlet was collected, CBQCA was added to the samples, and a plate reader was used to quantify the fluorescence, thereby reflecting the amount of protein present in the samples. By performing the CBQCA assay on samples over time during the cleaning procedure, the cross-contamination as a function of washing volume was determined (Figure 1d).

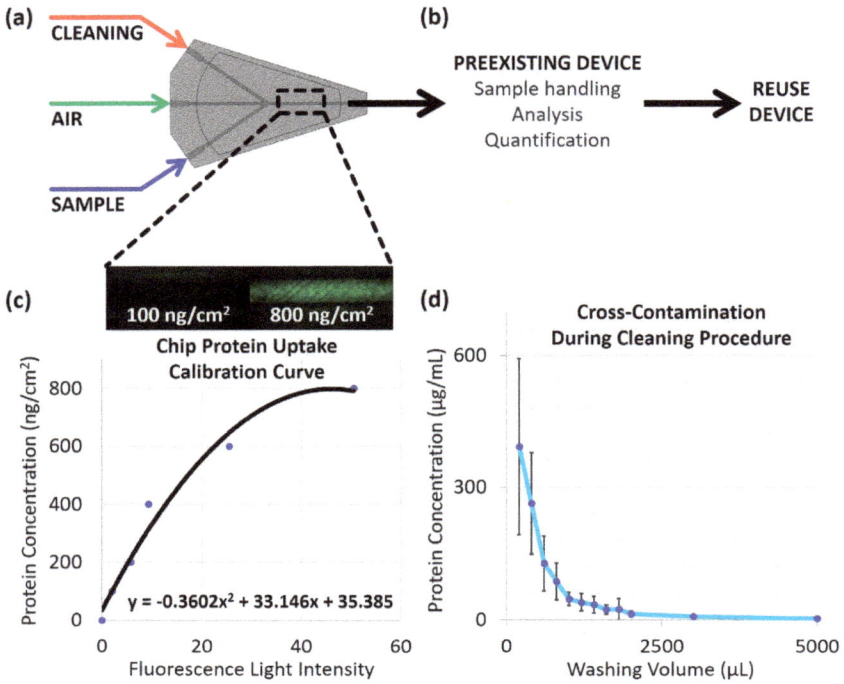

Figure 1. Design, core functionality, and experimental overview: (**a**) Generalized design of cleaning chip. The cleaning chip device consists of inlets for the sample, cleaning solution, and air, all connected to a single outlet. This design allows for a cleaning procedure to be performed on the chip itself and any external microfluidic device that is connected to the cleaning chip's outlet; (**b**) Core functionality of the cleaning chip. The cleaning chip may be used in conjunction with a pre-existing microfluidic device, which may itself perform sample handling, analysis, quantification, etc., in order to render the external device a reusability microfluidic device by connecting the outlet of the cleaning chip to the inlet of the pre-existing device; (**c**) Experimental preview of chip protein uptake. The 3D-printed cleaning chip was connected to a syringe pump, and the cleaning solution was collected from the outlet of the cleaning chip for protein cross-contamination quantification, while the chip itself was assessed for protein absorption. A fluorescence detection method was used to determine the concentration of protein present in the collected samples from the outlet and within the chip itself; the CBQCA assay used fluoresces green by excitation with blue light. Above, the two images shown provide an example of the fluorescent images with their respective protein concentrations. Below, the calibration curve used for determining protein uptake by the chip was constructed by injecting known concentrations of protein into channels of a chip and imaging the channels using a fluorescence microscope, where the raw light intensity (on a scale from 0 to 255) was used as the independent variable; (**d**) Experimental preview of cross-contamination results. Collected samples were quantified by using a plate reader to measure the fluorescence of the cleaning solution from the outlet. The observed trend showed the decreasing protein concentration with the increasing washing volume.

2. Materials and Methods

2.1. Design and Fabrication of 3D-Printed Microfluidic Cleaning Chip

Microfluidic cleaning chips were designed using SolidWorks CAD modeling software (SolidWorks 2017–2018, Dassault Systèmes SOLIDWORKS Corp, Waltham, MA, USA), as seen in Figure 2a. The chip consisted of four channels: a biological sample inlet, a cleaning solution inlet, an air inlet, and a universal outlet. For each iteration of the chip, the outer dimensions were slimmed down to reduce material waste, and a viewing window was added to the top and bottom of the chip to improve through-chip visibility, while the channel width was held constant at 1000 μm. The CAD models were then fabricated using the Formlabs Form 2 3D SLA printer (Form 2, Formlabs Inc., Somerville, MA, USA). This 3D printer was chosen due to its high-resolution capabilities. Furthermore, clear resin was used to fabricate the chips, so it would be possible to view the channels externally (Clear Resin GPCL04, Formlabs Inc., Somerville, MA, USA). The clear resin also allowed for fluorescence readings to be taken through the chip. After 3D-printed chips were fabricated, 1 mL of PDMS (Cell Guard Encapsulation Kit, ML Solar, Campbell, CA, USA) was deposited to the viewing windows on each side of the chip in order to improve clarity by filling pores of the surface roughness of the 3D-printed material (Figure 2a,b). The PDMS was degassed and then baked for 20–25 min at 80 °C. The refractive indices of PDMS and the clear resin have been reported as 1.4 and 1.5, respectively [45,46]. For reference, the refractive index of mineral oil, which is often used to improve optical clarity of a rough surface, is around 1.47; however, the bottom viewing window of the 3D-printed chip would necessitate the mineral oil-coated surface to be overturned, which could pose a risk to the microscope used for protein quantification [47]. Due to the very similar refractive indices of the PDMS and 3D-printed material used in the fabrication of the chip, there was negligible light interference at the interface between the two materials.

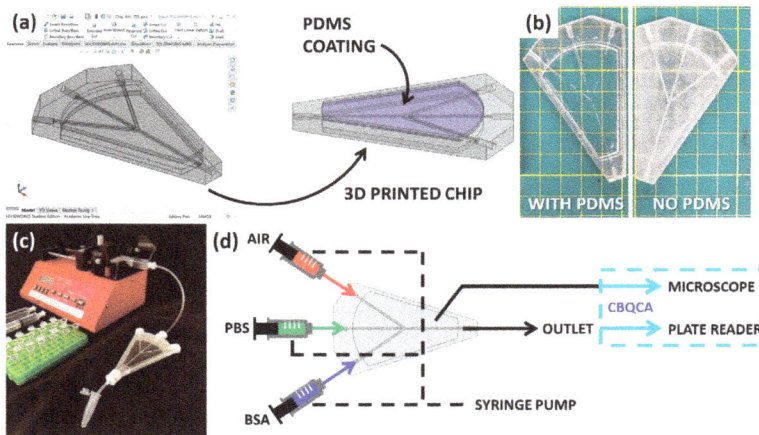

Figure 2. Overview of fabrication and experimental methods: (**a**) Microfluidic chip is first designed in SolidWorks, and then 3D-printed using a Formlabs Form 2 printer. Post-printing, a thin layer of PDMS is poured and cured on both faces of the chip; (**b**) Addition of the PDMS coating on the outside of the chip improves visibility. See a comparison of the chip with PDMS (on the left) to the chip without PDMS (on the right); (**c**) Image of experimental setup, showing the 3D-printed chip connected to a syringe pump. The sample from the outlet of the chip was collected in microcentrifuge tubes for fluorescence detection analysis; (**d**) Schematic representation of the experimental design and implementation of the 3D-printed microfluidic chip. The concentrated BSA solution, and phosphate buffered solution (PBS) as a cleaning solution, and air each have designated inlets which all converge to a single outlet. The sample from the outlet, collected at set intervals, and the chip itself are then analyzed for the protein concentration by the CBQCA quantitation method using a plate reader and microscope, respectively.

To determine the optimum channel height for the microfluidic chip, dimensions ranging from 100 μm to 1000 μm were tested. Printing success was assessed by measuring the length of each open channel and comparing this to the expected channel length. Using 500 μm as a benchmark channel height, which printed semi-reliably and of which success depended greatly on printing parameters, the ideal printing orientation was determined by analyzing the printing success, as described above, in relation to print orientation with respect to the build platform of the 3D printer. Ultimately, the chips were printed with an angled vertical orientation, such that the channels could self-drain excess resin during printing. Additionally, the Form 2 had higher resolutions in the x- and y-directions than in the z-direction. The highest-resolution layer height, 0.025 mm, was used; in the x- and y-directions, the laser could fabricate lines down to 140 microns and holes down to 25 microns. By orienting the chips vertically, the smallest feature size (the channels) benefited from the higher resolution of the printer. Furthermore, by angling the chips, the cross-sectional area of the features was effectively increased. The final chip design was assessed for printing accuracy by measuring the width of the printed channels using a desktop microscope and comparing this to the expected 1000 μm width.

2.2. Cleaning Procedure

The 3D-printed cleaning chip was designed to enable the reusability of microfluidic chips by implementing a simple cleaning procedure to reduce biofouling and cross-contamination between the samples. The cleaning procedure utilized a syringe pump and the four channels of the cleaning chip (Figure 2c,d). First, the biological sample inlet was used: a representative protein sample, 2 mL of bovine serum albumin (BSA) at a high concentration (20 mg/mL), was pumped through the chip, with the cleaning solution and air inlets sealed, using a syringe pump set to 1 mL/min (Thermo Fisher Scientific, Waltham, MA, USA). Following the BSA, air was pumped from the air inlet and through all three non-air channels individually by sealing the remaining two channels, with 0.5 mL of air for each channel, also set at 1 mL/min. The cleaning agent was then pumped, with the biological sample and air inlets sealed, for a total of 10 mL at 1 mL/min. Pure PBS and DI water were tested as the cleaning agent (Thermo Fisher Scientific, Waltham, MA, USA); PBS was determined to be the superior choice. The cleaning procedure ended with a second round of the air to dry all the channels of the chip.

2.3. Protein Cross-Contamination Quantification

While a cleaning agent was pumped through the chip during the cleaning procedure, 200 μL samples from the outlet were collected at 13 intervals: 200, 400, 600, 800, 1000, 1200, 1400, 1600, 1800, 2000, 3000, 5000, and 10,000 μL. The protein level of these samples passing through the device was quantified using a CBQCA fluorescence detection method (CBQCA Protein Quantitation Kit (C-6667), Molecular Probes, Inc., Eugene, OR, USA). Using this method, BSA protein as little as 10 ng can be detected. In the presence of protein, the CBQCA fluorescence green when excited by blue light. To measure the cross-contamination level of the cleaning solution after passing through the system, 5 μL of KCN (20 mM) and 10 μL of CBQCA reagent (5 mM) were added to 135 μL of each sample in a microplate. After allowing the samples in the microplate to incubate for 1 hour in darkness, the fluorescence level was measured using a BioTek Synergy H1 plate reader (Synergy H1 Hybrid Multi-Mode Reader, BioTek, Winooski, VT, USA).

2.4. Protein Uptake by Cleaning Chip Calibration and Quantification

In order to quantify the protein uptake by the cleaning chip, a calibration chip consisting of 10 channels of equal width and height as the channels of the cleaning chip was designed and 3D-printed. Each channel was filled with 135 μL of a BSA-in-PBS solution of known concentrations (1, 0.5, 0.25, 0.1, 0.08, 0.06, 0.04, 0.02, 0.01, and 0 mg/mL) mixed with 5 μL of KCN (20 mM) and 10 μL of CBQCA reagent (5 mM). After incubating for 1 hour in darkness, the channels of the chip were imaged using a Zeiss Observer z1 fluorescence microscope (Carl Zeiss Microscopy GmbH, Jena, Germany) on the 2.5x objective. The greater the concentration of protein present, the greater the intensity of the green fluorescence. Intensity was measured using ImageJ (v1.48k, National Institutes of Health, Bethesda, MD, USA).

The protein absorption by the chip—the same chip as the cross-contamination experiment after being exposed to the 20 mg/mL BSA solution—was then characterized by adding a mixture of 135 μL of PBS, 5 μL of KCN (20 mM), and 10 μL of CBQCA reagent (5 mM) to the chip's channels. After a 1-hour incubation period in darkness, the channels of the chip were imaged using the fluorescence microscope. The light intensity of the channels was compared to the intensity of the calibration chip to determine the amount of protein absorbed by the chip. The protein uptake was quantified for four cases: a single-use chip with DI, a single-use chip with PBS, a 6-cycle chip with PBS, and a 6-cycle chip with PBS and a post-cleaning treatment of 10 mL of 0.25% trypsin (Trypsin-EDTA (0.25%), phenol red, 25200-056, Thermo Fisher Scientific, Waltham, MA, USA). Trypsin is a protease which hydrolyses other proteins, and therefore reduces the levels of protein uptake by the cleaning chip significantly by digesting the adhered and absorbed proteins.

2.5. Longitudinal Protein Cross-Contamination

The longitudinal sample protein cross-contamination was characterized by repeatedly performing the same analysis as for protein cross-contamination. After completion of each round of pumping samples through the chip, the chip was cleaned following the prescribed cleaning process. This procedure of flowing the biological/protein sample, collecting the PBS samples from the outlet, and cleaning the chip was repeated to demonstrate the collective contamination as a result of repeated use of the chip.

2.6. Protein Quantification for Urine Processing

The reusability of the 3D-printed cleaning chip for real urine sample handling was validated by assessing the protein cross-contamination between samples and the protein uptake by the chip. Urine samples were collected from volunteers according to protocol # H17-043, approved by the University of Connecticut Institutional Review Board. All subjects provided informed consent. For studying cross-contamination, samples were pumped through the microfluidic device and collected in microcentrifuge tubes. Commercially available urine analysis dipsticks (Urinalysis Reagent Strips, HealthyWiser LLC., Los Angeles, CA, USA) were then dipped into the collected samples and allowed to incubate for 60 seconds before being imaged with a DSLR camera (REBEL T6, Canon, Melville, NY, USA) in the RAW image mode with flash for consistent lighting. The captured images were then converted to grayscale, such that the light intensity of the grayscale image could be correlated to a specific protein concentration. To facilitate this quantification, a calibration curve was constructed by least-squares curve fitting of the light intensities from images of the dipstick protein assay pads for various BSA concentrations (0, 0.15, 0.3, 1, 3, 10, and 20 g/L).

The cross-contamination was studied for an "extreme-case" scenario, in which pumped samples alternated between urine and high-concentration (20 g/L) BSA. A modified cleaning procedure was performed between each sample: in addition to pumping PBS (the cleaning solution) through the universal exit channel, PBS was also pumped through the sample inlet channel. For each cycle, four samples were collected and analyzed using the dipstick assay: urine, post-urine PBS, BSA, and post-BSA PBS. The protein uptake by the chip was also studied for this "extreme-case," in addition to exposure to pumped urine samples for 30 and 90 min. The total sample volume of the "extreme-case" was approximately equal to the volume of the 30-min exposure to urine. For each of the three urine experiments, protein uptake by the chip was quantified using the previously described CBQCA fluorescent detection method and fluorescent microscope.

3. Results

3.1. Design and Fabrication of 3D-Printed Microfluidic Cleaning Chip

An early chip design (the top-most design depicted in Figure S1a in Supplementary Materials) was fabricated with a 500 μm channel height by 3D printing at different orientations. That is, the chip was oriented, with respect to the print platform, at the (Formlabs Preform software, v2.17.1, Formlabs Inc., Somerville, MA, USA) default 45-degree angle, vertically, horizontally, and lied flat flush along the print platform. The channels of the printed chips were then injected with food dye to visualize the printed length of each channel. The printed length (i.e., the length colored by the food dye) was measured and reported as a fraction of the total expected length; these results are presented in Table 1. The outlet channel consistently had the lowest printing success, attributed to the fact that the outlet was the furthest from the print platform. As the chip was printed, the distance from the print platform increased, allowing for greater deflection that results in misalignment and misprints. The default orientation was chosen as the optimal orientation, despite the vertical orientation having slightly better success, since the default angled orientation allowed for a greater amount of supports. The final chip design (Figure S1b, Supplementary Materials) was also assessed for printing accuracy. A desktop microscope was used to image the width of the channels. The measured printed width and the percent error are reported in Table 2. It should also be noted that in both Tables 1 and 2, the left inlet exhibited greater printing success than the right inlet; this can be attributed to the slight rotation about the vertical axis used by the default orientation setting, which positions the left inlet was marginally closer to the print platform than the right inlet. The low average percent error in printing was deemed sufficient to not warrant a redesign of the chip's channels.

Table 1. Analysis and comparison of print orientations. A 500 µm channel height chip was printed using the four different possible orientations: the default angle, vertical, horizontal (along the chip's side), and flat (face-down). The 500 µm channel height was chosen for this quantification since it was at the threshold of printability. The channels of the chip were injected with red dye, and the fraction of the channel that was open was measured and recorded. For all values presented, there was measurement error of ±0.03 (1 mm measure per 38 mm channel length). The outlet channel consistently had inferior results due to its further distance from the build platform during printing. Vertical and the default angled orientations provide the best results, as evidenced by the highest average channel success values.

Print Orientation	Inlet (Left)	Inlet (Middle)	Inlet (Right)	Outlet	Average Printing Success
Default (45°)	0.83	0.57	0.57	0.03	0.5
Vertical	0.83	0.73	0.67	0.13	0.59
Horizontal	0.5	0.07	0.17	0.07	0.2025
Flat	0	0	0	0	0

Table 2. Measurements and analysis of printing success of the final 3D-printed microfluidic chip design. The expected width of all channels was 1000 µm. For all values presented, there was a measurement error of ±0.01 µm. The actual printed width and percent error are presented.

Channel	Printed Width (µm)	Percent Error (%)
Outlet	924.73	7.53
Inlet (Left)	946.25	5.38
Inlet (Middle)	913.98	8.60
Inlet (Right)	892.47	10.75
Average	919.36	8.07

The progression of the chip design process is depicted in Figure S1a (Supplementary Materials), from top to bottom. The initial chip was created with a specific focus on developing the interior channel pattern, as well as the channel inlet/outlet ports along the perimeter of the chip. As the design iterations continued, changes were made to the outer dimensions of the chip, reducing material usage and print time by removing excess solid portions (Figure S1e, Supplementary Materials). In addition, a viewing window was added, first to the top only and then to both faces of the chip, to increase through-chip visibility, which is particularly important for performing fluorescence imaging on the chip. The final chip design is presented in Figure S1b (Supplementary Materials), which features large viewing windows and an overall slim design with channels of 1000 μm wide by 750 μm high. To determine the ideal channel height, a parallel channel chip was designed, as seen in Figure S1c (Supplementary Materials), where each channel had a constant width of 1000 μm and the height was varied for 100, 200, 300, 400, 500, and 1000 μm. By eliminating channels that did not print (i.e., the channel was filled with cured resin), it was deduced that the optimal channel height was between 500 μm and 1000 μm. The final chip design was then printed with 500 μm, 625 μm, and 750 μm channel heights, and 750 μm was ultimately the dimension with the best and most reliable printability. Using the final channel dimensions, a calibration chip was printed (Figure S1d, Supplementary Materials), which consisted of ten channels; each channel was then filled with different concentrations of BSA and imaged using the fluorescence microscope.

3.2. Characterization of Protein Cleaning Procedure

The effectiveness of the cleaning process (explained graphically in Figure 3a) for reducing the degree of cross-contamination between samples and the protein absorption by the chip was assessed using CBQCA assay for various scenarios: after one cycle use of the chip with DI water as the cleaning agent (i.e., the BSA sample was applied and the channels were dried and washed with the cleaning agent), after one cycle use of the chip with PBS as the cleaning agent, after six cycles of repeated use of a chip with PBS as the cleaning agent, and a repeat of the six cycles of repeated use with PBS and an additional post-treatment of trypsin. For each experiment, 13 samples were collected from an outgoing cleaning agent at different time/volume points, as described above in Methods. Two cleaning agents, PBS and DI water, were used separately for comparison. For both of the cleaning reagents, the BSA concentration is high at first and then decreases with the more washing volume; this decreasing BSA cross-contamination trend follows an exponential decay as a function of washing volume. A negligible amount of BSA was seen in PBS samples after 3000 μL, while this value was 5000 μL for DI water (Figure 3b). These results showed that using PBS as a cleaning agent is over 4 times more effective than using DI water, in terms of cross-contamination.

In addition to cross-contamination, the amount of protein absorption by the chip was measured after one-cycle use and repeated cycles of sample introduction and cleaning. The fluorescent microscope was used to measure the amount of protein in the channels after adding CBQCA. The light intensities in the images were analyzed for each channel. The light intensities inside and outside the channel were calculated using ImageJ software (v1.48k, National Institutes of Health, Bethesda, MD, USA). The light intensity outside the channel was used as a control as there was no protein, and hence no illumination occurred in that area. Figure 3c shows the average amount of protein absorption after cleaning with PBS and DI water. The chip that was cleaned with DI water shows significantly more absorbed BSA than the chip cleaned with PBS, (794 ± 99) ng/cm^2 compared to (486 ± 104) ng/cm^2, respectively. This leads to the conclusion that cleaning with PBS is over 1.6 times as effective as cleaning with DI water, in terms of chip protein uptake. Therefore, in further experiments, PBS was chosen as the cleaning reagent.

The longitudinal cross-contamination was also investigated to validate the proposed 3D-printed microfluidic chip and cleaning process for application in a setting, where a single device could be used repeatedly. After flowing the BSA sample through the inlet, the channels were dried and washed by flowing air and PBS through the chip. This cycle was repeated six times, with the 13 samples collected for each cycle, and all the collected samples were analyzed with the CBQCA assay. Comparing the obtained plate reader fluorescence results, it was clear that the amount of protein in the collected samples is reduced by increasing the amount of washing volume (Figure 3d). The average BSA concentration in the first 200 μL of PBS collections of all cycles was (392 ± 200) μg/mL (standard deviation is depicted in Figure 3d using error bars) and this amount is decreased sharply within 1000 μL of washing with PBS and then dwindled to zero. The overall trend is similar to the results of one-cycle cross contamination assessment. As seen in Figure 3d, the standard deviation in the measured BSA concentration for the first four samples of each cycle is significant due to the accumulation of BSA in the chip after each cycle; in other words, the large error bars represent the variation between the increased number of cycles. For example, higher cycles would lie closer to the top of the error bars, while the beginning cycles would be towards the bottom. The error bars shorten afterwards, which demonstrated that no matter how many times the chip was used, after washing with as much as 5000 μL, all the possible BSA has been removed, or only negligible BSA is present afterwards. It should be noted that it does not mean that all the BSA has been washed out, as some of the BSA was absorbed and accumulated in the chip.

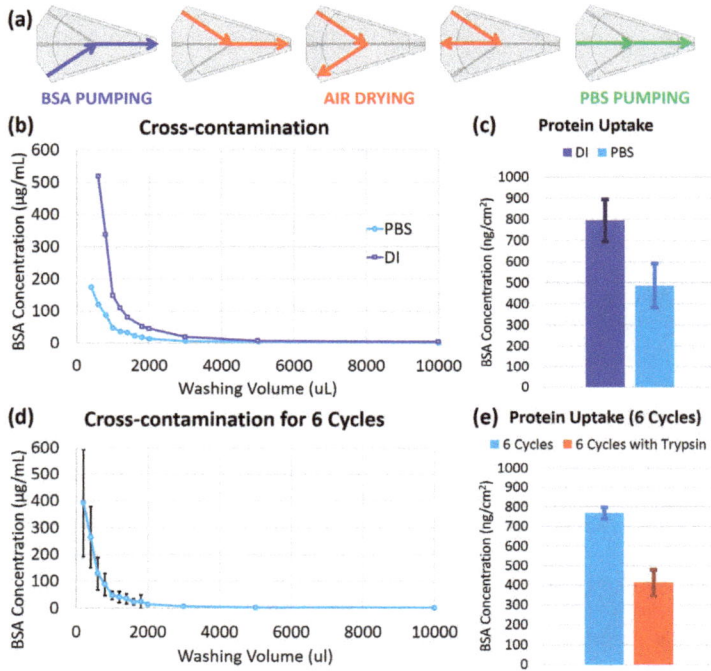

Figure 3. Protein cleaning characterization: (**a**) Graphical representation of one-cycle of the cleaning procedure. The PBS from the outlet was collected at set intervals for the cross-contamination results; (**b**) Cross-contamination of a single-use chip using deionized (DI) water compared to PBS as a cleaning solution with the BSA concentration in the output PBS or DI sample using a new chip. Cross-contamination is significantly reduced by using PBS, as compared to DI; (**c**) Protein absorption by a single-use 3D-printed chip after cleaning with DI compared to PBS. After cleaning with PBS, approximately half as much protein uptake remained, as compared to after cleaning with DI; (**d**) Cross-contamination of a repeated-use chip with the BSA concentration in the output PBS sample for 6 cycles using the same chip repeatedly, graphed as the mean with standard deviation error bars; (**e**) Protein absorption by the 3D-printed chip after 6 cycles of repeated use. By adding a post-cleaning treatment of trypsin after the 6 cycles of repeated use, the protein uptake is decreased by nearly half.

The protein uptake after six repeated cycles was shown to be (758 ± 30) ng/cm^2 (Figure 3e), which is approximately twice the amount of uptake after single use of the chip. This showed a promising trend that even after six repeated cycles of use, chip protein uptake less-than doubles, demonstrating a favorable relationship in that there was not a one-to-one relationship between cycles (and therefore exposure to BSA sample) and protein uptake. This can be taken to mean that the total protein uptake by the chip may approach a plateau after a relatively low number of cycles of use. Furthermore, by flowing 10 mL of trypsin after the six cycles of samples and cleaning, the protein uptake was reduced to (412 ± 66) ng/cm^2. These favorable results are due to the protease function of the trypsin, which digests a portion of the BSA on the chip. This means that by adding trypsin, the effectiveness of the cleaning procedure can be further improved.

3.3. Protein Quantification for Urine Processing

The 3D-printed cleaning chip was demonstrated for application to urine testing using commercially available dipstick protein assay pads. For handling multiple samples, the cleaning procedure, depicted graphically in Figure 4a, was slightly modified to include PBS cleaning of both the sample inlet channel and the universal outlet channels. Additionally, the volume of cleaning solution was reduced to 300 µL. Using this modified cleaning procedure, urine samples and high-concentration (20 g/L) BSA were alternatingly pumped through the chip. The resulting protein uptake by the chip was (657 ± 69) ng/cm^2 (Figure 4b). Comparatively, the protein uptake after 30 min of pumping urine (approximately the same total sample volume) with no intermediate cleaning was (364 ± 54) ng/cm^2. The difference between these values is attributed to the high-concentration BSA, which was used in the former experiment. When the exposure time to urine without cleaning was increased to 90 min, the protein uptake was (443 ± 70) ng/cm^2. This is a favorable result because for tripling the exposure time to urine, and tripling the volume of urine processed by the chip, the protein uptake only increased by 21.7%.

The cross-contamination between samples was also studied for the alternating urine and high-concentration (20 g/L) BSA. The protein assay pads of commercially available dipsticks were dipped in BSA samples of various concentrations (0, 0.15, 0.3, 1, 3, and 10 g/L). The resulting color change was converted into a change in grayscale light intensity. A calibration curve was created using least-squares curve fitting; the resulting curve, depicted on the left of Figure 4c, was found to be $y = -0.000002618x^4 + 0.001382x^3 + 0.2630x^2 + 20.8829x - 552.0138$ with a correlation coefficient of $R^2 = 0.963$. The calibration curve was then verified by comparing the protein concentration calculated via image analysis to the actual protein concentration of the samples, as shown on the right of Figure 4c.

Figure 4d shows the longitudinal quantification of protein concentration for 20 cycles of urine samples, post-urine PBS cleaning solution, BSA samples, and post-BSA PBS cleaning solution (a total of 81 pumping steps). The average protein concentration in the urine samples after pumping through the cleaning chip, represented by the orange dotted line, was 0.8072 g/L, as compared to the true value of 0.4958 g/L. This deviation is attributed to the low sensitivity of the urine dipstick assay for low protein concentrations; in fact, for this range of protein concentration, the color key provided by the dipstick manufacturer only discerns between 0.3 g/L and 1.0 g/L. The average protein concentration of the PBS, represented by the green dotted line, was 4.0715 g/L. Although there is no protein present in the PBS cleaning solution originally, the protein concentration after pumping through the chip is elevated as it washes out protein from the previous sample. Despite the cross-contamination between the high-concentration BSA and subsequent PBS cleaning solution, the cleaning and air-drying steps are sufficient for yielding consistent urine concentration results. Figure 4e shows images of the dipstick pads for the gray region of Figure 4d.

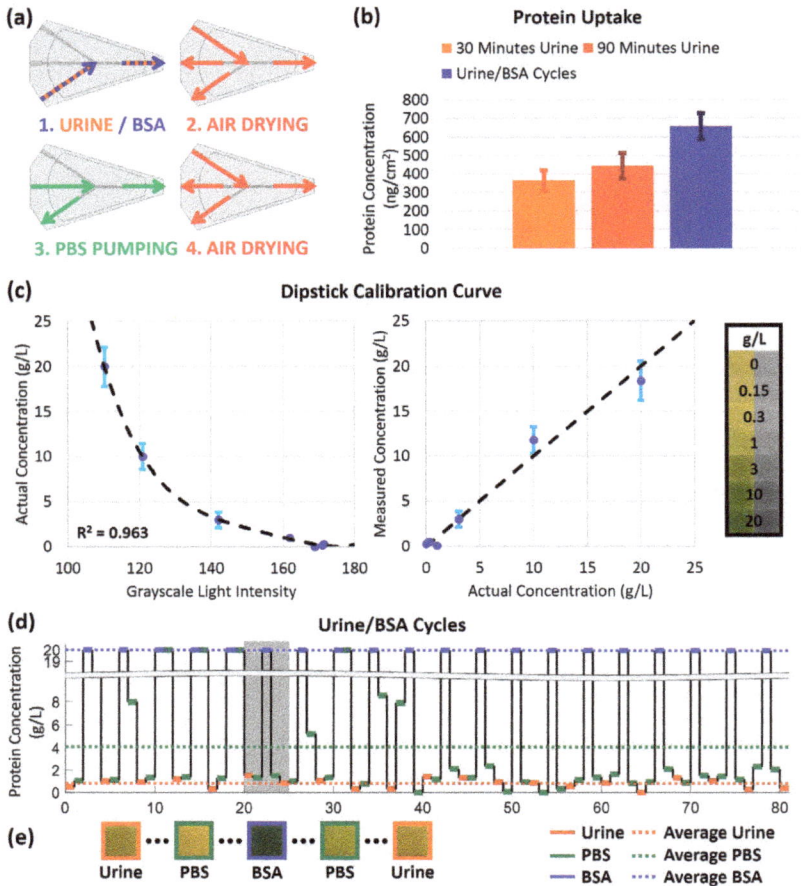

Figure 4. Performance of 3D-printed cleaning chip for urinary protein measurement: (**a**) Graphical representation of one-cycle of the modified cleaning procedure; (**b**) Protein absorption by the 3D-printed cleaning chip after 30 min of pumping urine (without cleaning), 90 min of pumping urine (without cleaning), and 20 cycles of alternating urine and high-concentration BSA samples (with cleaning between samples); (**c**) Left: calibration curve for the actual protein concentration versus the grayscale intensity of protein assay pads, calculated by least-squares curve fitting. Right: comparison of the calculated protein concentration using the calibration curve versus the actual concentration of the samples. Far right: photographs of the protein assay pads for 0, 0.15, 0.3, 1, 3, 10, and 20 g/L, shown in full color and grayscale; (**d**) Longitudinal protein concentration measurement for 20 cycles (81 pumping steps) of urine, PBS cleaning, high-concentration BSA, and a second round of PBS cleaning. Urine samples, PBS and BSA are represented by blue, green, and blue lines, respectively. Dotted lines represent the average protein concentration for the respective sample; (**e**) Photographs of the protein assay pads for Cycle 6, Steps 21–25.

4. Discussion

Reusability of microfluidic devices is critical for any devices that will go to market; however, fouling of the devices' channels can pose a significant challenge. For a microfluidic device capable of being commercialized, it must work reliably over a long period of time and for a wide variety of samples [38]. Characterization of the protein absorption of 3D-printed material, specifically Formlabs Clear Resin for SLA printing, was performed to validate the reusability of 3D-printed microfluidic chips. Furthermore, the cleaning chip and cleaning procedure proposed here is an efficient and cost- and time-effective solution for enabling the reusability of microfluidic devices. 3D printing is a low-cost, rapid-prototyping method for fabrication that is simpler, cheaper, and faster than soft lithography. The reusable design of this 3D-printed microfluidic chip also replaces the need for fabrication and disposal of many separate devices for the delivery and testing of multiple samples.

As the proposed cleaning chip design has shown favorable results, i.e., minimal protein cross-contamination and protein absorption has occurred, our work here sets a precedent for considering reusability when designing 3D-printed microfluidic devices. Future research directions may look to further assess other factors contributing to the reusability of devices. It is known that the nature of the sample, the materials from which the device is fabricated, and the flow of samples can all contribute to the fouling that occurs. Other 3D-printed materials may be studied to determine their respective absorption characteristics. BSA, a protein often found in urine, was chosen as a representative protein sample; however, different mixtures of proteins may exhibit different absorption characteristics based on charge, hydrophobicity/hydrophilicity, or other surface interactions. With different constituents in the mixture, proteins may unfold; for instance, a protein may be hydrophilic on the surface, but upon unfolding, it might expose a hydrophobic interior, which can affect its absorption behavior. Additionally, since bacteria can feed on residual samples left in the chip and can further contribute to the fouling problem, bacteria adsorption and cross-contamination studies may be performed. It is important to quantify and characterize the bacterial contamination as bacteria could negatively impact the results of some assays. Other future work may also consider the influence of flow rate. Finally, in order to achieve the goal of integrating the demonstrated cleaning chip into a functional microfluidic chip, in the future, the cleaning procedure should be automated and self-contained.

Supplementary Materials: The following is available online at http://www.mdpi.com/2072-666X/9/10/520/s1, Figure S1: Design iterations of the 3D-printed microfluidic chip for reusability assessment.

Author Contributions: S.T. designed the experiments. E.L. and R.A. performed the experiments. E.L. and S.T. wrote and edited the manuscript.

Acknowledgments: S.T. acknowledges the American Heart Association Scientist Development Grant (15SDG25080056), Connecticut Innovations Biopipeline Award, and the University of Connecticut Research Excellence Program award for financial support of this research. The authors acknowledge Julia Dugas, Patrick Madaus, and Farnoosh Saeedinejad for their assistance.

Conflicts of Interest: S.T. is a founder of, and has an equity interest in mBiotics, LLC, a company that is developing microfluidic technologies for point-of-care diagnostic solutions. S.T. and R.A. are a founder of, and has an equity interest in QRfertile, LLC, a company that is developing microfluidic technologies for fertility testing. S.T.'s and R.A.'s interests were viewed and managed in accordance with the conflict of interest policies. The authors have no other relevant affiliations or financial involvement with any organization or entity with a financial interest in or financial conflict with the subject matter or materials discussed in the manuscript, apart from those disclosed.

References

1. Wankhede, S.P.; Du, Z.; Berg, J.M.; Vaughn, M.W.; Dallas, T.; Cheng, K.H.; Gollahon, L. Cell detachment model for an antibody-based microfluidic cancer screening system. *Biotechnol. Prog.* **2006**, *22*, 1426–1433. [CrossRef] [PubMed]

2. Leoncini, E.; Ricciardi, W.; Cadoni, G.; Arzani, D.; Petrelli, L.; Paludetti, G.; Brennan, P.; Luce, D.; Stucker, I.; Matsuo, K.; et al. Lab-on-a-chip for oral cancer screening and diagnosis. *Head Neck* **2014**, *36*, 1391. [CrossRef]

3. Kim, L. Overview of the microfluidic diagnostics commercial landscape. In *Microfluidic Diagnostics. Methods in Molecular Biology (Methods and Protocols)*; Jenkins, G., Mansfield, C., Eds.; Humana Press: Totowa, NJ, USA, 2013; Volume 949.

4. Srinivasan, V.; Pamula, V.K.; Fair, R.B. An integrated digital microfluidic lab-on-a-chip for clinical diagnostics on human physiological fluids. *Lab Chip* **2004**, *4*, 310–315. [CrossRef] [PubMed]

5. Hsu, Y.-H.; Moya, M.L.; Hughes, C.C.W.; George, S.C.; Lee, A.P. A microfluidic platform for generating large-scale nearly identical human microphysiological vascularized tissue arrays. *Lab Chip* **2013**, *13*, 2990. [CrossRef] [PubMed]

6. Toh, Y.-C.; Lim, T.C.; Tai, D.; Xiao, G.; van Noort, D.; Yu, H. A microfluidic 3D hepatocyte chip for drug toxicity testing. *Lab Chip* **2009**, *9*, 2026. [CrossRef] [PubMed]

7. Yu, L.; Chen, M.C.W.; Cheung, K.C. Droplet-based microfluidic system for multicellular tumor spheroid formation and anticancer drug testing. *Lab Chip* **2010**, *10*, 2424. [CrossRef] [PubMed]

8. Chin, C.D.; Linder, V.; Sia, S.K. Commercialization of microfluidic point-of-care diagnostic devices. *Lab Chip* **2012**, *12*, 2118. [CrossRef] [PubMed]

9. Sia, S.K.; Kricka, L.J. Microfluidics and point-of-care testing. *Lab Chip* **2008**, *8*, 1982. [CrossRef] [PubMed]

10. Foudeh, A.M.; Fatanat Didar, T.; Veres, T.; Tabrizian, M. Microfluidic designs and techniques using lab-on-a-chip devices for pathogen detection for point-of-care diagnostics. *Lab Chip* **2012**, *12*, 3249. [CrossRef] [PubMed]

11. Yenilmez, B.; Knowlton, S.; Tasoglu, S. Self-Contained Handheld Magnetic Platform for Point of Care Cytometry in Biological Samples. *Adv. Mater. Technol.* **2016**, *1*, 1600144. [CrossRef]

12. Knowlton, S.; Joshi, A.; Syrrist, P.; Coskun, A.F.; Tasoglu, S. 3D-Printed Smartphone-Based Point of Care Tool for Fluorescence- and Magnetophoresis-Based Cytometry. *Lab Chip* **2017**, *17*, 2839–2851. [CrossRef] [PubMed]

13. Lepowsky, E.; Ghaderinezhad, F.; Knowlton, S.; Tasoglu, S. Paper-based assays for urine analysis. *Biomicrofluidics* **2017**, *11*, 051501. [CrossRef] [PubMed]

14. Maciosek, M.V.; Coffield, A.B.; Flottemesch, T.J.; Edwards, N.M.; Solberg, L.I. Greater use of preventive services in U.S. health care could save lives at little or no cost. *Health Aff.* **2010**, *29*, 1656–1660. [CrossRef] [PubMed]

15. Ashraf, M.W.; Tayyaba, S.; Afzulpurkar, N. Micro Electromechanical Systems (MEMS) based microfluidic devices for biomedical applications. *Int. J. Mol. Sci.* **2011**, *12*, 3648–3704. [CrossRef] [PubMed]

16. Jivani, R.R.; Lakhtaria, G.J.; Patadiya, D.D.; Patel, L.D.; Jivani, N.P.; Jhala, B.P. Biomedical microelectromechanical systems (BioMEMS): Revolution in drug delivery and analytical techniques. *Saudi Pharm. J.* **2016**, *24*, 1–20. [CrossRef] [PubMed]

17. Fujii, T. PDMS-based microfluidic devices for biomedical applications. *Microelectron. Eng.* **2002**, *61–62*, 907–914. [CrossRef]

18. Dario, P.; Carrozza, M.C.; Benvenuto, A.; Menciassi, A. Micro-systems in biomedical applications. *J. Micromech. Microeng.* **2000**, *10*, 235–244. [CrossRef]

19. Knowlton, S.; Yu, C.H.; Jain, N.; Ghiran, I.C.; Tasoglu, S. Smart-phone based magnetic levitation for measuring densities. *PLoS ONE* **2015**, *10*, 1–17. [CrossRef] [PubMed]

20. Knowlton, S.; Yenilmez, B.; Tasoglu, S. Towards Single-Step Biofabrication of Organs on a Chip via 3D Printing. *Trends Biotechnol.* **2016**, *34*, 685–688. [CrossRef] [PubMed]

21. Zhang, H.; Chiao, M. Anti-fouling coatings of poly(dimethylsiloxane) devices for biological and biomedical applications. *J. Med. Biol. Eng.* **2015**, *35*, 143–155. [CrossRef] [PubMed]

22. Picher, M.M.; Küpcü, S.; Huang, C.-J.; Dostalek, J.; Pum, D.; Sleytr, U.B.; Ertl, P. Nanobiotechnology advanced antifouling surfaces for the continuous electrochemical monitoring of glucose in whole blood using a lab-on-a-chip. *Lab Chip* **2013**, *13*, 1780. [CrossRef] [PubMed]

23. Lepowsky, E.; Tasoglu, S. Emerging Anti-Fouling Methods: Towards Reusability of 3D-Printed Devices for Biomedical Applications. *Micromachines* **2018**, *9*, 196. [CrossRef]

24. Amin, R.; Li, L.; Tasoglu, S. Assessing reusability of microfluidic devices: Urinary protein uptake by PDMS-based channels after long-term cyclic use. *Talanta* **2018**, *192*, 455–462. [CrossRef]

25. Shields, C.W.; Ohiri, K.A.; Szott, L.M.; López, G.P. Translating microfluidics: Cell separation technologies and their barriers to commercialization. *Cytom. Part B Clin. Cytom.* **2017**, *92*, 115–125. [CrossRef] [PubMed]

26. Amin, R.; Joshi, A.; Tasoglu, S. Commercialization of 3D-printed microfluidic devices. *J. 3D Print. Med.* **2017**, *1*, 85–89. [CrossRef]

27. Amin, R.; Knowlton, S.; Hart, A.; Yenilmez, B.; Ghaderinezhad, F.; Katebifar, S.; Messina, M.; Khademhosseini, A.; Tasoglu, S. 3D-printed microfluidic devices. *Biofabrication* **2016**, *8*, 022001. [CrossRef] [PubMed]

28. Beebe, D.J.; Moore, J.S.; Yu, Q.; Liu, R.H.; Kraft, M.L.; Jo, B.-H.; Devadoss, C. Microfluidic tectonics: A comprehensive construction platform for microfluidic systems. *Proc. Natl. Acad. Sci. USA* **2000**, *97*, 13488–13493. [CrossRef] [PubMed]

29. Rolland, J.P.; Van Dam, R.M.; Schorzman, D.A.; Quake, S.R.; DeSimone, J.M. Solvent-Resistant Photocurable "Liquid Teflon" for Microfluidic Device Fabrication. *J. Am. Chem. Soc.* **2004**, *126*, 2322–2323. [CrossRef] [PubMed]

30. Sugioka, K.; Cheng, Y. Femtosecond laser processing for optofluidic fabrication. *Lab Chip* **2012**, *12*, 3576–3589. [CrossRef] [PubMed]

31. Yalikun, Y.; Hosokawa, Y.; Iino, T.; Tanaka, Y. An all-glass 12 μm ultra-thin and flexible micro-fluidic chip fabricated by femtosecond laser processing. *Lab Chip* **2016**, *16*, 2427–2433. [CrossRef] [PubMed]

32. Wan, A.M.D.; Devadas, D.; Young, E.W.K. Recycled polymethylmethacrylate (PMMA) microfluidic devices. *Sens. Actuators B Chem.* **2017**, *253*, 738–744. [CrossRef]

33. Au, A.K.; Huynh, W.; Horowitz, L.F.; Folch, A. 3D-Printed Microfluidics. *Angew. Chem. Int. Ed.* **2016**, *55*, 3862–3881. [CrossRef] [PubMed]

34. Ho, C.M.B.; Ng, S.H.; Li, K.H.H.; Yoon, Y.-J. 3D printed microfluidics for biological applications. *Lab Chip* **2015**, *15*, 3627–3637. [CrossRef] [PubMed]

35. Waheed, S.; Cabot, J.M.; Macdonald, N.P.; Lewis, T.; Guijt, R.M.; Paull, B.; Breadmore, M.C. 3D printed microfluidic devices: enablers and barriers. *Lab Chip* **2016**, *16*, 1993–2013. [CrossRef] [PubMed]

36. Brennan, M.D.; Bokhari, F.F.; Eddington, D.T. Open design 3D-printable adjustable micropipette that meets the ISO standard for accuracy. *Micromachines* **2018**, *9*. [CrossRef]

37. Bhattacharjee, N.; Urrios, A.; Kang, S.; Folch, A. The upcoming 3D-printing revolution in microfluidics. *Lab Chip* **2016**, *16*, 1720–1742. [CrossRef] [PubMed]

38. Mukhopadhyay, R. When Microfluidic Devices Go Bad. *Anal. Chem.* **2005**, *77*, 429A–432A. [CrossRef] [PubMed]

39. Wong, I.; Ho, C.M. Surface molecular property modifications for poly(dimethylsiloxane) (PDMS) based microfluidic devices. *Microfluid. Nanofluidics* **2009**, *7*, 291–306. [CrossRef] [PubMed]

40. Goyanes, A.; Buanz, A.B.M.; Hatton, G.B.; Gaisford, S.; Basit, A.W. 3D printing of modified-release aminosalicylate (4-ASA and 5-ASA) tablets. *Eur. J. Pharm. Biopharm.* **2015**, *89*, 157–162. [CrossRef] [PubMed]

41. Leslie, D.C.; Waterhouse, A.; Berthet, J.B.; Valentin, T.M.; Watters, A.L.; Jain, A.; Kim, P.; Hatton, B.D.; Nedder, A.; Donovan, K.; et al. A bioinspired omniphobic surface coating on medical devices prevents thrombosis and biofouling. *Nat. Biotechnol.* **2014**, *32*, 1134–1140. [CrossRef] [PubMed]

42. Tu, Q.; Wang, J.C.; Liu, R.; He, J.; Zhang, Y.; Shen, S.; Xu, J.; Liu, J.; Yuan, M.S.; Wang, J. Antifouling properties of poly(dimethylsiloxane) surfaces modified with quaternized poly(dimethylaminoethyl methacrylate). *Colloids Surf. B Biointerfaces* **2013**, *102*, 361–370. [CrossRef] [PubMed]

43. Sunny, S.; Cheng, G.; Daniel, D.; Lo, P.; Ochoa, S.; Howell, C.; Vogel, N.; Majid, A.; Aizenberg, J. Transparent antifouling material for improved operative field visibility in endoscopy. *Proc. Natl. Acad. Sci. USA* **2016**, *113*, 201605272. [CrossRef] [PubMed]

44. Carve, M.; Wlodkowic, D. 3D-printed chips: Compatibility of additive manufacturing photopolymeric substrata with biological applications. *Micromachines* **2018**, *9*. [CrossRef]

45. Whitesides, G.M.; Tang, S.K.Y. Fluidic optics. *Proc. SPIE* **2006**, *6329*, 63290A. [CrossRef]

46. Refractive Index of Clear Resin. Available online: https://forum.formlabs.com/t/refractive-index-of-clear-resin/9282 (accessed on 14 October 2018).

47. International Gem Society Refractive Index List of Common Household Liquids. Available online: https://www.gemsociety.org/article/refractive-index-list-of-common-household-liquids/ (accessed on 14 October 2018).

MDPI

St. Alban-Anlage 66

4052 Basel

Switzerland

Tel. +41 61 683 77 34

Fax +41 61 302 89 18

www.mdpi.com

Micromachines Editorial Office

E-mail: micromachines@mdpi.com

www.mdpi.com/journal/micromachines

www.ingramcontent.com/pod-product-compliance
Lightning Source LLC
Chambersburg PA
CBHW051847210326
41597CB00033B/5800